Computational Imaging for

Scene Understanding

SCIENCES

Image, Field Director – Laure Blanc-Féraud

Sensors and Image Processing,
Subject Head – Cédric Demonceaux

Computational Imaging for Scene Understanding

Transient, Spectral, and Polarimetric Analysis

Coordinated by
Takuya Funatomi
Takahiro Okabe

WILEY

First published 2024 in Great Britain and the United States by ISTE Ltd and John Wiley & Sons, Inc.

ISTE Ltd
27-37 St George's Road
London SW19 4EU
UK

www.iste.co.uk

John Wiley & Sons, Inc.
111 River Street
Hoboken, NJ 07030
USA

www.wiley.com

Any opinions, findings, and conclusions or recommendations expressed in this material are those of the author(s), contributor(s) or editor(s) and do not necessarily reflect the views of ISTE Group.

Library of Congress Control Number: 2023942088

British Library Cataloguing-in-Publication Data
A CIP record for this book is available from the British Library
ISBN 978-1-78945-150-4

ERC code:
PE6 Computer Science and Informatics
 PE6_8 Computer graphics, computer vision, multi media, computer games
 PE6_11 Machine learning, statistical data processing and applications using signal processing (e.g. speech, image, video)
 PE6_12 Scientific computing, simulation and modelling tools

Contents

Chapter 5. Spectral Modeling and Separation
of Reflective-Fluorescent Scenes

Ying FU, Antony LAM, Imari SATO, Takahiro OKABE, and Yoichi SATO

Chapter 6. Shape from Water

Yuta ASANO, Yinqiang ZHANG, Ko NISHINO, and Imari SATO

Chapter 7. Far Infrared Light Transport Decomposition and Its Application for Thermal Photometric Stereo 161

Kenichiro TANAKA

Chapter 8. Synthetic Wavelength Imaging: Utilizing Spectral Correlations for High-Precision Time-of-Flight Sensing 187
Florian WILLOMITZER

Part 3. Polarimetric Imaging and Processing 219

Chapter 9. Polarization-Based Shape Estimation 221
Daisuke MIYAZAKI

Chapter 10. Shape from Polarization and Shading 241
Thanh-Trung NGO, Hajime NAGAHARA, and Rin-ichiro TANIGUCHI

Chapter 11. Polarization Imaging in the Wild Beyond the Unpolarized World Assumption 269
Jérémy Maxime RIVIERE

Chapter 12. Multispectral Polarization Filter Array 299
Kazuma SHINODA

Introduction

Takuya FUNATOMI[1] and Takahiro OKABE[2]

[1]*Division of Information Science, Nara Institute of Science and Technology, Japan*
[2]*Department of Artificial Intelligence, Kyushu Institute of Technology, Fukuoka, Japan*

In our physical world, light propagates from various positions towards various directions. Light is emitted from light sources such as the sun and lamps and reflected by surfaces such as walls and glasses. The amount of light flowing in every direction through every point is described as a light field. As readers know, light is an electro-magnetic wave that oscillates perpendicular to its traveling direction. Therefore, light is characterized both by the spatial period of oscillation, i.e. the wavelength, and by the direction of oscillation, i.e. the polarization state. Consequently, the light field is a high-dimensional function with respect to time t, position (x, y, z), direction $(\theta.\phi)$, wavelength λ and polarization state s of light.

Most cameras are inherently designed to mimic the human eye by having three channels of red, green and blue (RGB) color and achieving about 30 frames per second. Therefore, conventional cameras only capture a part of the modalities of a light field with limited spatial, temporal and spectral resolution. Some cameras are designed to capture other modalities, for example spectra from near UV to near IR rather than RGB, polarimetry and time of light travels. Such modalities are difficult to perceive, but provide much information about scenes.

This book focuses on emerging computer vision techniques known as computational imaging. These techniques include capturing, processing and analyzing light modalities for various applications used in scene understanding.

This book is divided into three parts corresponding to time of flight, spectral and polarimetric domains. Part 1 focuses on transient imaging, which enables us to

Computational Imaging for Scene Understanding,
coordinated by Takuya FUNATOMI and Takahiro OKABE.
© ISTE Ltd 2024.

capture events at a temporal resolution in the order of the speed of light. This field is rapidly growing with applications more than range-imaging, e.g. non-line-of-sight reconstruction. The key to rapid growth stands on imaging technologies with consumer cameras and analysis-by-synthesis techniques.

Part 1 has three chapters. Chapter 1 begins with an overview of transient imaging techniques. This chapter also separately addresses one of the most attractive topics in this field, non-line-of-sight (NLOS) imaging, which enables us to see scenes that are not directly visible. Chapter 2 introduces a transient imaging technique with a correlation image sensor used in consumer-level time-of-flight cameras. Chapter 3 reviews the time-of-flight/transient rendering techniques fundamental to the analysis-by-synthesis approach.

Part 2 focuses on spectral imaging and processing, which makes use of spectral information from near UV to IR rather than conventional RGB for scene understanding. Spectral imaging enables us not only to increase the spectral resolution of captured images but also to study wavelength-dependent phenomena such as refraction, scattering and absorption.

This part covers Chapters 4–8: it begins with the principles and architectures of hyperspectral camera in Chapter 4, and then introduces the emerging techniques based on spectral imaging. Chapter 5 addresses the absorption and emission due to fluorescent materials and shows that spectral imaging with a programmable illumination in the spectral domain is useful for separating reflective and fluorescent components in images. Chapter 6 exploits the fact that water absorbs light with near IR wavelengths and shows that the shape of an underwater scene can be recovered from near IR imaging. Chapter 7 studies the temporal transport of far IR light and heat and shows that thermal imaging enables shape recovery of challenging objects made of transparent and semi-transparent materials. Chapter 8 introduces a unique approach to time-of-flight and NLOS imaging based on spectral interferometry.

Part 3 focuses on polarimetric imaging and processing using the oscillating direction of light for scene understanding. The polarization state of light depends on the normal and refractive index of an object's surface on which light is reflected, and therefore polarimetric imaging is useful for estimating the geometry and material of the object. Polarization imaging is one of the actively studied areas as the imaging sensors with polarization filter arrays become widespread.

This part covers the fundamental theory of polarization and the emerging techniques based on polarimetric imaging. It consists of Chapters 9–12. Chapter 9 begins with the fundamental theory of polarization followed by various techniques for polarization-based shape recovery: the shape from the phase angle, the degree of linear polarization and the Stokes vector. Chapter 10 shows that integrating complementary clues, i.e. the polarimetric clue and shading clue, is useful for shape recovery.

Chapter 11 achieves shape recovery in the wild, in particular, by considering polarized illumination such as the sky. Finally, Chapter 12 is a bridge between multispectral imaging and polarimetric imaging. It achieves a multispectral polarization filter array using a photonic crystal and proposes a demosaicing algorithm for it.

Each part begins with an introductory chapter and reviews various achievements by several leading researchers in the field. We hope that such collections, ranging from the overview of fundamentals to cutting-edge technologies, inspire new directions of research and development in these modalities for various applications of scene understanding.

PART 1

Transient Imaging and Processing

1

Transient Imaging

Adrian JARABO

Graphics and Imaging Lab, University of Zaragoza, Spain

1.1. Introduction

In 1878, English photographer Eadweard Muybridge captured his famous *Horse in Motion* series, the first sequence of photographs able to capture a dynamic scene by using stop motion (Leslie 2001). His invention revolutionized photography, leading to ground-breaking innovations that ended up with the invention of the Lumière Cinématographe that gave birth to the cinema (Lumière 1936). But beyond that, this first sequence of photographs allowed people to analyze the dynamics of the horse at run, which was followed by many (up to 100,000 sequences) images of animals and humans in motion that allowed for better understanding on the dynamics of locomotion.

Being able to capture how the world evolves with time was a powerful tool for understanding the physical laws governing the world and how to carefully analyze experiments. Almost 90 years after Muybridge's invention, Harold Edgerton pushed these photographic analyses of the world dynamics further, capturing them at 10,000 frames per second. A famous example of Edgerton's captures is the *Bullet Through Apple* photograph, freezing in time ultrafast events such as the bullet penetrating through the apple (Bedi and Collins 1994). Seeing the world at that speed helped with understanding the mechanical behavior of fast dynamic systems such as liquid flows or explosions. However, while impressive on their own, all these high-speed

Computational Imaging for Scene Understanding,
coordinated by Takuya FUNATOMI and Takahiro OKABE.
© ISTE Ltd 2024.

photography techniques were not even close to capturing the fastest events in nature: the propagation of light.

Only recently, with the emergence of so-called *transient imaging*, it has been possible to capture events at a temporal resolution in the order of the speed of light, up to five orders of magnitude faster than Edgerton's stroboscope. While the seminal work by Abramson introduced the first light-in-flights visualization using holography (Abramson 1978, 1983) for coherent light in small scenes, it was not until 40 years later when the first macroscopic light-in-flight animations were demonstrated with the introduction of *femto-photography* (Velten et al. 2012b, 2013); for the first time, we were able to see light propagating at a scene at roughly a trillion frames per second (see Figure 1.1).

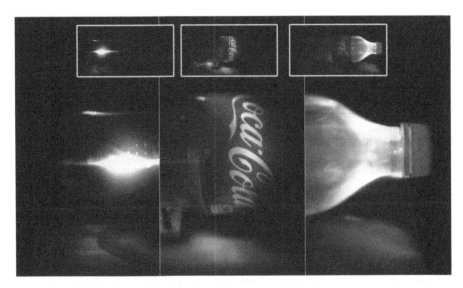

Figure 1.1. *Example capture of a light pulse propagating through a bottle filled with water using femto-photography (Velten et al. 2013), with an effective frame rate of 1 trillion frames per seconds. The large split image is a composite of the three complete frames shown in the insets (source: Jarabo et al. (2017)). The complete videos of this and other scenes captured with femto-photography can be downloaded from Velten et al. (2013a). For a color version of this figure, see www.iste.co.uk/ funatomi/computational.zip*

Capturing light transport at such temporal resolution has made it possible to *see the invisible*: from impressive animations of light in motion to a myriad of new computer vision and scene understanding techniques that leverage the enormous amount of information encoded in the temporal domain of light, enabling us, for example, to see around corners or through turbid media. In the rest of this chapter, we will introduce the mathematical properties of time-resolved light transport, the techniques developed for capturing light in the transient state, and some of the numerous scene understanding applications enabled by capturing light in motion. For a further in-depth review of transient imaging, we refer to monographs on the topic (Jarabo et al. 2017).

1.2. Mathematical formulation

One of the most general assumptions in computer vision and scene understanding is to consider the speed of light to be infinite, therefore assuming light transport in the steady state. This is a reasonable assumption since most of the existing imaging hardware is very slow compared to the speed of light. Even in the case of high-speed cameras, capable of capturing up to a few thousand frames per second, light travels tens of kilometers in each frame (Figure 1.2, left). With the emergence of ultrafast imaging devices, effective exposure times of nano- or pico-seconds have been made possible. At such exposure times, light travels a few centimeters per frame, effectively breaking the assumption of a steady state of light, requiring viewers to account for the time-resolved nature of light transport. The incoming light $L_t(\mathbf{x}, t)$ at the camera's pixel $\mathbf{x}$ is a function defined on the temporal domain t (Figure 1.2). This time-resolved radiance relates to its steady-state counterpart $L_s(\mathbf{x})$ as

$$L_s(\mathbf{x}) = \int_{-\infty}^{\infty} L_t(\mathbf{x}, t)dt. \qquad [1.1]$$

Figure 1.2. *Synthetic example of time-resolved light transport (data from Jarabo et al. (2014)). In the steady state (left), all light has finished propagating along the scene. However, in the transient state, the incoming radiance at the pixel is a function of time, and the light takes several nanoseconds to complete the propagation along the scene. The plot shows the temporal profile of the radiance in pixels (a), while the individual frames in the plot are shown as blue dashed lines. For a color version of this figure, see www.iste.co.uk/funatomi/computational.zip*

Mathematically speaking, the time-resolved radiance $L_t(x, t)$ relates to the impulse response of the scene $T(\mathbf{x}, t - \tau)$ and the temporal profile of the illumination $L_i(t)$ as

$$L_t(\mathbf{x}, t) = \int_{-\infty}^{\infty} T(\mathbf{x}, t - \tau) \, L_i(\tau)d\tau = (T(\mathbf{x}) *_t L_i)(t), \qquad [1.2]$$

with $*_t$ being the convolution operator on the temporal domain. In general, when we talk about *transient imaging*, we are interested on the impulse response of the scene. From equation [1.2], we can see that continuous illumination essentially reduces to equation [1.1] (see Figure 1.3); this illustrates the need for a temporally resolved illumination to capture transient light transport, ideally an ultrafast pulsed illumination that makes it possible to directly capture the impulse response of the scene, although as we will see in section 1.3, there are methods that make it possible to use continuous illumination for lower-cost transient imaging by coding the illumination in time.

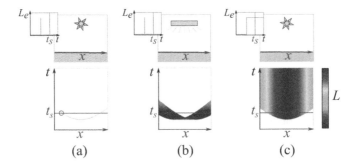

Figure 1.3. *Time-resolved incoming radiance (bottom) at the ground plane for different types of emission in a simple scene (top, the temporal profile of the emission is on the inset). The frequency in either the temporal or spatio-angular domain affects the frequency of the transient illumination. (a) A pulsed point light source generates a sharp signal on the temporal domain. By reducing the frequency in the spatial (b) or temporal (c) domains, by means of a pulsed area light or a non-Delta emission, respectively, the incoming light is no longer a differential pulse, and the high frequencies in the impulse response are lost. For a color version of this figure, see www.iste.co.uk/funatomi/computational.zip*

Transient light transport, as encoded in the impulse response matrix, also presents a cross-dimensional information transfer between the temporal and angular domains, as demonstrated by Wu et al. (2012) by using a Fourier analysis on the spatio-directional time-resolved light transport. This property has a myriad of potential applications, including, for example, bare-sensor imaging. In addition, it also shows that a loss on angular frequency translates into a loss on the temporal frequency of the integrated time-resolved signal, as illustrated in Figure 1.3 (c). Therefore, in order to get a narrow-banded impulse response of the scene, a light source with a very small surface is desired.

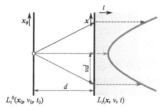

Propagation	Domain	Operator
Forward	Primal	Shear in x, hyperbolic curvature in t
	Frequency	Shear in ω_v, convolution in ω_v
Inverse	Primal	Tomographic reconstruction
	Frequency	Filtered backprojection

Figure 1.4. Left: *ray diagram illustrating the two-planes light field parametrization for an impulse point light source illumination. As the light propagates to distance d, the light field suffers from a hyperbolic curvature in the temporal domain.* **Right:** *overview of the operators for forward and inverse propagation of time-resolved light fields in both the primal and the frequency domains (source: adapted from Wu et al. (2012)). For a color version of this figure, see www.iste.co.uk/funatomi/computational.zip*

1.2.1. *Analysis of transient light transport propagation*

We have already seen one important property of the transient light transport encoded in $T(\mathbf{x}, t)$, which is the cross-dimensional transfer between the angular and the temporal domains. To study it further, and following Wu et al. (2012), let us first define the time-resolved light field as a time-dependent spatio-angular function $L_t(\mathbf{x}, v, t)$, where v encodes the direction of propagation of the light field. We can now define how a source light field $L_t^0(\mathbf{x}_0, v_0, t_0)$ changes as it propagates through the scene to a parallel plane at distance d (see Figure 1.4, left), as

$$L_t(\mathbf{x}, v, t) = L_t^0\left(\mathbf{x} - v\,d, v, t - \frac{d}{c}\sqrt{1 + v^2}\right),\qquad [1.3]$$

with $\mathbf{x}_0 = \mathbf{x} - v\,d$, $v_0 = v$ and $t_0 = t - \frac{d}{c}\sqrt{1+v^2}$, and c being the speed of light. Physically, light rays at more slanted angles (i.e. higher v) take longer to propagate. For a pulsed point light source at position $\mathbf{x}_0$, this means that the time-resolved light field has a hyperbolic curvature in the spatio-temporal domain. In essence, this means that the spatio-temporal light field suffers two transformations as it propagates: a shear on $\mathbf{x}$ and a hyperbolic curvature in t.

Now let us define the Fourier transform of the light field $L_t(\mathbf{x}, v, t)$ as $\mathcal{F}[L_t(\mathbf{x}, v, t)] = \tilde{L}_t(\omega_{\mathbf{x}}, \omega_v, \omega_t)$, with $\omega_{\mathbf{x}}, \omega_v,$ and ω_t being the frequency components in space, angle and time, respectively. As demonstrated by Wu et al. (we refer to the derivations and proofs on the original work (Wu et al. 2012)), the Fourier counterpart of equation [1.3] is

$$\tilde{L}_t(\omega_{\mathbf{x}}, \omega_v, \omega_t) = \tilde{L}_t^0(\omega_{\mathbf{x}}, \omega_v + \omega_{\mathbf{x}}\,d, \omega_t) *_{\omega_v} F_{\frac{\omega_t\,d}{c}}(\omega_v),\qquad [1.4]$$

with $F_{\frac{\omega_t\,d}{c}}(\omega_v)$ being a time-dependent convolution kernel on the domain of the angular frequencies. Mathematically, equation [1.4] models propagation on the frequency domain as a shear on the space-angular dimensions and a convolution with the kernel $F_{\frac{\omega_t d}{c}}(\omega_v)$ on the angular frequencies. This later convolution is a key implication on whether we can recover information from a time-resolved signal depending on its temporal frequency: only if the kernel parameter $|\frac{\omega_t d}{c}| > |\omega_v|$, then the information can be detected. Finally, while this mathematical analysis targets forward space-time propagation, its consequences apply also to inverse propagation for solving inverse problems. In particular, the primal operator for inverse propagation in time-resolved light fields results into a tomographic reconstruction; this problem can be efficiently solved using filtered backprojection. As we will see later in section 1.5, this observation is the base for most initial methods on non-line-of-sight reconstruction using transient imaging.

1.2.2. *Sparsity of the impulse response function* $T(\mathbf{x}, t)$

The impulse response function $T(\mathbf{x}, t)$ stores a vast amount of information about the light transport in a scene. However, in scenes with significant multiple scattering (either from surfaces and scattering media) such information might be difficult to extract directly from the input data. To overcome this challenge, and to extract potentially hidden information from the scene, a common approach is to leverage the sparsity of $T(\mathbf{x}, t)$. For that, several authors have proposed different bases to characterize the physical phenomena of light transport. The idea is that the impulse function can be approximated as (we remove the spatial dependence for clarity)

$$T(t) \approx \sum_{i=1}^{N} w_i\, g_i(t), \qquad\qquad [1.5]$$

with $g_i(t)$ being a functional base dependent on time, and w_i being a weight for $g_i(t)$.

The most simplistic basis is the K-sparse model representing the different scattering events of light before reaching the sensor. This basis is actually very intuitive, and is particularly useful for reasoning about the applications of transient light transport, as demonstrated in their seminar work (Raskar and Davis 2008). In a nutshell, the impulse response of a scene is a set of sparse scattering events, each represented as a delta function on the temporal domain $g_i(t) = \delta(t - t_i)$, with t_i being the time of flight of each scattering event as

$$T(t) \approx \sum_{i=1}^{N} w_i\, \delta(t - t_i), \qquad\qquad [1.6]$$

where w_i is directly related to the path contribution on the sensor (i.e. its radiance). The K-sparse model is accurate for the direct component of a pulsed illumination, as well as the case of sparse scenes or in surfaces with near-singular reflectance (Kadambi et al. 2013; Peters et al. 2015; Qiao et al. 2015). Unfortunately, it quickly fails for complex scenes, or under arbitrary reflectances. In the case of non-singular reflectance (e.g. Lambertian) and continuous surfaces, the time-resolved reflected radiance is continuous, which makes it impractical to approximate it with a large number of delta functions.

To avoid using a potentially infinite N, several works have proposed to use continuous ad hoc bases to represent different elements of light transport. These bases are, in general, relatively simple analytical functions, such as Gaussians or exponentially modified Gaussians, that resemble the temporal profile of different components of light transport. For example, Wu et al. (2014) propose to use Gaussians of varying mean and standard deviation to represent the indirect scattering of continuous surfaces. In a similar phenomenological manner, Freedman et al. (2014) proposed to use exponential distributions as a basis for representing the indirect illumination reflected out by Lambertian surfaces. These are, of course, approximations, since the temporal profile of indirect scattering is a complex function dependent of the surface's shape and reflectance, and closed-form exact solutions exist only for planar unoccluded Lambertian surfaces (Iseringhausen and Hullin 2020).

In the case of scenes with scattering media, and based on the observation that scattering in media follows an exponential decay in time (which can be rigorously derived by assuming Brownian motion of photons (D'Eon and Irving 2011)), Wu et al. (2014) proposed to use an exponential distribution to account for the temporal reflectance on translucent objects. Similarly, Heide et al. (2014b) proposed to use exponentially modified Gaussians to account for the different responses of light transport in the presence of participating media.

Finally, beyond these phenomenological models that aim to define simple bases for representing light transport, Liang et al. (2020) derived a set of low-dimensional data-driven bases for transient light transport by machine learning. All these works demonstrate that such mathematical space for sparsity exists on the impulse response function $T(\mathbf{x}, t)$; unfortunately, the relationship with the actual physical image formation process only exists so far as ad hoc models. Still, these models have been proven very useful for many different applications, including capture, denoising, compression or analysis of time-resolved light transport.

1.3. Capturing light in flight

Imaging light in flight is equivalent to capturing the impulse response of a scene along the temporal domain. For that, a camera sensor captures the time-resolved radiance (equation [1.2]) arriving to the sensor during an exposition time, as

$$I(t) = \int_{-\infty}^{\infty} W(\tau)(T *_t L_i)(t + \tau)d\tau = W *_t T *_t L_i, \qquad [1.7]$$

where $W(t)$ is the sensor's temporal response during the exposition time. Since the goal is to capture $T(t)$, an ideal transient imaging system would consist of delta illumination and sensor response. However, this is not possible for practical reasons; limiting the systems' pulsed illumination and ultrashort expositions are, in general, expensive. For that, different approaches have been proposed, offering a range of temporal resolution, generality and cost. These systems can be categorized in three main categories: (1) interferometry-based systems, (2) direct temporal recording and (3) time-of-flight (ToF) cameras.

Interferometry: interferometry was the first technology for imaging light in motion (Abramson 1978), based on imaging systems similar to the Michelson interferometer. However, it has a limited range of applicability, which has made its scope relatively narrow to small scenes under very controlled setups. The key idea is to leverage electromagnetic interference of coherent light to disambiguate between slightly different path lengths (i.e. times of flight). While the results into a very high temporal resolution (at the femtosecond scale, see, for example, Gkioulekas et al. (2015)), the imaging setup needs to be highly controlled, and only small scenes can be imaged. In addition, early interferometry-based methods did not capture incoherent light like indirect illumination, although more recent ones (Kotwal et al. 2020) do handle incoherent light by using sophisticated light transport probing.

Direct temporal recording: this is the conceptually simplest approach, and consists of using an ultrashort pulsed illumination in combination with ultrafast imaging hardware to capture equation [1.7]. Several different optical systems have been proposed, including time-gated systems (Busck and Heiselberg 2004), streak cameras (Velten et al. 2013), single-photon avalanche diodes (SPAD) (Gariepy et al. 2015) or dispersive optics (Nakagawa et al. 2014; Lei et al. 2016). These systems range in terms of cost, temporal resolution, operating time or spatial resolution. For example, streak cameras are expensive, difficult to operate, require a Galvo for scanning the y-axis and need long capture times, but have large spatial resolution and effective exposure times in the order of 2 picoseconds. SPADs are cheap and very sensitive to photons, which make them very adequate for operating in the wild, but they are currently limited by their spatio-temporal resolution. Given the widespread use of SPADs in current transient imaging systems, we will devote more space to them in section 1.3.1.

Phase-based time-of-flight cameras: the last type of imaging system used for transient imaging is phase-based time-of-flight (P-ToF) cameras, also called correlation-based time-of-flight (C-ToF) or simply ToF cameras (Hansard et al. 2012). These are the cheapest systems of the three, and were designed for low-cost time-of-flight-based range imaging with high spatial resolution. P-ToF cameras cross-correlate emitted time-modulated light $L_i(t)$ and the sensor response $W(t)$, generally following a periodic continuous wave. This correlation encodes the impulse response of a pixel modulated at the sensor. By using a small set of measurements with different phase shifts, it makes it possible to disambiguate light time-of-flight. Beyond range imaging, Heide et al. (2013) and Kadambi et al. (2013) demonstrated that by taking several measurements with different modulation frequencies and phase shifts, it was possible to reconstruct transient images using optimization, leading to low-budget transient imaging using easy-to-operate hardware. Key to this optimization was the choice of sparse basis for modeling $T(t)$, as described in section 1.2.2.

Later work (Lin et al. 2014) demonstrated that measuring the response of cross-correlated periodic continuous waves was analogous to measuring the Fourier transform of the impulse response $T(t)$. Follow-up work (Gupta et al. 2015) used these frequency-domain properties to dramatically improve the performance of ToF sensors on range imaging. Interestingly, despite the good frequency properties of using cross-correlated continuous waves, measuring in the Fourier domain still requires measuring for a large number of frequencies to avoid aliasing. In fact, the performance of ToF sensors can be significantly improved by using temporal codes carefully designed via optimization (Gupta et al. 2018; Gutierrez-Barragan et al. 2019) or machine learning (Marco et al. 2017; Su et al. 2018). Still, the price to pay for the low cost of these systems is its lower temporal resolution and dynamic range, compared with the two alternative approaches. This limits the applicability of these sensors when imaging multiply scattered light, significantly dimmer than the direct component.

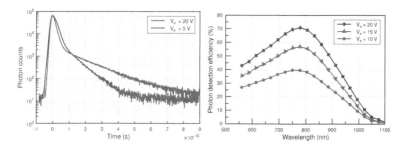

Figure 1.5. *Left: measured temporal impulse response of a 20-μm CMOS SPAD with excess voltages of 5 V and 20 V. **Right:** photon detection efficiency $E(\lambda, V_e)$ as a function of wavelengths for different excess voltages (Cova et al. 1996) (source Hernandez et al. (2017)). For a color version of this figure, see www.iste. co.uk/funatomi/computational.zip*

1.3.1. *Single-photon avalanche diodes (SPAD)*

Single-photon avalanche diodes are affordable photodetectors, which can capture extremely fast low-energy events due to their single-photon sensibility. This makes them suitable for time-of-flight-based imaging systems, without the large cost and power requirements of other direct recording systems (e.g. streak cameras), without a large loss of temporal resolution (on the order of picosecond resolution). SPADs detect signals by producing an avalanche current reaction when activated by a photon. They have been demonstrated useful in several fields including single molecule detection (Andreoni and Cubeddu 1984), quantum mechanics (Rarity and Tapster 1990) or measurements of fluorescent decays (Li and Davis 1993; Soper et al. 1993).

A SPAD detector consists of a reversed biased p-n junction (diode) above its breakdown voltage. The sensor is in a semi-stable state, in which a single photon is able to start an avalanche through the electrons inside the layers of semiconductors, generating a measurable electric current. Once the avalanche is triggered, the sensor needs to be quenched and restored to the original voltage to detect the next photon. The interval between quenching and restoration is the sensor's hold-off time t_o, in which the diode cannot detect other incoming photons. Special phenomena must be taken into account due to external and internal noise or material flaws (Renker 2006; Charbon 2007; Zappa et al. 2007). These might lead to signal degradation and are a physical bound on the temporal resolution of SPADs, as shown in Figure 1.5 (left), which shows the temporal response of a SPAD for an impulse signal. In the following, we describe the physical properties of SPADs, as well as the sources of signal degradation.

Photon detection efficiency: before detection, a photon must be absorbed by the diode, and then trigger the avalanche process. The probability of triggering the avalanche is modeled by the photon-detection efficiency E defined as

$$E(\lambda, V_e) = \eta(\lambda) \cdot P_T(V_e), \qquad\qquad [1.8]$$

where $\eta(\lambda)$ is the wavelength-dependent SPAD absorption efficiency, and $P_T(V_e)$ is the avalanche trigger probability (Savuskan et al. 2013). The absorption efficiency $\eta(\lambda)$ is related to the physical configuration of the sensor and its material properties (Zappa et al. 2007). On the contrary, the avalanche trigger probability $P_T(V_e)$ is a function of the excess voltage of the diode V_e. Figure 1.5 (right) shows different photon detection efficiency curves as a function of wavelength λ for different excess voltages V_e.

Time jitter: as photons hit the sensor, the SPAD records their time of arrival within an error margin. This error is called *time jitter*. The time jitter defines the temporal maximum resolution of the sensor (i.e. its point spread function). In general, time jitter t_j is modeled as a characteristic curve obtained with a time-correlated single-photon (TCSP) counting device (O'Connor 2012). Qualitatively, this characteristic curve has a Gaussian peak centered on the actual time of arrival, followed by an exponential tail (see Figure 1.5, left).

Quenching and pile-up distortions: once an avalanche is triggered, then the SPAD cannot detect other incoming photons. To restore the diode to operating levels, the avalanche must be quenched by lowering the bias below breakdown voltage. This process can be done passively or actively (Tisa et al. 2007). Passive circuits are relatively simple to implement on-chip, but are more prone to quenching errors which limit the temporal resolution. Active quenching, on the contrary, is implemented on more sophisticated electronics, forcing quenching in much shorter times. This is achieved by triggering a fast comparator that forces the avalanche to extinct. This makes the hold-off time t_o constant in every avalanche, making the system faster and more reliable.

A key limitation of the quenching and hold-off time t_o is that earlier photons are more likely to be detected than later ones, which results in a severe measurement skew known as *pile-up*. Unfortunately, pile-up cannot be corrected in hardware, and might result into gross time-stamping errors, orders of magnitude larger than the actual time precision of the sensor.

Noise: due to their high sensibility, SPADs suffer from errors in the form of sensor internal noise, which might trigger incorrect photon counts and therefore degrade the signal. This noise is visible as three effects mainly: dark counts, afterpulsing and crosstalk between neighbor pixels. Dark counts are photons generated by the sensor's thermal emission, which can trigger an avalanche process, resulting in a false photon count. Another source of false counts is afterpulsing, in which electrons got trapped on the depletion layer of the sensor and trigger an additional avalanche with significant delay. Finally, crosstalk between pixels in 1D and 2D SPAD arrays can also lead to avalanche interference between adjacent pixels.

1.3.1.1. *SPADs in transient imaging*

SPADs offer an affordable alternative for straight recording transient light transport via pulsed illumination, which given the high sensitivity make them very suitable even for low-power active illumination. However, they have some limitations compared to more expensive devices, especially in terms of spatio-temporal resolution. While Gariepy et al. (2015) demonstrated light-in-flight imaging using a 32×32 SPAD array, these are still not widespread. One-dimensional SPAD arrays are more common, but require using a Galvo and multiple measurements for recording a full 2D video (O'Toole et al. 2017). Still, most applications including range sensing or non-line-of-sight imaging use single-pixel SPADs, both in confocal (O'Toole et al. 2018) and non-confocal setups (Buttafava et al. 2015), requiring a large number of measurements. Large SPAD arrays would significantly reduce capture times in these applications, paving the way for real-time capture and analysis of transient imaging.

The second main limitation is temporal resolution, which is up to two orders of magnitude lower than, for example, streak cameras. This is a direct consequence of the sensor's pile-up, as well as the time jitter and other sources of noise. While these errors can be partially addressed by operating in a low-flux regime, this limits the applicability of SPADs in many in-the-wild applications. In order to fix this problem, inverse methods based on deconvolution (O'Toole et al. 2017) or statistical priors (Heide et al. 2018) have been proposed, increasing the resolution even up to sub-picosecond scale.

1.4. Applications

Access to time-resolved information of light transport has enabled a wide set of applications in computer vision and scene understanding. Novel robust solutions to long-standing problems in computer vision such as depth recovery or vision through turbid media have been proposed (Raskar and Davis 2008), while other exciting improved-vision capabilities have emerged, including material estimation, global illumination components separation, or most notable, non-line-of-sight imaging of hidden scenes. In the following, we briefly describe these applications; later, given its relevance, we devote an entire section for non-line-of-sight (NLOS) imaging (section 1.5).

1.4.1. *Range imaging*

Range imaging aims to recover the depth information from a scene, and is a key problem in computer vision and is a fundamental technology in robotics or autonomous driving. Several approaches have been proposed to tackle the problem (including stereo and multiview reconstruction or structured illumination). From these, time-of-flight-based range imaging is one of the most successful approaches. These systems emit a pulsed illumination, and measure the time (and therefore the distance) from the visible surface to the sensor. Range imaging has been the main application for transient imaging for decades, including LIDAR sensors (Mallet and Bretar 2009) or phase-based time-of-flight sensors (Hansard et al. 2012). Unfortunately, these techniques are sensitive to errors produced by **multipath interference** (MPI), where the light from multiple light paths affects the range measurement of direct illumination. This type of errors might be problematic in the case of indirect illumination from nearby geometry, but are completely pathological for imaging through turbid media, where MPI dominates. Capturing the full temporal response, beyond the single impulse one, makes it possible to disambiguate direct and indirect illumination, as well as the contribution of media. This has resulted into significantly more robust depth estimation techniques (see, for example, Wu et al. 2014; Gupta et al. 2015; Marco et al. 2017; Su et al. 2018), even for transparent objects (Kadambi et al. 2013) or in the presence of dense scattering media (Heide et al. 2014b; Raviv et al. 2014).

1.4.2. *Material estimation and classification*

Data of time-resolved light transport encodes significant information about the optical properties of matter that has interacted with light. In particular, the temporal profiles of the specific light–matter interactions describe the scattering in both surfaces and media. By analyzing the temporal profile of scattering in translucent materials, it is possible to observe the type of decay on the temporal domain, which gives direct information on the optical parameters of the media (Wu et al. 2014), and can be used to classify different materials (Su et al. 2016). Time-resolved light data is also a powerful tool for fluorescence lifetime imaging, allowing users to measure the reemission response of fluorescent materials to detect or categorize, which are important applications in medical imaging and remote sensing (see, for example, Satat et al. 2015). Finally, the cross-dimensional information transfer between temporal and angular domains (see section 1.2.1) can be used to demultiplex light paths, which has been demonstrated useful to measure and capture materials' reflectance in single-shot capture devices (Naik et al. 2011, 2014).

1.4.3. *Light transport decomposition*

As explained in section 1.2.2, transient light transport can be modeled as a sparse combination of physically meaningful bases describing different components of light transport. This decomposition is useful from a scene understanding perspective, and also allows several interesting applications directly related to range imaging or material estimation (see above). In particular, the sparse basis proposed by Wu et al. (2014) decomposed the illumination between its direct, indirect and subsurface components. A similar approach was proposed by Heide et al. (2014b) for disambiguating between surface and media scattering. An alternative approach for illumination components separation is leverage of spatially coded illumination (Nayar et al. 2006). O'Toole et al. (2014) combined this approach with transient imaging for obtaining a time-resolved direct–indirect separation, improving significantly the performance of time-of-flight sensors in scenes with complex light paths.

1.5. Non-line-of-sight imaging

Non-line-of-sight (NLOS) imaging aims to see scenes that are not directly in the line of sight of the observer, or that are partially or totally occluded. It is generally known as "looking around corners", although its range of applications goes beyond these scenarios. The key idea is to use the indirect reflections on a secondary visible wall to reconstruct the hidden scene (see Figure 1.6). There are several different approaches for NLOS imaging, including those based on passive and active illumination, or using steady-state or time-resolved data. Here, we focus on transient-based NLOS imaging, and refer to other sources for a wider view on the field (Maeda et al. 2019).

Using transient data for NLOS imaging was first proposed by Raskar and Davis (2008) and Kirmani et al. (2011), and later demonstrated by the seminal work by Velten et al. (2012a), who used a setup similar to femto-photography (Velten et al. 2013) for reconstructing hidden simple tabletop scenes. Since then, the problem has gained significant attention, with simpler and cheaper hardware (Heide et al. 2014a; Buttafava et al. 2015), faster performance (even real-time (O'Toole et al. 2018) in simple scenes) and generalizing to more complex and larger scenes (Liu et al. 2019). In the following, we first describe the most common approach based on backprojection of three-bounce paths. Then, we review alternative, more robust techniques for NLOS imaging.

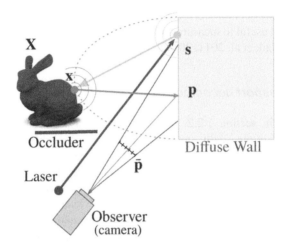

Figure 1.6. *Illustration of our reconstruction setup: a laser pulse is emitted towards a diffuse wall S (the relay wall), creating a virtual point light $s \in S$, illuminating the occluded scene. The light scattered by the occluded geometry is scattered towards a point $\mathbf{p}$ at a secondary surface P, and is imaged by the camera at pixel $\bar{\mathbf{p}}$. Usually S and P are the same surface, as shown in the figure (source: Arellano et al. (2017)). For a color version of this figure, see www.iste.co.uk/funatomi/computational.zip*

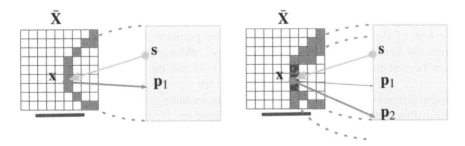

Figure 1.7. *Left: the propagation time from a point $\mathbf{x}$ at a hidden surface forms an ellipsoid with focal points s and $\mathbf{p}$. Right: the intersection of several of these ellipsoids defines a probability map for the occluded geometry (source: Arellano et al. (2017)). For a color version of this figure, see www.iste.co.uk/funatomi/computational.zip*

1.5.1. *Backprojection*

Techniques based on backprojection (Velten et al. 2012a) have been the first demonstrating NLOS imaging using time-resolved light transport information. They reconstruct a voxelization of the hidden scene $\mathbf{X}$. For that, these methods assume a relatively simple image formation model that can be inverted. By assuming that all lights at the sensor follow a three-bounce light path (see Figure 1.6) and no occlusions, the impulse response of the scene $T(\tilde{\mathbf{p}}, t|\mathbf{s})$ at pixel $\tilde{\mathbf{p}}$ for a laser shot at point $\mathbf{s}$ in the relay wall is

$$T(\tilde{\mathbf{p}}, t|\mathbf{s}) = \int_{\mathbf{X}} L(\mathbf{s} \to \mathbf{x}, t_0) f_r(\mathbf{s} \to \mathbf{x} \to \mathbf{p}) G(\mathbf{s} \to \mathbf{x} \to \mathbf{p}) d\mathbf{x}, \qquad [1.9]$$

with $\mathbf{x} \in \mathbf{X}$ being a point in the hidden scene, $\mathbf{p}$ being the point in the relay wall visible at pixel $\tilde{\mathbf{p}}$, $L(\mathbf{s} \to \mathbf{x}, t_0)$ is the reflected radiance at point $\mathbf{s}$ in the relay wall towards $\mathbf{x}$, $t_0 = t - (|\mathbf{x} - \mathbf{s}| - |\mathbf{p} - \mathbf{x}|)c^{-1}$ (note that for clarity, we ignore the distance between $\mathbf{p}$ and the camera and between $\mathbf{s}$ and the light source) and c being the speed of light, f_r is the BRDF at $\mathbf{x}$ and $G(\mathbf{s} \to \mathbf{x} \to \mathbf{p})$ is the geometric attenuation of the full path. For a single measurement $I(\tilde{\mathbf{p}}, t|\mathbf{s})$, the potential points contributing to pixel $\tilde{\mathbf{p}}$ are defined by the ellipsoid with locii at $\mathbf{p}$ and $\mathbf{s}$, and radius $t\,c$ (see Figure 1.7, left). By taking many measurements and assuming Lambertian reflectance, backprojection algorithms build a three-dimensional probability function $I(\mathbf{x})$ (Figure 1.7, right), in which the probability of point $\mathbf{x}$ being part of the occluded geometry is

$$I(\mathbf{x}) = \iiint T(\tilde{\mathbf{p}}, t|\mathbf{s}) \delta(t - (|\mathbf{x} - \mathbf{s}| - |\mathbf{p} - \mathbf{x}|)) G(\mathbf{s} \to \mathbf{x} \to \mathbf{p})^{-1} dt ds d\mathbf{p}$$

$$\approx \sum_{\tilde{\mathbf{p}}} \sum_{\mathbf{P}} T(\tilde{\mathbf{p}}, (|\mathbf{x} - \mathbf{s}| - |\mathbf{p} - \mathbf{x}|)|\mathbf{s}) G(\mathbf{s} \to \mathbf{x} \to \mathbf{p})^{-1}. \qquad [1.10]$$

This probability voxel map is later used to reconstruct the geometry by using some operator over $I(\mathbf{x})$, typically a non-linear three-dimensional edge filter such as the Laplacian filter (Velten et al. 2012a) or the Laplacian of Gaussian filter (Laurenzis and Velten 2014). Figure 1.8 shows the reconstruction of a simple hidden scene (left) using filtered backprojection (Velten et al. 2012a).

1.5.2. *Confocal NLOS and the light-cone transform*

Unfortunately, although fast backprojection implementations exist (Arellano et al. 2017), this approach scales poorly with resolution and number of measurements. To solve this issue, O'Toole et al. (2018) found that, by assuming confocal measurements (so that $\mathbf{s}$ and $\mathbf{p}$ lie at the same point), the higher-order light transport in equation [1.9] could be expressed as a convolution, using the so-called light-cone transform (LCT). The key benefit of this approach is that it makes it possible to exactly invert equation [1.9] simply as a three-dimensional deblurring problem with a known kernel.

By solving the deblurring in the Fourier domain, this allows a very fast reconstruction algorithm with low memory footprint, demonstrating even real-time performance in simple scenes. Further developments of this approach also demonstrated surface reconstruction by reconstructing normals using the so-called directional LCT (Young et al. 2020).

Figure 1.8. *Reconstruction result of hidden scene (a) consisting of a mannequin from the measurements on a visible surface, using the technique by Velten et al. (2012a). While the overall shape of the mannequin is reconstructed, it has little detail (middle) and lacks volume (right) (source: Arellano et al. (2017)). For a color version of this figure, see www.iste.co.uk/funatomi/computational.zip*

1.5.3. *Surface-based methods*

Both backprojection and the LCT reconstruct a voxelized representation (*volumetric albedo*) of the occluded scene, by attempting to invert the light transport model in equation [1.9]. Although this approach has proven effective and relatively efficient (especially the LCT), it is generally more sensitive to noise, assumes no occlusions, lacks information about the actual topology of the scene (it is limited to the voxel resolution, losing the information of the local topology, with the notable exception of the directional LCT [Young et al. 2020]) and its reflectance.

As an alternative approach, different works have proposed to optimize surfaces directly by using an analysis-by-synthesis approach (Tsai et al. 2016; Iseringhausen and Hullin 2020). In a nutshell, the idea is to use a gradient descent optimization of the hidden scene, including its geometry and/or materials. In each step of the optimization, the time-resolved response of an estimate of the hidden scene is synthesized (rendered); this simulation is then compared against the actual measurement, which makes it possible to compute the derivative of the error with respect to the parameters defining the scene. In each new step of the optimization, the parameters are updated following the gradient of the error. While potentially very powerful and flexible, these approaches suffer from the same limitations as inverse rendering methods. The optimization process is, in general, very inefficient, and there is no guarantee of finding the global minima.

An alternative approach of surface reconstruction, based on similar principles, is the so-called *Fermat paths method* (Xin et al. 2019). Fermat paths are light paths that satisfy Fermat's principle, and can be either specular paths or an object's boundaries. Based on the observation that in the transient domain the Fermat paths correspond to discontinuities in the transient measurements, and imposing constraints on the derivatives of these discontinuities, it is possible to reconstruct high-resolution complex geometry.

1.5.4. *Virtual waves and phasor fields*

While potentially very effective and efficient, methods consisting of inverting a light transport model are limited by the model itself. In most cases, this physical model is a relatively simple three-bounce light paths (equation [1.9]) and a particular reflectance model. Thus, these simplified models do not take into account multiple scattering, complex reflectances, occlusions or clutter in the hidden scene. An interesting observation is that on the line-of-sight (LOS) setting, conventional cameras do not need to invert any physical model. They are agnostic of the scene being captured, and just focus an incident field in the sensor. Virtual wave-based methods were developed following this line of thought: what if we could create a virtual camera at the visible relay wall, and transform the NLOS problem into a *virtual LOS one*? In principle, that would make it possible to image a hidden scene, without any assumptions on the underlying physical light transport model.

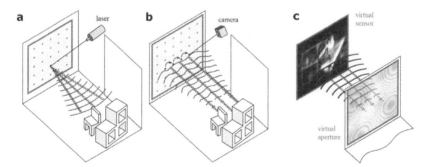

Figure 1.9. *The virtual imaging process. (a) A virtual light source at the relay wall emits a monochromatic field towards the scene, which is (b) scattered back towards the relay wall, potentially after multiple scattering events. (c) A virtual camera with aperture at the relay wall focuses the scattered field in the virtual sensor computationally, generating a sharp reconstruction of the hidden scene. For a color version of this figure, see www.iste.co.uk/funatomi/computational.zip*

To do so, Liu et al. (2019) proposed to use what they called phasor fields $\mathcal{P}(\mathbf{x}, t)$, a mathematical polychromatic wave field that behaves similarly to electromagnetic waves. This makes it possible to use well-established knowledge on Fourier optics to model the emission, propagation and focusing of these virtual waves, by using a virtual projector–camera pair system. The **virtual** imaging process works as shown in Figure 1.9: first (Figure 1.9a), a phasor field $\mathcal{P}(\mathbf{s}, t)$ is emitted from the virtual projector S towards the hidden scene. Then, by leveraging the linearity and time invariance of the impulse response function $T(\mathbf{p}, t|\mathbf{s})$, the incoming phasor field at the virtual sensor S in the relay wall is computed as (Figure 1.9b):

$$\mathcal{P}(\mathbf{p}, t) = \int_P [\mathcal{P}(\mathbf{s}, t) * T(\mathbf{p}, t|\mathbf{s})] \, d\mathbf{s}. \qquad [1.11]$$

Then, any point $\mathbf{x}$ in the hidden scene can be imaged (Figure 1.9c) by propagating $\mathcal{P}(\mathbf{p}, t)$ with an imaging operator $\Phi(\cdot)$ as

$$I(\mathbf{x}) = \Phi\left(\mathcal{P}(\mathbf{p}, t)\right). \qquad [1.12]$$

This imaging operator models a virtual lens and sensor system, and can be formulated in terms of a Rayleigh–Sommerfeld diffraction (RSD) propagator. This is the key insight of wave-based methods. As discussed by Liu et al. (2019), since the phasor field uses essentially the same propagator as electromagnetic carriers (the RSD), we can effectively design any *virtual* optical system using Fourier optics.

Thus, when propagating monochromatic phasors $\mathcal{P}_\omega(\mathbf{p}, t)$ of a single frequency ω, this image formation operator $\Phi\left(\mathcal{P}_\omega(\mathbf{p}, t)\right)$ becomes (Skolnik 2002)

$$I(\mathbf{x}) = \Phi\left(\mathcal{P}_\omega(\mathbf{p}, t)\right) \qquad [1.13]$$

where $\mathcal{L}_\omega(\cdot)$ is a complex operator that changes the phase of $\mathcal{P}_\omega(\mathbf{p}, t)$ effectively acting as a lens. Therefore, by setting the appropriate illumination and imaging operators, $\mathcal{L}_\omega(\cdot)$ and $\Phi(\cdot)$, respectively, we can define our virtual optical system. For example, a thin-lens camera focusing at $\mathbf{x}$ would be implemented as $\mathcal{L}_\omega(\mathbf{p}, \mathbf{x}) = e^{-ik|\mathbf{p}-\mathbf{x}|}$, with $k = \omega/c$ being the wave number. The supplementary material of Liu et al. (2019) (Tables S.2 and S.3) includes the exact definition of $\mathcal{L}_\omega(\cdot)$ and $\Phi(\cdot)$ for different example imaging systems.

Figure 1.10 shows a reconstruction using phasor fields of a complex hidden scene with arbitrary reflectances and significant multiple scattering, geometric complexity and clutter. Given that the phasor fields method operates as a virtual camera, no assumption is made with respect of the hidden scene. Note that phasor fields are not the only NLOS imaging method based on virtual waves. Lindell et al. (2019) proposed an alternative wave-based formulation based on the confocal capture setup obtaining results of comparable quality to phasor fields.

A very interesting consequence of phasor fields is that they make it possible to transfer most LOS imaging setups into hidden scenes by exploiting the virtual camera analogy. For example, Marco et al. (2021) proposed to extend the idea of the light transport matrix (Ng et al. 2003), which can be used for many applications including relighting (O'Toole and Kutulakos 2010) or global illumination decomposition (Nayar et al. 2006; O'Toole et al. 2012), to hidden scenes by creating virtual projector/camera pairs. However, we need to note that what is recorded by the virtual camera is in the wavelength of the phasor field $\lambda = \omega^{-1}$. This means that, as in any other imaging system, the resolution of the reconstruction is limited by the wavelength.

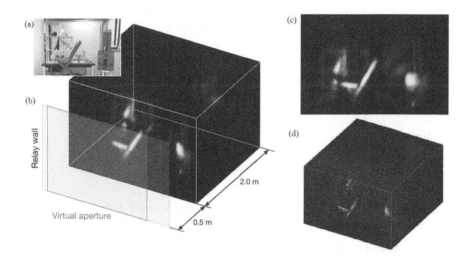

Figure 1.10. *Reconstruction result of a complex hidden scene (a) with significant geometric and reflectance complexity, multiple scattering and occlusions, using phasor fields. By using different imaging operators Φ phasor fields can image virtually three-dimensional (b) or two-dimensional (c) reconstructions of the hidden scene. In contrast, previous backprojection-based methods (d) struggle in the presence of multiple scattering and complex reflectances, which break the underlying physical model being inverted. For a color version of this figure, see www.iste.co.uk/funatomi/ computational.zip*

1.5.5. *Discussion*

We have briefly outlined some of the most notable applications of transient imaging; all of these applications are in different stages of development, but they have already proved their potential for game-changing imaging. Still, the rapid development of the field, together with the development of better (or cheaper) sensors, are likely to foster novel applications that might revolutionize the way we sense our world.

1.6. Conclusion

In this chapter, we have reviewed the emerging rapidly growing field of transient imaging. The main characteristic of this field with respect to other imaging modalities is that the assumption of infinite speed of light no longer holds, and by using ultrafast imaging devices, we can access the vast information encoded in the temporal domain of light transport. By accessing such, until now hidden, information we can easily tackle traditional problems in computer vision, while opening numerous new exciting problems of super-human vision. Along the chapter, we have given a historical perspective of ultrafast imaging, as well as discussed some of the most important mathematical properties of transient light transport. We have also categorized the different technologies proposed to capture transient imaging, with focus on the now-affordable SPADs, which are the most promising imaging technology for widespread transient imaging. Finally, we have discussed some of the emerging applications in the field, giving hints towards the most promising avenues of future work.

1.7. References

Abramson, N. (1978). Light-in-flight recording by holography. *Optics Letters*, 3(4), 121–123.

Abramson, N. (1983). Light-in-flight recording: High-speed holographic motion pictures of ultrafast phenomena. *Applied Optics*, 22(2), 215–232.

Andreoni, A. and Cubeddu, R. (1984). Photophysical properties of photofrin II in different solvents. *Chemical Physics Letters*, 108(2), 141–144 [Online]. Available at: http://www.sciencedirect.com/science/article/pii/0009261484857085.

Arellano, V., Gutierrez, D., Jarabo, A. (2017). Fast back-projection for non-line of sight reconstruction. *Optics Express*, 25(10), 11574–11583.

Bedi, J. and Collins, D. (1994). *Seeing the Unseen: Dr. Harold E. Edgerton and the Wonders of Strobe Alley*. MIT Press, Cambridge, MA.

Busck, J. and Heiselberg, H. (2004). Gated viewing and high-accuracy three-dimensional laser radar. *Applied Optics*, 43(24), 4705–4710.

Buttafava, M., Zeman, J., Tosi, A., Eliceiri, K., Velten, A. (2015). Non-line-of-sight imaging using a time-gated single photon avalanche diode. *Optics Express*, 23(16), 20997–21011.

Charbon, E. (2007). Will avalanche photodiode arrays ever reach 1 megapixel? *International Image Sensor Workshop*.

Cova, S., Ghioni, M., Lacaita, A., Samori, C., Zappa, F. (1996). Avalanche photodiodes and quenching circuits for single-photon detection. *Applied Optics*, 35(12), 1956–1976 [Online]. Available at: http://ao.osa.org/abstract.cfm?URI= ao-35-12-1956.

D'Eon, E. and Irving, G. (2011). A quantized-diffusion model for rendering translucent materials. *ACM Transactions on Graphics*, 30(4), 56:1–4.

Freedman, D., Smolin, Y., Krupka, E., Leichter, I., Schmidt, M. (2014). SRA: Fast removal of general multipath for ToF sensors. *European Conference on Computer Vision*.

Gariepy, G., Krstajić, N., Henderson, R., Li, C., Thomson, R.R., Buller, G.S., Heshmat, B., Raskar, R., Leach, J., Faccio, D. (2015). Single-photon sensitive light-in-fight imaging. *Nature Communications*, 6, 6021.

Gkioulekas, I., Levin, A., Durand, F., Zickler, T. (2015). Micron-scale light transport decomposition using interferometry. *ACM Transactions on Graphics*, 34(4), 1–14.

Gupta, M., Nayar, S.K., Hullin, M.B., Martin, J. (2015). Phasor imaging: A generalization of correlation-based time-of-flight imaging. *ACM Transactions on Graphics*, 34(5), 156:1–18.

Gupta, M., Velten, A., Nayar, S.K., Breitbach, E. (2018). What are optimal coding functions for time-of-flight imaging? *ACM Transactions on Graphics (TOG)*, 37(2), 13:1–18.

Gutierrez-Barragan, F., Reza, S.A., Velten, A., Gupta, M. (2019). Practical coding function design for time-of-flight imaging. *IEEE Computer Vision and Pattern Recognition*, pp. 1566–1574.

Hansard, M., Lee, S., Choi, O., Horaud, R.P. (2012). *Time-of-Flight Cameras: Principles, Methods and Applications*. Springer Science & Business Media, London.

Heide, F., Hullin, M., Gregson, J., Heidrich, W. (2013). Low-budget transient imaging using photonic mixer devices. *ACM Transactions on Graphics*, 32(4), 1–14.

Heide, F., Xiao, L., Heidrich, W., Hullin, M.B. (2014a). Diffuse mirrors: 3D reconstruction from diffuse indirect illumination using inexpensive time-of-flight sensors. *IEEE Computer Vision and Pattern Recognition*.

Heide, F., Xiao, L., Kolb, A., Hullin, M.B., Heidrich, W. (2014b). Imaging in scattering media using correlation image sensors and sparse convolutional coding. *Optics Express*, 22(21), 26338.

Heide, F., Diamond, S., Lindell, D.B., Wetzstein, G. (2018). Sub-picosecond photon-efficient 3D imaging using single-photon sensors. *Scientific Reports*, 8(1), 1–8.

Hernandez, Q., Gutierrez, D., Jarabo, A. (2017). A computational model of a single-photon avalanche diode sensor for transient imaging. *arXiv:1703.02635vi.*

Iseringhausen, J. and Hullin, M.B. (2020). Non-line-of-sight reconstruction using efficient transient rendering. *ACM Transactions on Graphics*, 39(1), 1–14.

Jarabo, A., Marco, J., Muñoz, A., Buisan, R., Jarosz, W., Gutierrez, D. (2014). A framework for transient rendering. *ACM Transactions on Graphics*, 33(6), 177:1–10.

Jarabo, A., Masia, B., Marco, J., Gutierrez, D. (2017). Recent advances in transient imaging: A computer graphics and vision perspective. *Visual Informatics*, 1(1), 65–79.

Kadambi, A., Whyte, R., Bhandari, A., Streeter, L., Barsi, C., Dorrington, A., Raskar, R. (2013). Coded time of flight cameras: Sparse deconvolution to address multipath interference and recover time profiles. *ACM Transactions on Graphics*, 32(6), 1–10.

Kirmani, A., Hutchison, T., Davis, J., Raskar, R. (2011). Looking around the corner using ultrafast transient imaging. *International Journal of Computer Vision*, 95(1), 13–28.

Kotwal, A., Levin, A., Gkioulekas, I. (2020). Interferometric transmission probing with coded mutual intensity. *ACM Transactions on Graphics (TOG)*, 39(4), 74:1–74:16.

Laurenzis, M. and Velten, A. (2014). Nonline-of-sight laser gated viewing of scattered photons. *Optical Engineering*, 53(2), 023102.

Lei, C., Guo, B., Cheng, Z., Goda, K. (2016). Optical time-stretch imaging: Principles and applications. *Applied Physics Reviews*, 3(1), 011102.

Leslie, M. (2001). The man who stopped time. *Stanford Magazine.*

Li, L. and Davis, L.M. (1993). Single photon avalanche diode for single molecule detection. *Review of Scientific Instruments*, 64(6), 1524–1529 [Online]. Available at: http://scitation.aip.org/content/aip/journal/rsi/64/6/10.1063/1.1144463.

Liang, Y., Chen, M., Huang, Z., Gutierrez, D., Muñoz, A., Marco, J. (2020). Compression and denoising of time-resolved light transport. *Optics Letters*, 45(7), 1986–1989.

Lin, J., Liu, Y., Hullin, M.B., Dai, Q. (2014). Fourier analysis on transient imaging with a multifrequency time-of-flight camera. *IEEE Computer Vision and Pattern Recognition.*

Lindell, D.B., Wetzstein, G., O'Toole, M. (2019). Wave-based non-line-of-sight imaging using fast fk migration. *ACM Transactions on Graphics*, 38(4), 1–13.

Liu, X., Guillén, I., La Manna, M., Nam, J.H., Reza, S.A., Le, T.H., Jarabo, A., Gutierrez, D., Velten, A. (2019). Non-line-of-sight imaging using phasor-field virtual wave optics. *Nature*, 572(7771), 620–623.

Lumière, L. (1936). The lumière cinematograph. *Journal of the Society of Motion Picture Engineers*, 27(6), 640–647.

Maeda, T., Satat, G., Swedish, T., Sinha, L., Raskar, R. (2019). Recent advances in imaging around corners. *arXiv:1910.05613*.

Mallet, C. and Bretar, F. (2009). Full-waveform topographic lidar: State-of-the-art. *ISPRS Journal of Photogrammetry and Remote Sensing*, 64(1), 1–16.

Marco, J., Hernandez, Q., Munoz, A., Dong, Y., Jarabo, A., Kim, M.H., Tong, X., Gutierrez, D. (2017). Deeptof: Off-the-shelf real-time correction of multipath interference in time-of-flight imaging. *ACM Transactions on Graphics (ToG)*, 36(6), 1–12.

Marco, J., Jarabo, A., Nam, J.H., Liu, X., Cosculluela, M.A., Velten, A., Gutierrez, D. (2021). Virtual light transport matrices for non-line-of-sight imaging. *Proceedings of the IEEE/CVF International Conference on Computer Vision (ICCV)*, pp. 2440–2449.

Naik, N., Zhao, S., Velten, A., Raskar, R., Bala, K. (2011). Single view reflectance capture using multiplexed scattering and time-of-flight imaging. *ACM Transactions on Graphics*, 30.

Naik, N., Barsi, C., Velten, A., Raskar, R. (2014). Estimating wide-angle, spatially varying reflectance using time-resolved inversion of backscattered light. *JOSA A*, 31(5), 957–963.

Nakagawa, K., Iwasaki, A., Oishi, Y., Horisaki, R., Tsukamoto, A., Nakamura, A., Hirosawa, K., Liao, H., Ushida, T., Goda, K. et al. (2014). Sequentially timed all-optical mapping photography (STAMP). *Nature Photonics*, 8(9), 695–700.

Nayar, S.K., Krishnan, G., Grossberg, M.D., Raskar, R. (2006). Fast separation of direct and global components of a scene using high frequency illumination. *ACM Transactions on Graphics*, 25(3).

Ng, R., Ramamoorthi, R., Hanrahan, P. (2003). All-frequency shadows using non-linear wavelet lighting approximation. *ACM Transactions on Graphics*, 22(3).

O'Connor, D. (2012). *Time-Correlated Single Photon Counting*. Academic Press, London.

O'Toole, M. and Kutulakos, K.N. (2010). Optical computing for fast light transport analysis. *ACM Transactions on Graphics*, 29(6), 164.

O'Toole, M., Raskar, R., Kutulakos, K.N. (2012). Primal-dual coding to probe light transport. *ACM Transactions on Graphics*, 31(4).

O'Toole, M., Heide, F., Xiao, L., Hullin, M.B., Heidrich, W., Kutulakos, K.N. (2014). Temporal frequency probing for 5D transient analysis of global light transport. *ACM Transactions on Graphics*, 33(4).

O'Toole, M., Heide, F., Lindell, D.B., Zang, K., Diamond, S., Wetzstein, G. (2017). Reconstructing transient images from single-photon sensors. *Proceedings of the IEEE Conference on Computer Vision and Pattern Recognition*, pp. 1539–1547.

O'Toole, M., Lindell, D.B., Wetzstein, G. (2018). Confocal non-line-of-sight imaging based on the light-cone transform. *Nature*, 555(7696), 338–341.

Peters, C., Klein, J., Hullin, M.B., Klein, R. (2015). Solving trigonometric moment problems for fast transient imaging. *ACM Transactions on Graphics*, 34(6).

Qiao, H., Lin, J., Liu, Y., Hullin, M.B., Dai, Q. (2015). Resolving transient time profile in ToF imaging via log-sum sparse regularization. *Optics Letters*, 40(6), 918–921.

Rarity, J.G. and Tapster, P.R. (1990). Experimental violation of Bell's inequality based on phase and momentum. *Physical Review Letters*, 64, 2495–2498 [Online]. Available at: http://link.aps.org/doi/10.1103/PhysRevLett.64.2495.

Raskar, R. and Davis, J. (2008). 5D time-light transport matrix: What can we reason about scene properties? Technical report, MIT [Online]. Available at: http://hdl.handle.net/1721.1/67888.

Raviv, D., Barsi, C., Naik, N., Feigin, M., Raskar, R. (2014). Pose estimation using time-resolved inversion of diffuse light. *Optics Express*, 22(17), 20164–20176.

Renker, D. (2006). Geiger-mode avalanche photodiodes, history, properties and problems. *Nuclear Instruments and Methods in Physics Research Section A: Accelerators, Spectrometers, Detectors and Associated Equipment*, 567(1), 48–56.

Satat, G., Heshmat, B., Barsi, C., Raviv, D., Chen, O., Bawendi, M.G., Raskar, R. (2015). Locating and classifying fluorescent tags behind turbid layers using time-resolved inversion. *Nature Communications*, 6, 1–8.

Savuskan, V., Brouk, I., Javitt, M., Nemirovsky, Y. (2013). An estimation of single photon avalanche diode (SPAD) photon detection efficiency (PDE) nonuniformity. *IEEE Sensors Journal*, 13(5), 1637–1640.

Skolnik, M. (2002). *Introduction to Radar Systems*, 3rd edition. McGraw-Hill Education, Maidenhead.

Soper, S.A., Mattingly, Q.L., Vegunta, P. (1993). Photon burst detection of single near-infrared fluorescent molecules. *Analytical Chemistry*, 65(6), 740–747 [Online]. Available at: http://dx.doi.org/10.1021/ac00054a015.

Su, S., Heide, F., Swanson, R., Klein, J., Callenberg, C., Hullin, M., Heidrich, W. (2016). Material classification using raw time-of-flight measurements. *IEEE Computer Vision and Pattern Recognition*.

Su, S., Heide, F., Wetzstein, G., Heidrich, W. (2018). Deep end-to-end time-of-flight imaging. *Proceedings of the IEEE Conference on Computer Vision and Pattern Recognition*, pp. 6383–6392.

Tisa, S., Zappa, F., Tosi, A., Cova, S. (2007). Electronics for single photon avalanche diode arrays. *Sensors and Actuators A: Physical*, 140(1), 113–122.

Tsai, C.-Y., Veeraraghavan, A., Sankaranarayanan, A.C. (2016). Shape and reflectance from two-bounce light transients. *IEEE International Conference on Computational Photography*.

Velten, A., Willwacher, T., Gupta, O., Veeraraghavan, A., Bawendi, M.G., Raskar, R. (2012a). Recovering three-dimensional shape around a corner using ultrafast time-of-flight imaging. *Nature Communications*, 3(1), 1–8.

Velten, A., Wu, D., Jarabo, A., Masia, B., Barsi, C., Lawson, E., Joshi, C., Gutierrez, D., Bawendi, M.G., Raskar, R. (2012b). Relativistic ultrafast rendering using time-of-flight imaging. *ACM SIGGRAPH 2012 Talks*.

Velten, A. et al. (2013a). Femto-photography – project page [Online]. Available at: http://giga.cps.unizar.es/~ajarabo/pubs/femtoSIG2013/.

Velten, A., Wu, D., Jarabo, A., Masia, B., Barsi, C., Joshi, C., Lawson, E., Bawendi, M., Gutierrez, D., Raskar, R. (2013b). Femto-photography: Capturing and visualizing the propagation of light. *ACM Transactions on Graphics*, 32(4), 44:1–8.

Wu, D., Wetzstein, G., Barsi, C., Willwacher, T., O'Toole, M., Naik, N., Dai, Q., Kutulakos, K., Raskar, R. (2012). Frequency analysis of transient light transport with applications in bare sensor imaging. *European Conference on Computer Vision*.

Wu, D., Velten, A., O'Toole, M., Masia, B., Agrawal, A., Dai, Q., Raskar, R. (2014). Decomposing global light transport using time of flight imaging, *International Journal of Computer Vision*, 107(2), 123–138.

Xin, S., Nousias, S., Kutulakos, K.N., Sankaranarayanan, A.C., Narasimhan, S.G., Gkioulekas, I. (2019). A theory of Fermat paths for non-line-of-sight shape reconstruction. *IEEE Computer Vision and Pattern Recognition (CVPR)*, pp. 6800–6809.

Young, S.I., Lindell, D.B., Girod, B., Taubman, D., Wetzstein, G. (2020). Non-line-of-sight surface reconstruction using the directional light-cone transform. *Proc. CVPR*.

Zappa, F., Tisa, S., Tosi, A., Cova, S. (2007). Principles and features of single-photon avalanche diode arrays. *Sensors and Actuators A: Physical*, 140(1), 103–112.

2

Transient Convolutional Imaging

Felix HEIDE

Princeton Computational Imaging Lab, Princeton University, USA

2.1. Introduction

Inspired by biological sonar, optical time-of-flight sensors measure the delay between light emitted into a scene and received by a co-located sensor. It was the invention of the laser in 1960, as well as accurate timing electronics, that allowed for a commercially successful application of this principle in Lidar systems. While delivering high accuracy and range, a major drawback is that distance is only measured to a single point, and hence, scanning is required. Recently, correlation image sensors (Lange 2000) have been revolutionizing the 3D depth imaging market by temporally convolving a scene with amplitude-modulated flood illumination, which no longer requires scanning. Correlation image sensors convolve an incoming temporal signal with a reference signal. From this convolutional measurement, the time-of-flight can be estimated with high accuracy. As such time-of-flight systems do not require a baseline or lack of textured objects, they outperform structured light or stereo matching techniques by a significant margin for scenes with large range. Correlation image sensor technology has been developed since the early 2000s, and adoption has been rapidly accelerated by sensor fabrication in the CMOS process. CMOS technology has been driven by the microprocessor industry and is therefore inexpensively available for mass-market production.

Adding computation to these convolutional correlation measurements, in the form of Bayesian inference, allows us to overcome limitations of existing conventional

Computational Imaging for Scene Understanding,
coordinated by Takuya FUNATOMI and Takahiro OKABE.
© ISTE Ltd 2024.

imaging systems. Relying on temporally and spatially convolutional structure in correlation measurements, we extract a novel image modality that was essentially "invisible" before: a new temporal, transient dimension of light propagation, obtained using consumer correlation time-of-flight cameras.

2.2. Time-of-flight imaging

Typical time-of-flight (ToF) range imaging systems use either impulse modulation or continuous wave modulation. In impulse ToF imaging, including classical Lidar range imaging, a pulse of light is emitted into the scene and an accurate timer is started synchronously. The light pulse travels from the illumination source to the object and is directly reflected back to the sensor, assuming direct reflection at the object surface. Once the sensor detects the reflected light pulse, the timer is stopped, measuring the round-trip time τ. Given the speed of light $c = 299,792,458$ m/s and assuming collocated sensor and illumination, we can then calculate the depth d of the object point reflecting the pulse as $d = c\tau/2$.

Impulse time-of-flight imaging: impulse illumination has the advantage that a high amount of energy is transmitted in a short time window. Therefore, a high signal-to-noise ratio (SNR) is achieved with regard to ambient background illumination (Lange 2000), while the mean optical power stays at low eye-safe levels. The high SNR allows for high-accuracy measurements over a wide range of distances. A further advantage is that multi-path interference (MPI) can be reduced simply by detection of the first peak and temporal gating.

The drawbacks of impulse illumination are that due to varying ambient illumination, attenuation in scattering volumes and varying scene depth, both the peak intensity of the pulse and the ambient illumination offset change. Thus, the sensors require a large dynamic range to detect pulses, where the detection is usually implemented by thresholding. Most impulse-illuminated systems are either point-scanning or line-scanning systems because of the challenges in developing low-cost high-power laser diodes and the difficulty in implementing high-accuracy pixel-level timers. To illustrate the required precision, note that for a distance resolution of 1 mm, light pulses need to be separated with a time resolution of 6.6 ps. Only recently, advances in implementing large arrays of single-photon avalanche detectors (SPADs) in standard CMOS enable efficient scannerless pulsed range imaging. However, SPADs implemented in silicon technology at room temperature are not able to reach the timing resolutions necessary for sub-mm resolution. Therefore, either simple averaging over many periodic pulse trains in time-correlated single-photon counting (TCSPC) or continuous wave modulation schemes are adopted to improve the timing uncertainty.

Continuous wave time-of-flight imaging: in continuous wave ToF imaging, rather than using impulse illumination, the light source is modulated. The electromagnetic waves present in the illumination itself, for example, in the case of laser illumination, only with a single wavelength, represent the carrier wave that gets modulated. Different modulation methods exist. While frequency modulation for ToF imaging has been explored, by far the most popular modulation technique is amplitude modulation due to its comparatively simple hardware implementation. Using amplitude-modulated continuous wave (AMCW) illumination rather than pulsed illumination has the advantage that the light source does not need to provide a short pulse-width at high rates; thus, in practice, inexpensive LED illumination can be used (Lange 2000). Furthermore, accurate per-pixel timing is not necessary for correlation sensors or time-gated photon counters. As the fundamental principle of these sensors is based on temporal correlation (or convolution), we dub these sensors *Correlation Image Sensors* in the following.

In correlation range, imaging the intensity of the light source is modulated with a periodic modulation signal. Rather than directly measuring the ToF, as in impulse ToF imaging, the phase difference θ between a reference signal (often the transmitted signal) and the received signal is measured. θ can be estimated by cross-correlation between the reference and received signal. Given the modulation frequency f of the periodic illumination, it is $\theta = 2\pi f\tau$, and thus, the depth is $d = c\tau/2 = c/(4\pi f)\theta$. One of the first correlation image sensors was introduced in Lange (2000).

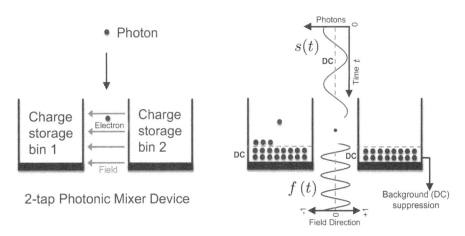

Figure 2.1. *Working principle of a PMD pixel. Photo-generated charges are directed towards two bins using an electric field between two electrodes, i.e. the CCD transport principle (left). The intensity g of the light source and the electric field f is modulated during integration (right). The sensor correlates the signal g with f. Unmodulated light is equally accumulated in both buckets and can be removed during or after integration. For a color version of this figure, see www.iste.co.uk/funatomi/computational.zip*

2.2.1. *Correlation image sensors*

Sensors for correlation imaging are based on two working principles: correlation during or after exposure. The most popular class of sensors performs the correlation of the incoming modulated signal *during exposure*. This is done in photonic mixer device (PMD) pixels that direct the photo-generated charge towards multiple taps (commonly two), as illustrated in Figure 2.1. The charge accumulation and transport are achieved using the CCD principle (Boyle and Smith 1970). The main difference from a common CCD image sensor is that the charge transfer between the group of accumulation sites occurs during the exposure. The direction of the transfer can be modeled as a scalar between -1 (maximum field strength towards the left bin) and +1 (fully towards the right bin) according to a temporally varying modulation function f. The left bin corresponds then to negative values, while the right models correspond to positive values. Assuming a periodic modulation function, integrating over a large number of periods (ca. 10^4–10^5) means that the sensor essentially *correlates* (or convolves) an illumination function g with the sensor modulation function f; see the right side of Figure 2.1. The correlation value is the subtraction of the left from the right (as we assigned negative values to the left bin). Using a zero-mean sensor modulation function f, unmodulated ambient light is removed in this subtraction as a constant bin offset after or even during integration. This is also illustrated on the right side of Figure 2.1. An alternative approach is to do the correlation *after exposure* on the electronic signal generated by the photosensitive pixel during measurement. The correlation can be done with analog electronics using standard photodiodes as sensors. However, this sensing approach requires complex electronics per pixel, has high-power consumption and suffers from noise in the signal processing. Using emerging SPAD sensors in the Geiger mode, the correlation can be performed in the digital domain after read-out.

2.2.2. *Convolutional ToF depth imaging*

Depth ToF imaging commonly assumes the *direct illumination* of a *single object point*, illustrated in the top of Figure 2.2. Here, the correlation ToF camera measures the time-of-flight along a single, direct light path connecting source, reflector and sensor. Modeling illumination amplitude as a periodic, intensity-modulated signal g_ω with frequency ω, we obtain the incident signal

$$s_\omega(t) = I + \alpha a \cdot g_\omega(t + \tau). \tag{2.1}$$

where I is the DC component of the light source plus any ambient illumination, a is the modulation amplitude for the light source, α is an attenuation term due to surface reflectance and distance-based intensity fall-off, and τ is the ToF from the light source to the object point and then to a given correlation pixel.

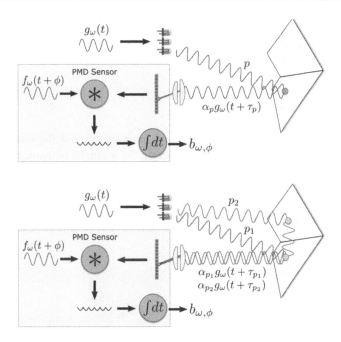

Figure 2.2. *Top: Operating principle of a conventional correlation PMD sensor. Light from a modulated light source arrives at a pixel sensor via a single light path p with time delay τ. The PMD sensor modulates the incident light with a reference signal f_ω and integrates the resulting modulated exposure to obtain a distance-dependent correlation. Bottom: in the presence of global illumination, the incident illumination is a superposition of light paths, for example path p_1 and p_2, with different phase shifts. For a color version of this figure, see www.iste.co.uk/funatomi/computational.zip*

On the correlation image sensor pixel, the incident signal s_ω is correlated with the periodic reference f_ω with the identical frequency ω, assuming a homodyne measurement setup, and a programmable temporal offset ϕ. The modulated exposure $b_{\omega,\phi}$ measured by the sensor for a given frequency/phase pair for a number of periods N becomes

$$b_{\omega,\phi} = N \int_0^T s_\omega(t) f_\omega(t+\phi) \, dt = \alpha a \cdot \underbrace{N \int_0^T g_\omega(t+\tau) f_\omega(t+\phi) \, dt}_{=c_{\omega,\phi}(\tau)}, \quad [2.2]$$

with a correlation coefficient c introduced for the attenuation and intensity-independent factors. Note that the offset I disappears as the sensor reference function is zero-mean.

These correlation coefficients can either be analytically determined for specific f_ω, g_ω, such as sinusoids, or they can be calibrated using objects at different known distances d using the relationship $\tau = \omega/c \cdot d$, where c is here the speed of light. For traditional range imaging, a minimum of two measurements $b_{\omega,0/\omega}$ and $b_{\omega,\pi/(2\omega)}$ are obtained for per-pixel range estimation. Assuming perfect sinusoidal modulation, the correlation coefficients become cosines

$$c_{\omega,\phi} = \frac{1}{2}a \, \cos((\tau - \phi)\omega)$$ [2.3]

and it is then possible to solve for the pixel intensity αa and the time-of-flight τ of the object visible at that pixel. In general, for a single frequency, correlation cameras acquire multiple measurements b_{ω_i,ϕ_i} with different reference offsets $\phi_i = i \cdot 2\pi/(N\omega)$ for $i \in \{0, \ldots, N - 1\}$, effectively performing a frequency analysis of b by using the discrete Fourier transform. Having finally computed the estimate τ_{est} of τ, the distance d can be computed from the time of flight using $d_{est} = \frac{c}{2} \cdot \tau_{est}$.

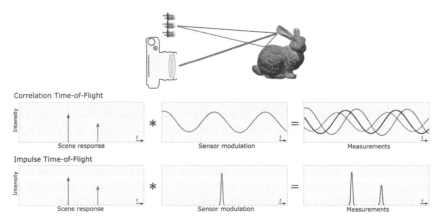

Figure 2.3. *In the scene on the top, contributions from direct and indirect light paths (direct blue and indirect green) are superimposed in the measurement. In correlation ToF imaging, the sinusoids for each path result in another measured sinusoid of a different phase (black), indistinguishable of a single reflector at a different depth. Impulse ToF can separate the components, in principle. However, the sensor requires extremely high-temporal resolution and almost continuous read-out. For a color version of this figure, see www.iste.co.uk/funatomi/computational.zip*

2.2.3. *Multi-path interference*

Above, we assumed a single reflector in the scene, and hence light transport only along a single, direct path. In non-trivial scenes, however, light can be scattered many times at different scene points until returned to the sensor, and thus, multiple returns along different paths can be superimposed at the sensor (see Figure 2.3). Resolving this MPI problem is critical in correlation ToF imaging for two reasons. First, the scene is flood-illuminated, creating significantly more potential MPI than with a scanned focused beam illumination. Second, when sine modulation is used, multiple superimposed sinusoids of the same frequency result in another sinusoid with the same frequency but different phase.

This means that even for a very low number of mixed paths, resolving MPI is hopeless with a single frequency. In contrast, impulse ToF can untangle a low number of separated paths using simple thresholding.

2.3. Transient convolutional imaging

Essentially, it is global light transport, i.e. light paths with multiple scattering events at non-local points in the scene, which causes the MPI (and is ignored in direct ToF). Once an illumination source is turned on, this global light transport does not occur in an instantaneous process, but it has a temporal dimension, because the speed of light (propagation) is finite. Transient imaging is a new imaging modality that captures this temporal dimension, thereby untangling all path contributions that are mixed together in conventional imaging systems. To illustrate this, imagine an illumination source emitting a short pulse of light. If we had a camera with ultra-high-temporal resolution, these pulses could be observed "in flight" as they traverse a scene and before the light distribution achieves a global equilibrium. A *transient image* is then the rapid sequence of frames capturing this scene response, hence representing the *impulse response of a scene*. As light transport is a linear time-invariant system that is fully described by its impulse response, such a transient image fully describes the light transport in the scene. This illustrates the potential of this new image modality.

The original concept of transient imaging goes back to work performed in the late 1970s by Abramson (1978) under the name "light-in-flight recording". Abramson created holographic recordings of scenes illuminated by picosecond lasers, from which it was possible to optically reconstruct an image of the wavefront at a specific time. However, the complexity of the captured scenes was severely limited by technical constraints of the holographic setup. Recently, interest in transient imaging has been rekindled by the development of femtosecond and picosecond lasers, allowing for ultra-short impulse illumination, as well as ultra-fast camera technologies. Velten et al. (2011) were the first to combine both technologies to capture transient images, which allows for a simplified setup compared to the holographic approach, as well as significantly more general scene geometries. Transient imaging has many exciting applications. Starting with the pilot experiments by Kirmani et al. (2009), several research groups have proposed using transient images as a means of reconstructing 3D geometry that is not directly visible to either the camera or the light sources (Velten et al. 2012), to capture surface reflectance, to perform lens-less imaging or to visualize light transport in complex environments to gain a better understanding of optical phenomena.

Figure 2.4. *Left: our capture setup for transient images (from left: computer, signal generator, power supply, modulated light source and correlation camera). Center: a disco ball with many mirrored facets. Right: the same sphere as seen by our transient imager when illuminated from the left, colored according to the time offset of the main intensity peak. For a color version of this figure, see www.iste.co.uk/funatomi/ computational.zip*

Unfortunately, impulse transient imaging relies on expensive custom hardware, i.e. a femtosecond laser as a light source, and a streak camera for the image capture as in Velten et al. (2011). While capturing high-resolution transient images, these components amount to costly equipment – i.e. multiple hundreds of thousands of dollars – that is bulky, extremely sensitive, difficult to operate, potentially dangerous to the eye and slow. For example, a streak camera measures only a single scanline of a transient image in each measurement. To obtain a full transient image, it is, therefore, necessary to mechanically scan the scene. Due to the very limited amount of light in a femtosecond pulse, averaging multiple measurements, complicated calibration and noise suppression algorithms are required to obtain good image quality. Hence, capture times of an hour or more have been reported for a single transient image (Velten et al. 2011).

To enable fast, simple and inexpensive transient imaging, we replace direct, impulse-based acquisition methods with a correlation ToF camera plus computation. We build on the convolutional correlation image sensor model from the previous sections to derive the relationship between transient imaging and traditional ToF imaging using PMDs. The recovery of transient images from correlation measurements is then formulated as an inverse problem solved computationally post-capture. Relying on this novel reconstruction method, we capture transient images using a consumer PMD camera, only modified to allow for the acquisition of a sweep of different modulation frequencies and phases. The system is inexpensive, portable, eye-safe and insensitive to background light, and acquires data at a much higher rate than streak cameras, enabling transient imaging even outside laboratory environments.

2.3.1. *Global convolutional transport*

We now relax the requirement from before that light only travels along a single light path and develop a full global illumination model instead, illustrated in Figure 2.2 on the bottom. We encapsulate all path components of the same length in I, which is a pixel of the transient image, i.e.

$$I(\tau) = \int_{\mathcal{P}} \delta(|p| - \tau)\alpha_p \, dp, \qquad\qquad [2.4]$$

with $\mathcal{P}$ being the space of all ray paths connecting camera and light source. The factors α_p and $|p|$ are the light attenuation and the travel time along such a path p, respectively. The delta function makes sure that path contributions with the same travel time τ are integrated for a given time offset τ. Note that we model here measurements for a single pixel, while a regular grid of transient pixels is a *transient image*; hence, it is three-dimensional, with two spatial and one time coordinate. Its definition immediately reveals that a transient image is the impulse response of a scene, i.e. the intensity profile as a function of time when the scene is illuminated by a short pulse of light. With this transient definition in hand, we can next model the measurement in the presence of global transport as

$$b_{\omega_i,\phi_i} = \int_0^\infty \int_0^{NT} (I + I(\tau) \cdot ag_{\omega_i}(t + \tau)) \, f_\omega(t + \phi_i) \, dt \, d\tau. \qquad [2.5]$$

Recognizing that I does not depend on the integration variable t, we can reformulate it using the correlation coefficients from section 2.2.2 and Figure 2.2, i.e.

$$b_{\omega_i,\phi_i} = a \int_0^\infty I(\tau)c_{\omega_i,\phi_i}(\tau)d\tau. \qquad\qquad [2.6]$$

Hence, the convolutional model allows for separating the transient component from the correlation on the correlation sensor. Discretizing τ in m temporal bins $\{\tau_1, \ldots, \tau_m\}$ results in the linear model

$$\mathbf{b} = \mathbf{Ci} \quad \text{with} \quad \mathbf{C}_{i,j} = c_{\omega_i,\phi_i}(\tau_j), \qquad\qquad [2.7]$$

and the transient image being $\mathbf{i} \in \mathbb{R}^m$. The measurement $\mathbf{b} \in \mathbb{R}^n$ denotes the vector of all n modulated exposures observed at all pixels in the image. The correlation coefficients c_{ω_j,ϕ_j} are arranged in a matrix $\mathbf{C}$ that maps a transient image to the vector of observations $\mathbf{b}$.

2.3.2. *Transient imaging using correlation image sensors*

Unfortunately, the transform $\mathbf{C}$ in the linear image formation model is ill-conditioned. This becomes obvious when approximating the modulation wavefronts with sinusoidal waveforms, resulting in the coefficient matrix becoming a truncated Fourier transform. Its maximum frequency is limited by the sensor and light source and usually lies around 100 MHz. This means the correlation coefficients vary slowly with distance for the range of feasible modulation frequencies and scene scales. Due to the ill-conditioned, truncated coefficient matrix and the measurement noise, the inverse problem of recovering a transient image i from measurements b is ill-posed; hence, we cannot directly solve for i as a direct solve of the linear system.

Following a Bayesian approach, we solve this inverse problem as a maximum a posteriori (MAP) estimation, relying on a spatial gradient prior and a temporal sparsity prior. Specifically, we assume that the spatial and temporal gradients are assumed to follow a Laplacian distribution, encouraging outliers at object boundaries which are sparsely distributed in the spatial dimension (for each individual time slice), and resulting in sparse reflection behavior along the temporal dimension. The resulting MAP problem is a convex optimization problem that we efficiently solve with a proximal algorithm (Chambolle and Pock 2011), which is equivalent to a linearized version of the alternate direction method of multipliers (ADMMs) (Parikh and Boyd 2013).

Our experimental acquisition setup is shown in Figure 2.4. We use a PMD Technologies CamBoard nano platform that is equipped with a PMD PhotonICs 19k-S3 image sensor. Since our technique is based on the use of a wide range of modulation frequencies, we modified the camera board and intercepted the on-board modulation signal and replaced it with our own input from an external source and added a trigger output that signals the start of the integration phase. As a light source, we use a bank of six 650 nm laser diodes with a total average output power of 2.4 W. Note that these modifications do not increase the component or manufacturing cost substantially, and are merely vertical replacements of existing illumination components and signal control circuits.

We captured various scenes with characteristic light transport. The individual time slices of three additional data sets are shown in Figure 2.5. The left column of Figure 2.5 shows a wavefront propagating through a scene with a mirrored disco ball placed in the corner of a room. In the first frame, the wavefront has just reached the front of the ball. In the second frame, the ball is now fully illuminated, and we see the wavefront propagating along the left wall. The third frame shows the first caustics generated by reflections in the mirror. More caustics appear for longer light paths near the top and bottom of the fourth image, and the direct illumination is now approaching the back of the corner from both sides. First, indirect illumination in the floor is visible. In the last frame, the caustics have disappeared, and the indirect illumination is now lighting up the shadow of the ball.

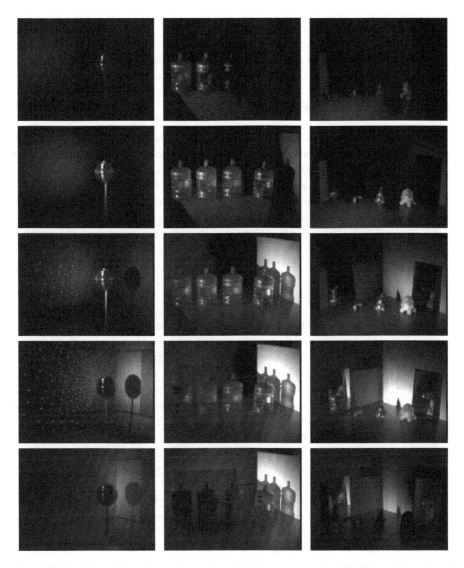

Figure 2.5. *Time slices from three transient images captured with our setup and reconstructed with the proposed method. Each of the three columns shows a result for the test scenes shown in Figure 2.4, the disco ball scene and two further scenes with filled water bottles and a set of mirrors, respectively. A variety of global illumination effects can be observed. Please see the text for detailed descriptions of the different effects we capture in each example*

The second column of the figure shows a scene with several bottles filled with water and a small amount of milk to create scattering. In the first frame, the wavefront has just reached the front of the leftmost bottles and is reflecting off their surface. In the second frame, scattering effects are becoming visible in the bottles. Next, the light reaches the far wall, showing caustics of the light transport through the bottles. Indirect illumination of the back wall from the light scattered in the bottles appears in the fourth frame. This light continues to illuminate the back wall even after the bottles themselves have darkened (last frame).

The last column of Figure 2.5 shows a scene with several foreground objects and two mirrors. We first see initial reflections coming off the foreground objects. As the light propagates, the foreground objects are now fully illuminated, and the wavefront reaches the back walls, but the mirrors remain dark. In the third frame, reflections of the foreground objects are starting to become visible in both the left and the right mirror. In the fourth frame, the left mirror shows a reflection of an object in the right mirror. This reflection lingers in the last frame, even after the wavefront has passed by the foreground objects.

2.3.3. *Spatio-temporal modulation*

Above, we only rely on temporal modulation. However, the approach can also be extended with spatial modulation. A number of works, starting with Debevec et al. (2000) and Sen et al. (2005), analyze or capture light transport through spatial modulation using projector–camera systems, where the transport is again assumed to be instantaneous. Adding spatial modulation leads to a straightforward extension of our convolutional model, as the model separable in each correlation measurement per pixel. Hence, we can simply code the illumination by using a projector instead of our flood light source. For example, we can replace the RGB light source of an off-the-shelf DLP LightCrafter projector with a collimated version of the light source described above. The additional spatial modulation can then, for example, be used to only selectively capture direct/retroreflective and indirect components. Since direct paths always satisfy the epipolar constraint, simply randomly turning pixels corresponding to epipolar lines on the projector (with probability 0.5) will suppress indirect paths from the projector to the camera. The direct/retroreflective component can then be estimated by capturing an unmasked measurement and subtracting the indirect-only component from it (O'Toole et al. 2014). Figure 2.6 shows results of transient imaging combined with this approach. The added spatial modulation aids the transient reconstruction by separating the direct and indirect components, hence significantly simplifying the unmixing. Note that the direct pulse of light traveling along the wall and the caustic light reflected by the mirror have high-temporal resolution in our reconstructions, whereas they appear broader when reconstructed without transport decomposition.

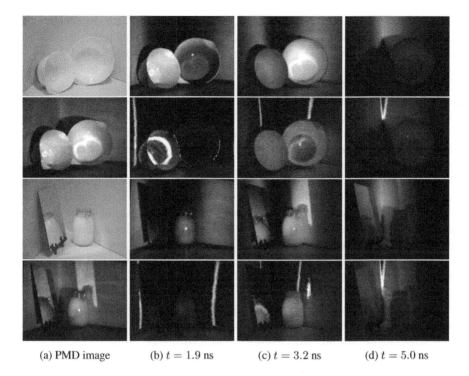

| (a) PMD image | (b) $t = 1.9$ ns | (c) $t = 3.2$ ns | (d) $t = 5.0$ ns |

Figure 2.6. *Transient imaging in combination with spatial modulation. (a) Scene captured with a normal camera under ambient illumination (rows 1 and 3) and correlation camera under projector illumination (rows 2 and 4). (b–d) Frames from the transient image reconstructed as before (rows 1 and 3) and with added spatial modulation (rows 2 and 4). Note the direct component is a sharp impulse traveling along the walls. For a color version of this figure, see www.iste.co.uk/funatomi/ computational.zip*

2.4. Transient imaging in scattering media

Next, we demonstrate an application of the novel temporal image dimension described in the previous section: imaging through scattering and turbid media. Imaging through scattering media has recently received significant attention. While many approaches have focused on microscopic settings, such as imaging in biological tissue, we consider here the macroscopic problem, ultimately targeting applications such as underwater imaging or imaging through fog.

The existing impulse-based methods image individual laser pulses with fast-gated cameras. However, this approach suffers from low SNR because ambient illumination can easily overpower the pulse amplitude. Mullen and Contarino (2000) propose a system that combines gated imaging with microwave amplitude modulation per pulse. While improving on conventional gated imaging techniques, this hybrid method directly samples the modulated incident scene response, which contains substantial ambient light. Hence, averaging many repeated captures is necessary to achieve sufficient SNR. In contrast, correlation ToF cameras allow for the integration over many (approximately 10^4–10^5) pulses in a single capture while performing adaptive in-pixel background suppression *during* the integration. As discussed for the working principle of correlation sensors, the unmodulated background component of photo-electrons is equally measured in both buckets and can be drained during exposure (potentially multiple times). This operating mode allows correlation sensors to amplify the signal component independently of the ambient light by increasing the exposure time, while saturation due to ambient light no longer occurs. Hence, correlation image sensors promise to be an efficient detector technology for imaging in scattering and turbid media. However, for imaging in scattering media, multi-path contributions are even stronger than they are for regular ToF range imaging. At a considered pixel, a mixture of path lengths is measured, where only a few ballistic photons directly hit objects submerged in the scattering media, and a large number of indirect paths are created by scattering events in the entire volume. This strong scattering, which makes traditional imaging very challenging, can only be handled if multi-path contributions are effectively removed. The convolutional transient imaging approach from the previous section allows us to solve this problem.

Figure 2.7. *Example of imaging in scattering media using our approach. Left: original scene with objects submerged in water-filled glass tank. Center: 160 ml milk added. Right: objects that are "invisible" due to strong scattering, like the structured object on the top right or parts of the circular object become detectable using our approach. For a color version of this figure, see www.iste.co.uk/funatomi/computational.zip*

Figure 2.7 shows imaging results using the correlation image sensor prototype from Figure 2.4 and the reconstruction method from the previous section. Here, we placed a water tank with a glass hull at a distance of 1.2 m in front of our camera so that the optical axis intersects the center of the largest planar side. The light source is placed at a slight offset to the right to eliminate direct reflections. We filled the tank with 80 liters of water and submerged objects at different positions in the water. In conventional captures of the scene that record the sum of all light path components, the submerged objects become "invisible" due to strong scattering, such as the structured object on the top right or parts of the circular object. These hidden objects are recovered using transient correlation imaging that resolves the temporal mixing of light transport components, while efficiently suppressing ambient scattering from other sources as background during the capture. Moreover, the proposed correlation imaging method not only increases the ability to detect objects in highly turbid solutions but also allows for a simultaneous estimation of distance.

2.5. Present and future directions

The proposed methods make steps towards the vision of inverting light transport under arbitrary conditions. However, to make this problem tractable at all, we also had to make strong assumptions on the scene, and we only analyzed part of the light transport. For example, we relied on paths corresponding to ballistic photons for imaging in scattering media. Hence, a promising direction for future research would be to investigate inverse methods that use a more accurate forward model and include more complex light paths.

However, even with improved inverse methods, light transport inversion is fundamentally limited by the resolution of the transient measurements, both in the temporal and spatial domains. A large benefit of using ToF depth cameras is that they have already become widely adopted consumer products, with emerging applications in gaming, virtual reality and robotics. While the spatial resolution of these sensors is rapidly increasing, higher temporal modulation frequencies, however, cannot be expected in consumer products in the immediate future. Due to the CCD charge transfer, at higher modulation frequencies, power consumption drastically increases and modulation contrast decreases. While this represents a severe limitation of the correlation image sensor, it is important to note that this limitation does not exist on the illumination side, and we only relied on amplitude modulation in this work. Combining the proposed methods with interference-based techniques could result in a system that relies on the amplitude modulation for low frequencies and the carrier for high-frequency sampling.

2.6. References

Abramson, N. (1978). Light-in-flight recording by holography. *Optics Letters*, 3(4), 121–123.

Boyle, W.S. and Smith, G.E. (1970). Charge coupled semiconductor devices. *Bell System Technical Journal*, 49(4), 587–593.

Chambolle, A. and Pock, T. (2011). A first-order primal-dual algorithm for convex problems with applications to imaging. *Journal of Mathematical Imaging and Vision*, 40(1), 120–145.

Debevec, P., Hawkins, T., Tchou, C., Duiker, H.-P., Sarokin, W., Sagar, M. (2000). Acquiring the reflectance field of a human face. *ACM SIGGRAPH*, pp. 145–156.

Kirmani, A., Hutchison, T., Davis, J., Raskar, R. (2009). Looking around the corner using transient imaging. *Proc. ICCV*, pp. 159–166.

Lange, R. (2000). 3D time-of-flight distance measurement with custom solid-state image sensors in CMOS/CCD-technology. Dissertation, Department of Electrical Engineering and Computer Science, University of Siegen.

Mullen, L.J. and Contarino, V.M. (2000). Hybrid lidar-radar: Seeing through the scatter. *IEEE Microwave Magazine*, 1(3), 42–48.

O'Toole, M., Heide, F., Xiao, L., Hullin, M.B., Heidrich, W., Kutulakos, K.N. (2014). Temporal frequency probing for 5D transient analysis of global light transport. *ACM Transactions on Graphics (SIGGRAPH)*, 33(4).

Parikh, N. and Boyd, S. (2013). Proximal algorithms. *Foundations and Trends in Optimization*, 1(3), 123–231.

Sen, P., Chen, B., Garg, G., Marschner, S., Horowitz, M., Levoy, M., Lensch, H.P.A. (2005). Dual photography. *ACM SIGGRAPH*, pp. 745–755.

Velten, A., Raskar, R., Bawendi, M. (2011). Picosecond camera for time-of-flight imaging. *OSA Imaging Systems and Applications*.

Velten, A., Willwacher, T., Gupta, O., Veeraraghavan, A., Bawendi, M., Raskar, R. (2012). Recovering three-dimensional shape around a corner using ultrafast time-of-flight imaging. *Nature Communications*, 3, 745.

3

Time-of-Flight and Transient Rendering

Adithya Kumar PEDIREDLA

Dartmouth College, Hanover, USA

3.1. Introduction

Time-of-flight rendering (Jarabo et al. 2014; Marco et al. 2019; Pediredla et al. 2019b, 2020) refers to generating an image or sequence of images that a time-of-flight camera captures for a given scene. Typically, a time-of-flight render, capable of time-of-flight rendering, is realized by replacing intensity cameras in a physics-based rendering engine with time-of-flight cameras. There are several time-of-flight cameras on the market, such as the Swiss ranger (Cazorla et al. 2010), Kinect (Smisek et al. 2013), intensified charged coupled device (ICCD) (Cester et al. 2019), single-photon avalanche diode detector (Villa et al. 2014) and Hamamatsu streak cameras. They are employed for depth sensing, fluorescence lifetime imaging, transient imaging, non-line-of-sight imaging and imaging through a scattering medium. At an abstract high level, the time-of-flight measurements are equal to the weighted sum of photon intensities with a weight ($|w| \leq 1$) that depends on the photon's *time-of-travel*[1] and the design of the time-of-flight camera. More details on this weighting function are in section 3.2.1. A time-of-flight renderer simulates a time-of-flight camera by

---

1. Time-of-travel is defined as the total time taken by a photon for the round trip travel, i.e. from the light source to the scene and back to the camera.

Computational Imaging for Scene Understanding,
coordinated by Takuya FUNATOMI and Takahiro OKABE.
© ISTE Ltd 2024.

implementing the weighting function accurately and for scenes with arbitrary scene geometry, material parameters and camera optics.

Time-of-flight rendering finds several applications in sensor design, machine learning and computer vision. For example, sensor designers are faced with a myriad of choices for any time-of-flight camera. All of these choices may vary for each pixel, or over time, and the correct choice tends to be application-dependent. Having access to efficient physically accurate renderers for time-of-flight cameras can greatly simplify the process of iterating through all these design permutations and combinations. Sensor designs can be evaluated virtually without building new sensors, which is a very expensive task. This results in significant acceleration and reduction in costs. In addition, rendering algorithms can help accelerate the development of computer vision algorithms that use data from time-of-flight cameras. For example, applications are emerging where deep learning algorithms use data from time-of-flight cameras for object recognition (Hagebeuker et al. 2007), shape estimation (Tsai et al. 2016), pose estimation (Hong et al. 2012), depth estimation (Honnungar et al. 2016; Horaud et al. 2016; Raghuram et al. 2019) and non-line-of-sight imaging (Velten et al. 2012). Efficient time-of-flight rendering algorithms could generate the large amounts of data (Chen et al. 2020; Gutierrez-Barragan et al. 2021) required to train and test these deep learning algorithms. The time-of-flight renderers are differentiable, and hence, we could infer scene parameters such as material properties and geometry with analysis-by-synthesis techniques (Tsai et al. 2019; Wu et al. 2021).

The goal of this book chapter is to familiarize the readers with time-of-flight rendering. We will present a high-level mathematical formulation of these cameras, followed by some unique challenges and path sampling solutions for time-of-flight cameras. We will then proceed to discuss some of the open-source rendering engines that the readers could integrate into their workflow. Finally, we will show some successful applications that use time-of-flight rendering. We will conclude with the current limitations of these renderers and possible future directions, both for improving the renderers and the applications with the help of the time-of-flight rendering. The mathematical formulations, derivations, results, and part of the text of this chapter is adapted from the author's work (Pediredla et al. 2017a, 2017b, 2019a, 2019b, 2020; Raghuram et al. 2019).

3.2. Mathematical modeling

We will first understand rendering scenes with intensity-only sensors before proceeding onto rendering with time-of-flight cameras. In computer graphics, we can simulate scenes with intensity-only sensors using rasterization rendering, radiosity rendering and the Monte Carlo rendering techniques. Among these, Monte Carlo rendering algorithms render scenes with high fidelity, and hence, in this chapter, we will focus on these techniques.

Path integral formulation (Veach 1998) is the foundation of modern Monte Carlo rendering algorithms. According to this formulation, the image I recorded by any intensity-only sensor is given by the path integral of the form

$$I = \int_{\mathcal{P}} f(\bar{x}) d\mu(\bar{x}). \tag{3.1}$$

where $\bar{x} = x_0 \rightarrow \ldots \rightarrow x_{B+1}$ is an ordered sequence of $B + 2$ three-dimensional points, for any $B \geq 0$, and represents a path photons may follow from a light source (path vertex x_0) to the sensor (path vertex x_{B+1}). Figure 3.1 shows an example path for a simple scene. The function $f(\bar{x})$ is the *radiance throughput* of the path, which depends on the geometry and material of the scene the light path has interacted with, the spatio-angular emission of light sources and the spatio-angular sensitivity (or importance) of the sensor. Finally, $\mathcal{P}$ is the space of all possible light paths ($\bar{x}$), and μ is its Lebesgue measure.

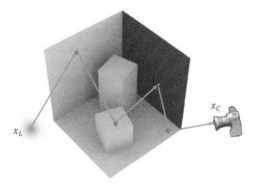

Figure 3.1. *An example light path for surface only mesh topology. For a color version of this figure, see www.iste.co.uk/funatomi/computational.zip*

In general, the integral in equation [3.1] is analytically intractable, except for small and simple scenes. Therefore, we evaluate this integral with Monte Carlo techniques, for which we first sample random light paths and approximate the integral with the summation:

$$I \approx \frac{1}{N} \sum_{n=1}^{N} \frac{f(\bar{x}_n)}{p(\bar{x}_n)}, \tag{3.2}$$

where $p(\bar{x}_n)$ is the probability of sampling the random path $\bar{x}_n$.

3.2.1. *Mathematical modeling for time-of-flight cameras*

As mentioned earlier, a time-of-flight camera weights photons based on their total time-of-travel, which, in the case of path integral formulation translates to weighting the contribution of each path with *pathlength importance* term. The path integral formulation for time-of-flight sensors would be

$$I = \int_{\mathcal{P}} W_\tau(\|\bar{x}\|) f(\bar{x}) d\mu(\bar{x}),$$ [3.3]

where $\|\bar{x}\|$ is the optical pathlength of the path. The optical pathlength is equal to the sum of the time delays caused at each point $(x_i; i = \{0, 1, \cdots, B + 1\};)$ of the path $(\bar{x})$, possibly due to fluorescence emission, and the time delays caused due to optical propagation of light from x_i to $x_{i+1}; i = \{0, 1, \cdots, B\}$. The pathlength importance function W_τ is independent of the scene; instead, it is uniquely determined by the technology, hardware design and parameters of the time-of-flight sensor.

We will broadly distinguish between three time-of-flight imaging modalities, which can be modeled using equation [3.3] with different functions W_τ. In Figure 3.2, we show renderings of various time-of-flight cameras.

Time-gated cameras: a time-gated camera accepts photons whose time-of-travel is in a narrow range and rejects all other photons. In terms of path integral formulations, this behavior translates to measuring the radiance of light paths whose optical pathlength is within some narrow range, and ignoring all other light paths. If the pathlength range is $[\tau - \frac{\Delta\tau}{2}, \tau + \frac{\Delta\tau}{2}]$, then such a camera can be modeled using a pathlength importance function:

$$W_\tau(\|\bar{x}\|) = \text{rect}\left(\frac{\|\bar{x}\| - \tau}{\Delta\tau}\right).$$ [3.4]

In this context, the pathlength importance function is also known as the *gate* of the camera. In practice, such cameras have imperfect gating mechanisms, and therefore may be better modeled with lower frequency but narrow pathlength importance functions; for example, a Gaussian with mean τ and standard deviation $\Delta\tau$:

$$W_\tau(\|\bar{x}\|) = \text{Gaussian}\left(\|\bar{x}\|; \tau, \Delta\tau\right).$$ [3.5]

Examples of time-gated cameras include Stanford ICCD (Cester et al. 2019), gated SPAD (Burri et al. 2014; Morimoto et al. 2020), Brightway Vision's brightway and VISDOM cameras (Gruber et al. 2019).

Transient cameras: a transient camera records a *sequence* of images $I(\tau), \tau \in \{\tau_{min}, \ldots, \tau_{max}\}$, with the entire sequence often termed a *transient image*. Each frame $I(\tau)$ in the sequence records contributions only from photons accepted by a narrow time gate centered around τ. For an *ideal* transient camera, this gate only permits photons with time-of-flight exactly τ, corresponding to a *per-frame* pathlength importance function that is a Dirac delta:

$$W_\tau(\|\bar{x}\|) = \delta(\|\bar{x}\| - \tau).$$ [3.6]

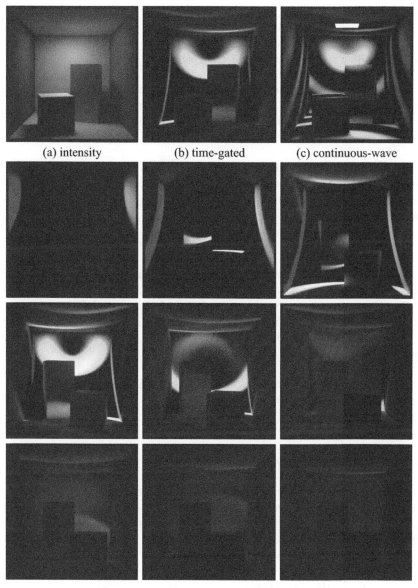

(a) intensity   (b) time-gated   (c) continuous-wave

(d) sequence of images captured in a transient

Figure 3.2. *Various time-of-flight rendering tasks: for the Cornell box scene, we show the renderings of (a) intensity camera (b), time-gated camera, (c) continuous-wave time-of-flight camera and (d) transient camera (we show frames of transient rendering). For a color version of this figure, see www.iste.co.uk/funatomi/ computational.zip*

In practice, the per-frame time gates can be imperfect, and are sometimes better modeled as in equations [3.4] and [3.5]. We note that a transient camera captures measurements equivalent to those captured by a time-gated camera with multiple gates, each with a gate centered at a different pathlength τ.

Examples of transient cameras include streak cameras (Velten et al. 2013) and SPAD cameras (O'Toole et al. 2017). By sweeping the time gates, a time-gated camera could also behave like a transient camera, but not all transient cameras are time-gated cameras.

Continuous-wave time-of-flight (CW-ToF) cameras: a CW-ToF camera uses illumination with temporally modulated amplitude, coupled with a sensor with temporally modulated sensitivity. This corresponds to a pathlength importance function (Lin et al. 2014):

$$W_\tau(\|\bar{x}\|) = C(\|\bar{x}\|), \qquad\qquad [3.7]$$

where C is the cross-correlation between the illumination and sensor modulation functions. Typically, C is a sinusoid, with frequency in the range of tens to hundreds of MHz. However, cross-correlations of trapezoidal (Gupta et al. 2018) and narrow rectangular (Tadano et al. 2015) shapes have also been demonstrated. In the latter case, the camera effectively reduces to a time-gated camera for some narrow pathlength importance function W_τ.

Examples of CW-ToF cameras include Kinect (Smisek et al. 2013), Swiss ranger (Cazorla et al. 2010) and photonic mixer devices (Kadambi et al. 2013, 2016; Bhandari et al. 2014).

3.3. How to render time-of-flight cameras?

Rendering measurements of a time-of-flight sensor require evaluating the integral of equation [3.3]. As in the intensity-only case ($W_\tau(\|\bar{x}\|) = 1$), analytic integration is impossible except for trivial scenes. Instead, Monte Carlo rendering algorithms (Veach 1998; Dutre et al. 2006) attempt to approximate the integral using Monte Carlo integration, forming an estimate of the form

$$\tilde{I} = \frac{1}{N}\sum_{n=1}^{N}\frac{W_\tau(\|\bar{x}_n\|)f(\bar{x}_n)}{p(\bar{x}_n)}, \qquad\qquad [3.8]$$

where each $\bar{x}_n$ is a randomly sampled path, and $p(\bar{x}_n)$ is the probability of sampling this path. When the path sampling distribution $p(\bar{x})$ has non-zero measure for all paths $\bar{x}$, estimators of this form are both consistent ($\lim_{N\to\infty}\tilde{I} = I$) and unbiased ($\mathbb{E}_{\{\bar{x}_n\}}\left[\tilde{I}\right] = I$).

Owing to the similarity between the path-space integral for intensity-only cameras and time-of-flight cameras, we could directly use the path sampling techniques such as path tracing, bidirectional path tracing, Metropolis light transport and its variants, designed to generate paths $(\bar{x}_n)$ for intensity-only cameras to time-of-flight cameras as well.

3.3.1. *Challenges and solutions in time-of-flight rendering*

The efficiency of Monte Carlo estimates critically depends on the path sampling techniques. Efficient path sampling techniques generate paths with distribution $p(\bar{x}_n)$ proportional to the integrand $W_\tau(\|\bar{x}_n\|)f(\bar{x}_n)$. Trivially using path sampling techniques developed for intensity-only rendering results in inefficient time-of-flight rendering. For example, consider rendering a transient camera. We bin the light paths based on their optical pathlength into various bins. As the path sampling techniques developed for intensity cameras do not have any control on the pathlength, some of the bins of the transient renderer will have no corresponding paths, and hence, zero intensity. This problem becomes more severe for rendering time-gated cameras as all the sampled paths might fall outside the narrow time-range of interest. Next, we will explain some specialized path sampling techniques developed for time-of-flight cameras.

Path reuse for transient rendering: to improve the convergence rate of transient renderers, Jarabo et al. (2014) proposed *temporal path reuse*, where every randomly sampled path contributes not just to one bin of the transient renderer but to all the temporal bins. Jarabo et al. achieve this with kernel-based temporal density estimation. The temporal sequence measured by the transient camera is

$$\tilde{I}(\tau) = \sum_{n=1}^{N} \mathcal{K}_\mathcal{T}(\tau - \|\bar{x}_n\|)\frac{f(\bar{x}_n)}{p(\bar{x}_n)}, \qquad\qquad [3.9]$$

where $\mathcal{K}_\mathcal{T}$ is the kernel with temporal bandwidth $\mathcal{T}$. Evidently, the kernel density estimation blurs the transient and hence cannot capture the high-frequency transient content and introduces bias. The error of the Monte Carlo approximation in equation [3.9] would be due to both bias and variance of the estimate. While the variance decays with the number of samples, the bias would not vanish similarly. To mitigate this problem, Jarabo et al. proposed to progressively reduce the kernel's temporal bandwidth with each sample as $\mathcal{T}_{n+1} = \mathcal{T}_n(n + \alpha)/(n + 1)$, which would ensure that the kernel density estimation is asymptotically unbiased. For interested readers, we also recommend reading the follow-up work: transient photon beams (Marco et al. 2017) and progressive transient photon beams (Marco et al. 2019).

Path sampling with pathlength constraint: we (Pediredla et al. 2019b) proposed an unbiased path sampling technique that is well suited for time-gated rendering and rendering time-of-flight cameras that accept light paths within a narrow pathlength range including delta sampling in path space. Instead of using kernel-based techniques, we proposed to sample light paths with an exact pathlength constraint. This sampling procedure ensures that we only sample paths which are accepted by a time-gated camera.

Our procedure for path sampling time-gated cameras works in two steps. First, we sample a target pathlength ($\|\bar{x}\|$) by importance-sampling the camera's weighting term (W_τ). Then, we generate a path that exactly matches this target length. The starting point for this second step is bidirectional path tracing (BDPT), a common technique used in computer graphics. In standard BDPT, we generate two sub-paths, one starting from the light source ($x_L \rightarrow \cdots \rightarrow x_l$ in Figure 3.3) and another from the camera ($x_C \rightarrow \cdots \rightarrow x_c$ in Figure 3.3). Then, we form complete paths by directly connecting vertices of these sub-paths. Because of this direct connection, we have no control over the pathlength of the resulting path. To overcome this, we modify BDPT so that, instead of directly joining sub-pathends, we join them through a new intermediate vertex (x_e). By controlling the location of this connecting vertex, we could control the pathlength of the resulting path. In particular, we need to select the connecting vertex so that the sum of its distances from the light and camera sub-path vertices is equal to the difference between the target pathlength and the total lengths of the light and camera sub-paths:

$$\|x_e \rightarrow x_l\| + \|x_e \rightarrow x_c\| = \|x\| \qquad [3.10]$$

This geometric constraint is the definition of an ellipsoid, whose focal points are the vertices of the light and camera sub-path, and its major axis length is equal to the left-over pathlength constraint. Depending on the location of this ellipsoid, we end up with two main scenarios as shown in Figure 3.3. If the ellipsoid is contained inside a volume, then we can use any point on it as the connecting vertex to form a complete path. Otherwise, we need to intersect the ellipsoid with the scene surfaces. Then, the connecting vertex must be on the curves resulting from the ellipsoid-scene intersection. The details of how to do this intersection analytically and fast are beyond the scope of this book chapter, but interested readers can find them in the manuscript (Pediredla et al. 2019b). We refer to these connections as *ellipsoidal connections*, and we will next compare them with direct connections (standard BDPT) for a few example scenarios.

In the first example scenario, we consider the case of transient imaging of dynamic scenes with time-gated cameras such as ICCDs or Kerr gates. As mentioned earlier, transient images could be obtained with time-gated cameras by capturing sequences of time-gated images, one at a time. The integration time to obtain a time-gated image is typically in the order of a few milliseconds even though each time-gate would be in the order of tens or hundreds of picoseconds. Therefore, in dynamic scenes, each time-gated frame of the transient sequence effectively images a different scene due to scene motion. Consequently, simulating this situation requires performing a sequence of time-gated rendering operations. We demonstrate this in Figure 3.4, where we compare the relative performance of BDPT with direct and ellipsoidal connections for this transient rendering task. We show a few frames of transient sequences from a dynamic scene, rendered for a transient camera with a per-frame gate width of 200 ps, typical of ICCDs (Cester et al. 2019). We observe that, at all frames, ellipsoidal connections produce renderings that are considerably less noisy. Note that, when transient rendering a static scene, path sampling with pathlength constraint would not offer a performance advantage over direct connections. In that case, temporal path reuse means that all paths sampled by BDPT with direct connections will have high importance, and therefore, it is not necessary to use ellipsoidal connections.

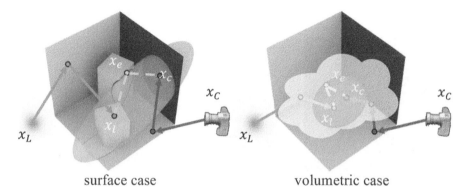

surface case   volumetric case

Figure 3.3. *Path sampling with pathlength constraint: in bidirectional path tracing, we trace sub-paths from light source and sensor and join them with a straight line, which results in paths with no pathlength constraint. Instead, we propose to join sub-paths with an additional vertex (x_e). For a given pathlength constraint $\|\bar{x}\|$, we show that the locus of possible locations for x_e will be an ellipsoid with focal points x_l and x_c and a major axis length equal to the leftover pathlength. Based on where this ellipsoid is, we have two cases: (a) surface case and (b) volumetric case. In the volumetric case, we randomly sample a point on the ellipsoid, and in the surface case, we intersect the surface geometry with the ellipsoid (shown in blue) and select one point on the intersection randomly to form the path. For a color version of this figure, see www.iste.co.uk/funatomi/computational.zip*

In the second example, we show the rendering of a proximity sensor for automobiles. Fixed time-gated sensors are increasingly used as proximity sensors in automobiles. Simulating these sensors is a time-gated rendering task, for which ellipsoidal connections can be beneficial. We show this in Figure 3.5, where we simulate a vehicle traveling through a road scene, equipped with a time-gated proximity camera. The figure shows a few frames of the rendered sequence, where we observe that ellipsoidal connections outperform direct connections.

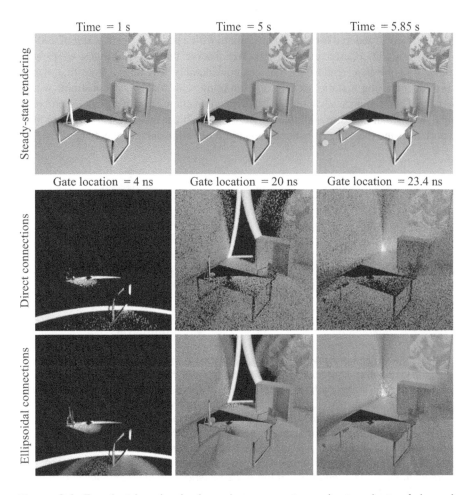

Figure 3.4. *Transient imaging in dynamic scenes: to render transients of dynamic scenes as captured by gated transient cameras, we have to render a sequence of time-gated frames, each for a slightly different scene due to motion. From top to bottom, we show steady-state renderings, transient renderings with direct connections and transient renderings with ellipsoidal connections. Different columns are different frames of the video. All images are rendered for* $15\,\mathrm{s}$, *with a gate width* $\Delta\tau = 200\,\mathrm{ps}$ *(1.74% of scene size) and per-gate exposure of* $50\,\mathrm{ms}$. *For a color version of this figure, see www.iste.co.uk/funatomi/computational.zip*

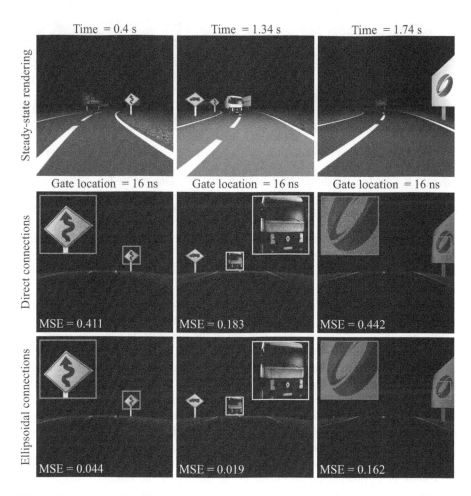

Figure 3.5. *Proximity detection camera: we simulate measurements from such a camera, equipped on a traveling automobile. From top to bottom, we show steady-state renderings, time-gated renderings with direct connections and time-gated renderings with ellipsoidal connections. Different columns are different frames of the video. All images are rendered for $10\,\mathrm{s}$, with gate width $\Delta\tau = 200\,\mathrm{ps}$ (1.14% of scene size). The MSE improvement for the shown frames is $9.25\times$, $9.39\times$ and $2.73\times$. For a color version of this figure, see www.iste.co.uk/funatomi/computational.zip*

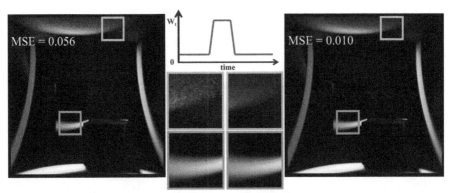

(a) BDPT with direct connections   (b) BDPT with ellipsoidal connections

Figure 3.6. *CW-ToF depth-selective camera: we simulate a CW-ToF camera using modulation codes resulting in depth-selectivity (Tadano et al. 2015). The inset at the top shows the corresponding pathlength importance function, with a high-importance area of width $\Delta\tau = 20\,\text{ps}$ (1.28% of scene size). All images are rendered for $60\,\text{s}$. The MSE improvement is 5.49×. For a color version of this figure, see www.iste.co.uk/funatomi/computational.zip*

In the third example, we show the rendering of a continuous-wave depth selective camera, first introduced by Tadano et al. (2015). Continuous-wave depth selective camera uses cross-correlation codes that are in the shape of an isosceles trapezium with a programmable base length. Therefore, these cameras produce an output similar to a time-gated camera even though the hardware is a CW-ToF camera. In Figure 3.6, we compare the renderings produced with direct connections and ellipsoidal connections. The function W_τ, in this case, has low (but non-zero) values for all pathlengths, except for a narrow trapezoidal area of width $\Delta\tau = 20\,\text{ps}$ (1.28% of scene size). Therefore, unlike the case of a time-gated sensor, here the contributions of all paths are non-zero, albeit in most cases very small. With ellipsoidal connections, we observe that the ability to importance sample the pathlength importance function W_τ results in significant improvement in the quality of the rendered image. This example demonstrates that ellipsoidal connections work well not just for time-gated cameras but also for time-of-flight cameras that have pathlength importance term (W_τ) with narrow support.

3.4. Open-source implementations

MitsubaToFRenderer (Pediredla 2019) is an implementation that we have built on top of Mitsuba (Jakob 2010) renderer (Mitsuba 0.5.0). Mitsuba renderer supports multiple light sources, cameras, materials, volumes and integrators. We have added support for transient, time-gated and continuous-wave time-of-flight cameras. In addition, we also added support for programmable projectors and cameras by integrating with *MitsubaCLTRenderer* (Sun and Gkioulekas 2019). Thanks to the Mitsuba implementation, the source code is scalable across multiple CPUs and also on a cluster of machines. However, the source code does not scale to GPUs. For easy, operating system independent usage, we also provided a docker implementation (at adithyapedireda/mitsubatofrenderer). Once a time-of-flight image or sequence of images is rendered, they can be viewed or processed in MATLAB or python using the support files we have provided.

Another open-source implementation that could render continuously varying refractive media is *MitsubaER* (Pediredla 2020; Pediredla et al. 2020), which also supports all time-of-flight rendering tasks. However, this renderer does not support ellipsoidal connections.

Jarabo et al. (2014) have open-sourced their implementation (Jarabo 2014) of temporal path reuse techniques. In addition to scalar light, their code also supports polarized light based on Stokes vectors (Jarabo and Arellano 2018).

3.5. Applications of transient rendering

Transient renderers find applications in validating a hardware design and evaluating the system performance, train deep neural networks for computer vision and imaging applications, and for analysis-by-synthesis algorithms to reconstruct geometry and materials of a scene. Below, we highlight some recent literature that uses transient rendering.

Optimal imaging system design: we used the renderer for choosing optimal parameters for scanning-based non-line-of-sight imaging (SNLOS). Non-line-of-sight imaging refers to reconstructing objects beyond the line-of-sight, around the corner as shown in Figure 3.7. While several imaging techniques for non-line-of-sight exist (Klein et al. 2016; Chandran and Jayasuriya 2019; Saunders et al. 2019; Faccio et al. 2020; Henley et al. 2020; Metzler et al. 2020; Willomitzer et al. 2021), the time-of-flight imaging-based solutions are most commonly employed (Pediredla et al. 2017a, 2017b; O'Toole et al. 2018; Liu et al. 2019, 2020; Xin et al. 2019; Rapp et al. 2020) for NLOS imaging, and they result in better reconstructions compared to the other imaging modalities. SNLOS uses time-gated cameras, and hence, the rendering technique presented in section 3.3.1 is well suited for optimal parameter selection for SNLOS.

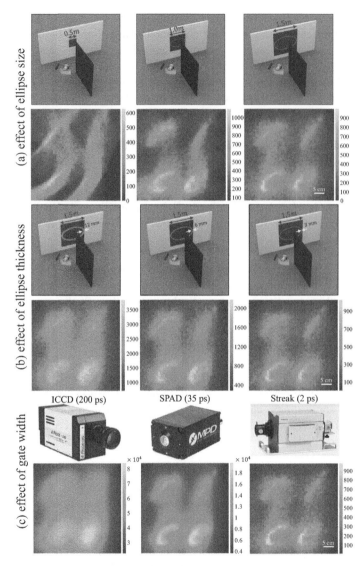

Figure 3.7. *Optimal imaging system design for scanning-based non-line-of-sight imaging (SNLOS): time-of-flight rendering helps in imaging system design. In this example, we study the effect of various design choices for SNLOS imaging systems for non-line-of-sight imaging. SNLOS uses a pulsed laser and illuminates the visible wall with an elliptic pattern and also images an elliptical ring on the wall with a single-pixel time-gated camera. We note that the resolution improves by increasing the ellipse size, decreasing the ellipse thickness and reducing the gate width. These choices not only enhance the resolution of the NLOS image but also increase photon noise. For a color version of this figure, see www.iste.co.uk/funatomi/computational.zip*

We will next present SNLOS at a high level and explain the optimal parameter selection procedure. SNLOS scans the hidden NLOS scene voxel-by-voxel. Hence, it is well suited to scan sparse scenes around the corner. To scan a voxel around the corner, SNLOS illuminates the visible wall with a pulsed laser and delays the light rays such that the total time of travel to the hidden voxel is a predetermined constant. On the imaging side, the light rays arriving back to the camera are also similarly delayed before they are integrated by the camera. This temporal delay of the rays during illumination and imaging makes all the rays that interacted with the hidden voxel focus temporally and other rays defocus temporally. By using a time-gated camera that measures only the focused rays, SNLOS scans a voxel around the corner. To scan a new voxel, only the delay pattern has to be changed.

Experimentally implementing the delay pattern is difficult as the delays required are in the order of nanoseconds, which would require bulky systems almost as large as the scanned hidden volume. In Pediredla et al. (2019a), we show that temporal focusing could also be achieved by illuminating and imaging an ellipse on the wall without the need for bulky temporal delay hardware. This suggests multiple design choices for implementing the SNLOS system: choose large ellipses versus small ellipses, thick ellipses versus thin ellipses and what time-of-flight cameras to use. Using the renderer, we show in Figure 3.7 that larger and thinner ellipses and high-temporal resolution time-of-flight cameras are better for high-resolution imaging around the corner but these choices result in noisier reconstruction as well. Such an analysis of the hardware would have taken several weeks and cost hundreds of thousands of dollars to try different time-of-flight hardware. Instead, this analysis took less than a day and less than a hundred dollars on AWS machines.

In a different work, Tsai (2019) reconstructed the shape of a specular concave object with the time-of-flight of two-bounce light paths. If we use an intensity camera to reconstruct the shape of a specular concave object, due to multiple light bounces, the information of both the shape and reflectance functions of multiple parts of the scene is merged in the intensity measurement. Tsai (2019) reasoned that the pathlength of light (and hence, the time-of-flight) is unaffected by the reflectance function and developed an algorithm to reconstruct specular concave objects using time-of-flight information of two-bounce light paths. The details of the algorithm are beyond the scope of this chapter, and the interested reader can find the details at (Tsai 2019; Tsai et al. 2016). To validate the algorithm, Tsai (2019) used a time-of-flight renderer. While their algorithm assumes only two-bounce light paths, multiple-bounce light paths reach the sensor. However, their algorithm is robust to higher-order bounce information. In this application of time-of-flight rendering, we could also render only two-bounce light paths to validate the algorithm, and then add higher-order bounces to quantify the degradation with each additional bounce. Such convenience is not possible in real hardware where all the light bounces are measured simultaneously.

Training neural networks: Gutierrez-Barragan et al. (2021) propose iToF2dToF, a data-driven technique to convert the data captured by a continuous-wave time-of-flight camera (iToF) to transients (dToF). They use sinusoidal codes and capture the CW-ToF images for two frequencies (20 MHz and 100 MHz) and two phases (0, $\frac{\pi}{2}$). Using a U-net with skip connections, ReLU activations and learned upsampling, they interpolate and extrapolate more frequency information (20 MHz to 600 MHz in steps of 20 MHz). By inverse Fourier transforming the frequency data, they obtain high-resolution transients. To train the U-net, they need a lot of data, especially at high frequencies (600 MHz) to learn extrapolation. Commercially available CW-ToF cameras have a maximum frequency of 100 MHz. To solve this training dataset problem, Gutierrez-Barragan et al. have employed the time-of-flight renderer (Pediredla et al. 2019b) to create a dataset.

Chen et al. (2020) use a transient camera and a deep learning framework for reconstructing and classifying objects around the corners. They use the data generated from a transient renderer to train the deep nets. Surprisingly, the resulting deep network generalizes to real data for reconstructing and classifying objects around the corner even though the network is not trained on any real data. The deep net reconstructions of real data are far superior to previous techniques that are based on physics-based inversion. The superior performance of renderer trained deep nets show that the transient renderers are good at accurately modeling the time-of-flight cameras.

Analysis-by-synthesis: it refers to estimating the scene parameters (lighting, geometry, textures, BRDFs) based on the images captured with single or multiple camera parameters. Camera parameters typically refer to multiple views, exposure durations, aperture settings and focal distances for intensity-only cameras. For time-of-flight cameras, in addition to these parameters, we can also vary the pathlength weighting function ($W_\tau(\|\bar{x}\|)$). To reconstruct the scene parameters, we solve the optimization problem:

$$\arg\min_{\Pi} \sum_{n=1}^{N} \ell(I_n - I(\Pi; W_n, \theta_n)), \qquad [3.11]$$

where Π is the set of unknown scene parameters, I_n are the measurements, W_n is the pathlength weighting function for the nth measurement, and θ_n are the scene and camera parameters, which are generally fixed ($\theta_n = \theta$). We commonly use gradient descent techniques for optimizing [3.11], which would require us to compute the derivate $\frac{dI(\Pi; W_n, v_n)}{d\Pi}$. We could compute this derivative using finite difference methods (Iseringhausen and Hullin 2020), with automatic differentiation (Nimier-David et al. 2019) or analytically (Tsai et al. 2019; Yi et al. 2021; Wu et al. 2021).

Tsai et al. (2019) proposed to solve non-line-of-sight imaging with analysis-by-synthesis. They assumed only third bounce photons and built a fast time-of-flight renderer. To compute the derivative term, they explicitly derived the derivative term. Around the same time, Iseringhausen and Hullin (2020) also proposed to solve non-line-of-sight imaging with analysis-by-synthesis. To compute the derivative term, they used finite differences. Both techniques restrict themselves to third bounce photons, and extending them to higher-order bounces is non-trivial. Recently, Yi et al. (2021) and Wu et al. (2021) have proposed differential transient rendering to compute the derivative term analytically for any number of light bounces.

3.6. Future directions

Modeling wave phenomenon: the path-space integral in equation [3.3] models only the ray behavior of light and not the wave behavior. Imaging systems such as optical coherence tomography (OCT) are time-of-flight imaging systems, but they exploit the wave nature of light to compute time-of-flight. Due to the wave nature of light, the images captured by the OCT have speckle in them which would be missing if we solve equation [3.3]. Recent techniques on rendering speckle (Bar et al. 2019; Alterman et al. 2021) could be integrated with the time-of-flight camera to render imaging systems such as OCT faithfully.

Modeling fluorescence: fluorescence is a non-linear light phenomenon, where a photon is absorbed in one wavelength and emitted at a different (higher) wavelength by a fluorescent molecule. The photon is emitted after a random delay that depends on the lifetime distribution of the fluorescent molecule. This lifetime is a function of environmental parameters such as pH, ion or oxygen concentration, and hence, fluorescence lifetime imaging (FLIM) technique is used for functional imaging. Rendering FLIM for various media finds applications in imaging through scattering media and is an interesting forward direction.

Adaptive sampling and Metropolis light transport for ellipsoidal connections: adaptive sampling techniques have been successful for steady-state rendering, with the key idea being that we can allocate more samples to spatial pixels that are difficult to render (e.g. edges, textures, caustics) (Zwicker et al. 2015). BDPT with ellipsoidal connections provides a way to adapt these techniques to time-of-flight rendering tasks, by making it possible to allocate more samples to temporal frames $I(\tau)$ that require them (e.g. high-frequency parts of a transient (Wu et al. 2014)).

Ellipsoidal connections are also promising within the framework of path-space Metropolis light transport (Veach and Guibas 1997): given a path, we can perturb it by removing one of its vertices and replacing it with a new ellipsoidal vertex selected so that the new path has the same length as the original one. Using ellipsoidal connections as pathlength-preserving perturbation operations opens up the possibility of using Metropolis light transport algorithms to efficiently explore path manifolds defined by pathlength constraints (Jakob and Marschner 2012).

3.7. References

Alterman, M., Bar, C., Gkioulekas, I., Levin, A. (2021). Imaging with local speckle intensity correlations: Theory and practice. *ACM Transactions on Graphics (TOG)*, 40(3), 1–22.

Bar, C., Alterman, M., Gkioulekas, I., Levin, A. (2019). A Monte Carlo framework for rendering speckle statistics in scattering media. *ACM Transactions on Graphics (TOG)*, 38(4), 1–22.

Bhandari, A., Kadambi, A., Whyte, R., Barsi, C., Feigin, M., Dorrington, A., Raskar, R. (2014). Resolving multipath interference in time-of-flight imaging via modulation frequency diversity and sparse regularization. *Optics Letters*, 39(6), 1705–1708.

Burri, S., Maruyama, Y., Michalet, X., Regazzoni, F., Bruschini, C., Charbon, E. (2014). Architecture and applications of a high resolution gated SPAD image sensor. *Optics Express*, 22(14), 17573–17589.

Cazorla, M., Viejo, D., Pomares, C. (2010). Study of the sr4000 camera. *Proceedings of XI Workshop of Physical Agents Fisicos*, Valencia.

Cester, L., Lyons, A., Braidotti, M.C., Faccio, D. (2019). Time-of-flight imaging at 10 ps resolution with an ICCD camera. *Sensors*, 19(1), 180.

Chandran, S. and Jayasuriya, S. (2019). Adaptive lighting for data-driven non-line-of-sight 3D localization and object identification. *arXiv:1905.11595*.

Chen, W., Wei, F., Kutulakos, K.N., Rusinkiewicz, S., Heide, F. (2020). Learned feature embeddings for non-line-of-sight imaging and recognition. *ACM Transactions on Graphics (TOG)*, 39(6), 1–18.

Dutre, P., Bekaert, P., Bala, K. (2006). *Advanced Global Illumination*. AK Peters/CRC Press.

Faccio, D., Velten, A., Wetzstein, G. (2020). Non-line-of-sight imaging. *Nature Reviews Physics*, 2(6), 318–327.

Gruber, T., Julca-Aguilar, F., Bijelic, M., Heide, F. (2019). Gated2depth: Real-time dense lidar from gated images. *Proceedings of the IEEE/CVF International Conference on Computer Vision*, pp. 1506–1516.

Gupta, M., Velten, A., Nayar, S.K., Breitbach, E. (2018). What are optimal coding functions for time-of-flight imaging? *ACM Transactions on Graphics (TOG)*, 37(2), 1–18.

Gutierrez-Barragan, F., Chen, H., Gupta, M., Velten, A., Gu, J. (2021). itof2dtof: A robust and flexible representation for data-driven time-of-flight imaging. *arXiv:2103.07087*.

Hagebeuker, D.-I.B. et al. (2007). *A 3D time of flight camera for object detection*. PMD Technologies GmbH, Siegen, 2.

Henley, C., Maeda, T., Swedish, T., Raskar, R. (2020). Imaging behind occluders using two-bounce light. *European Conference on Computer Vision*, Springer, pp. 573–588.

Hong, S., Ye, C., Bruch, M., Halterman, R. (2012). Performance evaluation of a pose estimation method based on the SwissRanger SR4000. *2012 IEEE International Conference on Mechatronics and Automation*. IEEE, pp. 499–504.

Honnungar, S., Holloway, J., Pediredla, A.K., Veeraraghavan, A., Mitra, K. (2016). Focal-sweep for large aperture time-of-flight cameras. *2016 IEEE International Conference on Image Processing (ICIP)*. IEEE, pp. 953–957.

Horaud, R., Hansard, M., Evangelidis, G., Ménier, C. (2016). An overview of depth cameras and range scanners based on time-of-flight technologies. *Machine Vision and Applications*, 27(7), 1005–1020.

Iseringhausen, J. and Hullin, M.B. (2020). Non-line-of-sight reconstruction using efficient transient rendering. *ACM Transactions on Graphics (TOG)*, 39(1), 1–14.

Jakob, W. (2010). Mitsuba renderer [Online]. Available at: https://www.mitsuba-renderer.org/index_old.html.

Jakob, W. and Marschner, S. (2012). Manifold exploration: A Markov chain Monte Carlo technique for rendering scenes with difficult specular transport. *ACM Transactions on Graphics (TOG)*, 31(4), 1–13.

Jarabo, A. (2014). A framework for transient rendering [Online]. Available at: http://webdiis.unizar.es/ajarabo/pubs/transientSIGA14/code/.

Jarabo, A. and Arellano, V. (2018). Bidirectional rendering of vector light transport. *Computer Graphics Forum*, 37(6), 96–105.

Jarabo, A., Marco, J., Munoz, A., Buisan, R., Jarosz, W., Gutierrez, D. (2014). A framework for transient rendering. *ACM Transactions on Graphics (ToG)*, 33(6), 1–10.

Kadambi, A., Whyte, R., Bhandari, A., Streeter, L., Barsi, C., Dorrington, A., Raskar, R. (2013). Coded time of flight cameras: Sparse deconvolution to address multipath interference and recover time profiles. *ACM Transactions on Graphics (ToG)*, 32(6), 1–10.

Kadambi, A., Zhao, H., Shi, B., Raskar, R. (2016). Occluded imaging with time-of-flight sensors. *ACM Transactions on Graphics (ToG)*, 35(2), 1–12.

Klein, J., Peters, C., Martín, J., Laurenzis, M., Hullin, M.B. (2016). Tracking objects outside the line of sight using 2D intensity images. *Scientific Reports*, 6(1), 1–9.

Lin, J., Liu, Y., Hullin, M.B., Dai, Q. (2014). Fourier analysis on transient imaging with a multifrequency time-of-flight camera. *Proceedings of the IEEE Conference on Computer Vision and Pattern Recognition*, pp. 3230–3237.

Liu, X., Guillén, I., La Manna, M., Nam, J.H., Reza, S.A., Le, T.H., Jarabo, A., Gutierrez, D., Velten, A. (2019). Non-line-of-sight imaging using phasor-field virtual wave optics. *Nature*, 572(7771), 620–623.

Liu, X., Bauer, S., Velten, A. (2020). Phasor field diffraction based reconstruction for fast non-line-of-sight imaging systems. *Nature Communications*, 11(1), 1–13.

Marco, J., Jarosz, W., Gutierrez, D., Jarabo, A. (2017). Transient photon beams. *ACM SIGGRAPH 2017 Posters, SIGGRAPH '17*. Association for Computing Machinery.

Marco, J., Guillén, I., Jarosz, W., Gutierrez, D., Jarabo, A. (2019). Progressive transient photon beams. *Computer Graphics Forum*, 38(6), 19–30.

Metzler, C.A., Heide, F., Rangarajan, P., Balaji, M.M., Viswanath, A., Veeraraghavan, A., Baraniuk, R.G. (2020). Deep-inverse correlography: Towards real-time high-resolution non-line-of-sight imaging. *Optica*, 7(1), 63–71.

Morimoto, K., Ardelean, A., Wu, M.-L., Ulku, A.C., Antolovic, I.M., Bruschini, C., Charbon, E. (2020). Megapixel time-gated SPAD image sensor for 2D and 3D imaging applications. *Optica*, 7(4), 346–354.

Nimier-David, M., Vicini, D., Zeltner, T., Jakob, W. (2019). Mitsuba 2: A retargetable forward and inverse renderer. *ACM Transactions on Graphics (TOG)*, 38(6), 1–17.

O'Toole, M., Heide, F., Lindell, D.B., Zang, K., Diamond, S., Wetzstein, G. (2017). Reconstructing transient images from single-photon sensors. *Proceedings of the IEEE Conference on Computer Vision and Pattern Recognition*, pp. 1539–1547.

O'Toole, M., Lindell, D.B., Wetzstein, G. (2018). Confocal non-line-of-sight imaging based on the light-cone transform. *Nature*, 555(7696), 338–341.

Pediredla, A. (2019). Mitsubatofrenderer [Online]. Available at: https://github.com/cmu-ci-lab/MitsubaToFRenderer.

Pediredla, A. (2020). Mitsubaer [Online]. Available at: https://github.com/cmu-ci-lab/MitsubaER.

Pediredla, A.K., Buttafava, M., Tosi, A., Cossairt, O., Veeraraghavan, A. (2017a). Reconstructing rooms using photon echoes: A plane based model and reconstruction algorithm for looking around the corner. *2017 IEEE International Conference on Computational Photography (ICCP)*, IEEE, pp. 1–12.

Pediredla, A.K., Matsuda, N., Cossairt, O., Veeraraghavan, A. (2017b). Linear systems approach to identifying performance bounds in indirect imaging. *2017 IEEE International Conference on Acoustics, Speech and Signal Processing (ICASSP)*. IEEE, pp. 6235–6239.

Pediredla, A., Dave, A., Veeraraghavan, A. (2019a). SNLOS: Non-line-of-sight scanning through temporal focusing. *2019 IEEE International Conference on Computational Photography (ICCP)*. IEEE, pp. 1–13.

Pediredla, A., Veeraraghavan, A., Gkioulekas, I. (2019b). Ellipsoidal path connections for time-gated rendering. *ACM Transactions on Graphics (TOG)*, 38(4), 1–12.

Pediredla, A., Chalmiani, Y.K., Scopelliti, M.G., Chamanzar, M., Narasimhan, S., Gkioulekas, I. (2020). Path tracing estimators for refractive radiative transfer. *ACM Transactions on Graphics (TOG)*, 39(6), 1–15.

Raghuram, A., Pediredla, A., Narasimhan, S.G., Gkioulekas, I., Veeraraghavan, A. (2019). Storm: Super-resolving transients by oversampled measurements. *2019 IEEE International Conference on Computational Photography (ICCP)*. IEEE, pp. 1–11.

Rapp, J., Saunders, C., Tachella, J., Murray-Bruce, J., Altmann, Y., Tourneret, J.-Y., McLaughlin, S., Dawson, R.M., Wong, F.N., Goyal, V.K. (2020). Seeing around corners with edge-resolved transient imaging. *Nature Communications*, 11(1), 1–10.

Saunders, C., Murray-Bruce, J., Goyal, V.K. (2019). Computational periscopy with an ordinary digital camera. *Nature*, 565(7740), 472–475.

Smisek, J., Jancosek, M., Pajdla, T. (2013). 3D with kinect. In *Consumer Depth Cameras for Computer Vision*, Fossati, A., Gall, J., Grabner, H., Ren, X., Konolige, K. (eds). Springer, London.

Sun, J. and Gkioulekas, I. (2019). Mitsuba CLT renderer [Online]. Available at: https://github.com/cmu-ci-lab/mitsuba_clt.

Tadano, R., Pediredla, A.K., Veeraraghavan, A. (2015). Depth selective camera: A direct, on-chip, programmable technique for depth selectivity in photography. *Proceedings of the IEEE International Conference on Computer Vision*, pp. 3595–3603.

Tsai, C.-Y. (2019). Shape from multi-bounce light paths. PhD Thesis, Carnegie Mellon University.

Tsai, C.-Y., Veeraraghavan, A., Sankaranarayanan, A.C. (2016). Shape and reflectance from two-bounce light transients. *2016 IEEE International Conference on Computational Photography (ICCP)*. IEEE, pp. 1–10.

Tsai, C.-Y., Sankaranarayanan, A.C., Gkioulekas, I. (2019). Beyond volumetric albedo – A surface optimization framework for non-line-of-sight imaging. *Proceedings of the IEEE/CVF Conference on Computer Vision and Pattern Recognition*, pp. 1545–1555.

Veach, E. (1998). *Robust Monte Carlo Methods for Light Transport Simulation*. Stanford University.

Veach, E. and Guibas, L.J. (1997). Metropolis light transport. *Proceedings of the 24th Annual Conference on Computer Graphics and Interactive Techniques*. ACM Press/Addison-Wesley Publishing Co., pp. 65–76.

Velten, A., Willwacher, T., Gupta, O., Veeraraghavan, A., Bawendi, M.G., Raskar, R. (2012). Recovering three-dimensional shape around a corner using ultrafast time-of-flight imaging. *Nature Communications*, 3(1), 1–8.

Velten, A., Wu, D., Jarabo, A., Masia, B., Barsi, C., Joshi, C., Lawson, E., Bawendi, M., Gutierrez, D., Raskar, R. (2013). Femto-photography: Capturing and visualizing the propagation of light. *ACM Transactions on Graphics (ToG)*, 32(4), 1–8.

Villa, F., Lussana, R., Bronzi, D., Tisa, S., Tosi, A., Zappa, F., Dalla Mora, A., Contini, D., Durini, D., Weyers, S. et al. (2014). CMOS imager with 1024 SPADs and TDCs for single-photon timing and 3-D time-of-flight. *IEEE Journal of Selected Topics in Quantum Electronics*, 20(6), 364–373.

Willomitzer, F., Rangarajan, P.V., Li, F., Balaji, M.M., Christensen, M.P., Cossairt, O. (2021). Fast non-line-of-sight imaging with high-resolution and wide field of view using synthetic wavelength holography. *Nature Communications*, 12(1), 1–11.

Wu, D., Wetzstein, G., Barsi, C., Willwacher, T., Dai, Q., Raskar, R. (2014). Ultra-fast lensless computational imaging through 5D frequency analysis of time-resolved light transport. *International Journal of Computer Vision*, 110(2), 128–140.

Wu, L., Cai, G., Ramamoorthi, R., Zhao, S. (2021). Differentiable time-gated rendering. *ACM Transactions on Graphics (TOG)*, 40(6), 1–16.

Xin, S., Nousias, S., Kutulakos, K.N., Sankaranarayanan, A.C., Narasimhan, S.G., Gkioulekas, I. (2019). A theory of Fermat paths for non-line-of-sight shape reconstruction. *Proceedings of the IEEE/CVF Conference on Computer Vision and Pattern Recognition*, pp. 6800–6809.

Yi, S., Kim, D., Choi, K., Jarabo, A., Gutierrez, D., Kim, M.H. (2021). Differentiable transient rendering. *ACM Transactions on Graphics (TOG)*, 40(6), 1–11.

Zwicker, M., Jarosz, W., Lehtinen, J., Moon, B., Ramamoorthi, R., Rousselle, F., Sen, P., Soler, C., Yoon, S.-E. (2015). Recent advances in adaptive sampling and reconstruction for Monte Carlo rendering. *Computer Graphics Forum*, 34(2), 667–681.

PART 2

Spectral Imaging and Processing

4

Hyperspectral Imaging

Nathan HAGEN

Department of Optical Engineering, Utsunomiya University, Japan

4.1. Introduction

Spectral imaging systems measure the spectral irradiance $I(x, y, \lambda)$ of a scene, collecting a 3D dataset typically called a datacube (see Figure 4.1). Since datacubes are of higher dimensionality than the 2D detector arrays available, system designers must either measure time-sequential 2D slices of the cube, or divide the datacube into multiple 2D elements that can be recombined into a cube in post-processing. These are typically described here as scanning and snapshot approaches.

In this chapter, we give a broad survey of scanning and snapshot instrumentation for doing spectral imaging and explain the advantages and disadvantages of various techniques, in order to explain why system designers choose different techniques depending on the needs of each application.

The terms *spectral imaging*, *imaging spectrometry* (or *imaging spectroscopy*), *hyperspectral imaging* and *multispectral imaging* used to describe the field are often used interchangeably. However, we generally find that in applications where the imaging aspect of the measurement is the focus of attention, authors generally prefer (multi-, hyper-)spectral imaging, whereas in applications where the spectrum is the focus of attention, imaging spectrometry/spectroscopy is more common.

Computational Imaging for Scene Understanding,
coordinated by Takuya FUNATOMI and Takahiro OKABE.
© ISTE Ltd 2024.

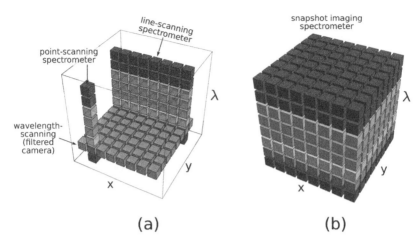

Figure 4.1. *The portions of the datacube collected during a single detector integration period for (a) scanning and (b) snapshot devices (source: from Hagen and Kudenov (2013)). For a color version of this figure, see www.iste.co.uk/funatomi/computational. zip*

Many researchers make a distinction between multispectral and hyperspectral imaging systems, but there are two competing, and sometimes conflicting, definitions of the difference. The first definition distinguishes the two by the number of spectral bands, by saying that *multispectral* is for systems with up to 10 spectral bands (or 20, or 100, depending on the author), whereas *hyperspectral* systems contain more bands than that limit. The second definition is largely limited to the field of remote sensing, in order to distinguish between the type of data obtained by the Landsat series of Earth-observing satellites and data obtained by systems based on grating spectrometers. Figure 4.2 provides an example of the type of measurements provided by these two. The filter-based Landsat system has gaps between spectral channels and overlapping channels. While grating spectrometers certainly have overlaps between their wavelength samples, their discrete data provide an approximate form of the underlying continuous spectrum. This makes it easier, for example, to single out a particular spectral line feature and integrate across it to get the total strength of emission or absorption from a narrow spectral line, even when the spectrometer resolution varies, as we can see from Figure 4.2(b).

In order to compare the strengths and weaknesses of different spectral imaging architectures, researchers generally discuss the following instrument properties:

– spectral range, resolution, resolving power;

– image field of view, resolution and resolving power;

– F-number, light collection efficiency;

– frame rate;

– size, weight and power consumption (SWAP);

– robustness;

– dynamic range.

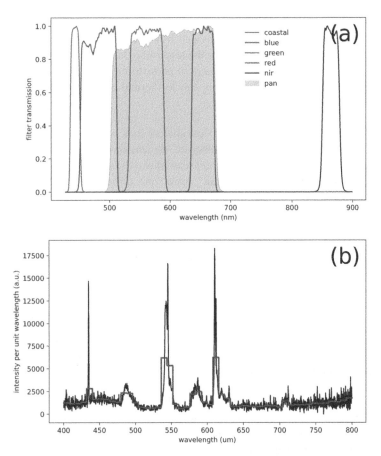

Figure 4.2. *(a) The transmission spectra for six of Landsat 8's spectral filters. (b) A fluorescent lamp emission spectrum obtained with a grating spectrometer, shown at a resolving power of 912 (black line) and 25 (red line) – i.e.* $N_w = 1824$ *and 50, respectively. For a color version of this figure, see www.iste.co.uk/funatomi/ computational.zip*

In many cases, the resolving power in both imaging and spectral dimensions is abbreviated by referring to the number of voxels in the datacube produced by the system. If we use N_x, N_y and N_w to enumerate the number of sample elements along the (x, y) spatial and λ spectral axes, respectively, then the total number of datacube voxels will be given by $N_x N_y N_w$. The term *resolution* is an idealized quantity giving the full-width at half-maximum (FWHM) width of the point spread function (PSF). For non-computational systems, the resolution typically varies only a small amount at different points within the datacube, so that it can be approximated as a scalar. For computational systems, however, the resolution typically varies within the datacube, and depends also on the scene under measurement. The *resolving power R* of a system is given by the width of the domain divided by the Nyquist-limited PSF. Thus, the resolving power of a spectrum would be the lesser of (a) the PSF width divided by the spectral range, or (b) $2N_w$. The *robustness* of an instrument refers to its immunity to external perturbations such as vibration, shock, thermal changes, humidity changes, etc.

The different types of imaging spectrometers can be divided into three basic architectures, determined by the number of dimensions over which the systems scan. Raster-scanning point spectrometers scan in two dimensions; slit spectrometers, tunable filter spectrometers, etc., scan across one dimension; snapshot imaging spectrometers require no scanning. Each of these architectures has a number of different implementations, as we detail below.

In order to use a point spectrometer for imaging spectrometry, the scene is scanned across the instrument's input pinhole aperture with the use of two orthogonal galvo mirrors (raster scanning) or just one mirror if the instrument platform is itself moving (linear whiskbroom scanning), allowing the collection of a full 3D datacube (see Figure 4.3). Note that linear whiskbroom scanning may seem an odd choice, since it creates strangely diagonally oriented voxels, but it is actually a good architecture choice for systems that require adjusting the swath/track width. This is particularly useful in airborne systems, in which the height and speed of the aircraft can be altered to compensate for changes in the swath width.

The second measurement mode is "pushbroom" scanning. If the slit of a line-imaging spectrometer is oriented orthogonal to the motion of the platform, then no moving components are needed – the readout rate of the spectrometer and the speed of platform motion are matched to capture the spectrum at each pixel of the scene.

The third measurement style is the "staring" mode, in which the device captures a 2D image of a scene at one wavelength channel. After adjusting the spectral filter, a second image is captured and so on until the full scene datacube is collected. Because the platform will be moving during this time, the image capture rate must be timed carefully with the platform speed, and the dataframes must be registered to one another in post-processing. Figure 4.3 shows views of each of these scanning modes in use by airborne and satellite imaging spectrometers.

Image acquisition modes

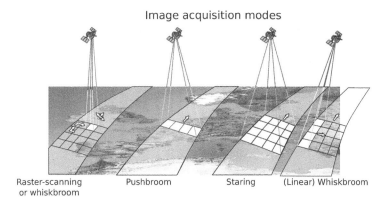

Raster-scanning   Pushbroom   Staring   (Linear) Whiskbroom
or whiskbroom

Figure 4.3. *Swath is cross-track; track is the platform motion direction (figure adapted from http://wray.eas.gatech.edu/remotesensing2015/). For a color version of this figure, see www.iste.co.uk/funatomi/computational.zip*

4.2. 2D (raster scanning) architectures

Spectrometers designed without imaging in mind are "point spectrometers". The most commonly found implementation of this is a fiber-optic-coupled spectrometer, in which the field stop is the fiber core itself, as shown in Figure 4.4. In this case, a fiber placed at an image plane will collect light from a field of view, referred to as an "instantaneous field of view (IFOV)" when considering a single patch within the scene, where the IFOV angle is determined by the focal length of the fore-optics and the size of the field stop. While the fiber core is generally small, it is typically much larger than the width of the slit, so that the light input into the spectrometer is confined to a long thin region emitted from the fiber output.

Because the extent of the object measured by a fiber-coupled spectrometer is limited to a small spatial region, it is easy to see that one approach to constructing an imaging spectrometer is to scan the system point-by-point across the scene, either by using galvo mirrors to scan the image across the face of the fiber, or by using translation stages to scan the fiber face across the image plane. Because collecting light into a fiber is lossy, users may be tempted to place the spectrometer's entrance slit directly at the fore-optics' image plane. While this improves the light throughput, in many grating spectrometers it also limits the ability to scan along the long dimension of the slit, since these spectrometers suffer from blurring of the light along the slit's long dimension. Thus, for use in *imaging* spectroscopy, these instruments require a pinhole aperture rather than a narrow slit, causing a significant loss in light throughput. At the detector array, a point spectrometer's spectrum is dispersed in a line across a row of pixels (the detector's x dimension). For most instruments, the sensor used is actually a 2D detector array, where the y dimension (the direction along the slit) is used to capture additional light rather than to perform imaging. As a result, the y-dimension data is commonly binned together to produce a single value at each point along the spectrum, allowing for fast readout rates.

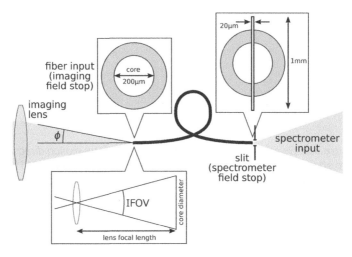

Figure 4.4. *A fiber spectrometer accepts light focused from the fore-optics onto the fiber input (the core diameter is exaggerated here for clarity). A slit is placed at the fiber output to limit the size of the field in the horizontal direction, while allowing all of the light through in the vertical direction. The core diameter (200 µm) and slit dimensions (20 µm × 1 mm) are example values, indicating typical relative sizes. For a color version of this figure, see www.iste.co.uk/funatomi/computational.zip*

While there are many different types of point spectrometers, we survey several of the most prominent designs:

– Czerny–Turner grating spectrometers;

– transmissive grating/prism spectrometers;

– coded aperture spectrometers;

– echelle spectrometers.

4.2.1. *Czerny–Turner grating spectrometers*

The Czerny–Turner (often abbreviated as CZT) spectrometer design uses two spherical mirrors and one plane grating to image a dispersed slit across the detector array (Czerny and Turner 1930). Figure 4.5(a) shows a standard $f/10$ Czerny–Turner design but adds a cylindrical lens in front of the detector array to correct for the inherent astigmatism in the design (Lee et al. 2010). The two spot diagrams underneath the optical layout show the uncorrected (astigmatic) and corrected spot diagrams from a single wavelength at the image plane. Whereas the uncorrected CZT design produces a PSF that is 672 μm long, the cylindrical lens manages to correct this so that the spot becomes nearly diffraction-limited (10 μm spot radius). For non-imaging applications, the cylindrical lens is unnecessary and so is generally not used.

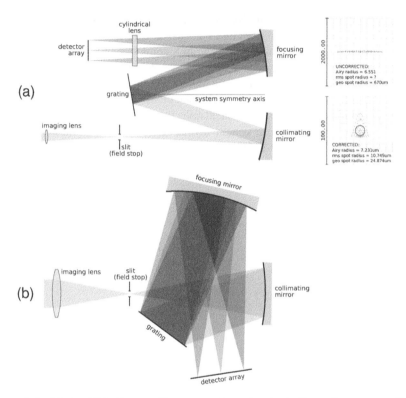

Figure 4.5. *(a) An $f/10$ Czerny–Turner spectrometer, including a (non-conventional) cylindrical lens for astigmatism correction. (b) An $f/2.8$ crossed Czerny–Turner spectrometer. For a color version of this figure, see www.iste.co.uk/funatomi/ computational.zip*

One of the advantages of a CZT design is that because the plane grating is relatively inexpensive and is used in reflection, some spectrometer manufacturers offer using a turret of gratings, so that users can readily switch between gratings with different blaze angles or different line spacings.

Figure 4.5(b) shows a folded implementation of the CZT design, in which the light paths are allowed to overlap one another in order to make the system more compact. The tradeoff with this approach is that it becomes more difficult to insert stray-light reducing baffles into the system (Pribram and Penchina 1968).

4.2.2. Transmission grating/prism spectrometers

For each reflective grating system geometry, there is a corresponding unfolded layout that can make use of transmission gratings. Historically, reflective gratings have been more common than transmissive, since they could be manufactured directly on metal substrates, by cutting grooves into the surface with a ruling engine. However, with the advent of modern lithographic and holographic manufacturing techniques, transmissive gratings are now as easy to make as reflective ones are, and bring their own system design advantages. These include the ability to use an unfolded optical layout, a reduced alignment sensitivity and a lack of Wood's anomalies (the sharp dips in the efficiency spectrum of a metal-coated reflective grating) (Ibsen Photonics 2017). Figure 4.6 shows two example system layouts employing transmission gratings, paralleling the CZT and crossed-CZT layouts used with reflective gratings.

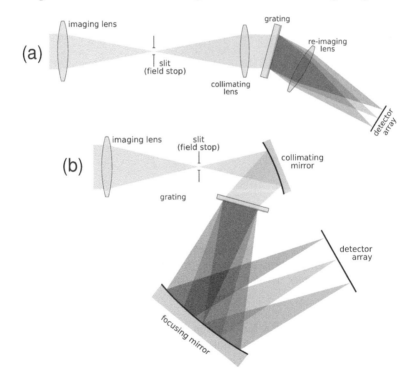

Figure 4.6. (a) An $f/2.8$ transmission grating spectrometer. (b) An $f/2.5$ folded transmission grating spectrometer. For a color version of this figure, see www.iste.co.uk/funatomi/computational.zip

Prisms are a third alternative available to the system designer. They have much less dispersive power than gratings do, though this can be compensated by the use of compound prisms and high angles of incidence (Hagen and Tkaczyk 2011). Their biggest advantage is that they can have a high efficiency over a very wide spectral range, allowing for spectrometers to capture data over wider bandwidths spanning multiple octaves. Compound prisms such as the double Amici type can also be used to provide a direct-view geometry, in which the optical path maintains a straight line, while providing the dispersive power needed for spectrometry.

For many researchers, the dispersion nonlinearity is a severe downside for prisms. However, for applications such as spectroscopic ellipsometry that work primarily in the wavenumber domain, the nonlinearity can be beneficial, by providing improved sampling in the short-wavelength region of the spectrum, where it is needed most.

A final alternative to system designers is to combine prisms and gratings, producing "grisms". Like the Amici prism, a grism disperser can be designed to not bend the beam at all, so that a direct-view geometry is possible, as shown in Figure 4.7.

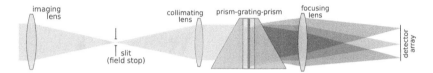

Figure 4.7. *A direct-view* $f/2.3$ *grism spectrometer. For a color version of this figure, see www.iste.co.uk/funatomi/computational.zip*

4.2.3. *Coded aperture spectrometers*

In a conventional slit spectrometer, the slit is typically quite narrow – often 5 to 50 µm wide – so that the amount of light captured in the spectrum is limited. While widening the slit allows more light into the system, it also reduces the spectral resolution, directly demonstrating the throughput-resolution tradeoff. In order to break the tradeoff, a coded aperture replaces the narrow slit with a wide slit containing a binary pattern of holes (Harwit and Sloane 1979). We can consider each column of the coded aperture as an effective slit, but each slit will then be overlapped with its neighbors on the detector array. If the aperture code is properly designed, then each column code can be made orthogonal to every other column, so that they can each be extracted from one another in post-processing.

The primary requirements for making this work well are that the aperture codes should be imaged well (i.e. not severely blurred or distorted) and the mask should be illuminated uniformly. The requirement for good imaging quality places a limit on the minimum size of the mask columns, since it is difficult to achieve good broadband imaging for mask elements smaller than about 5 µm. In order to fulfill the requirement for uniform illumination, the mask is generally placed at the Fourier plane of the objective lens. This works well if the source being measured does not have strong angular variations in intensity or spectrum.

Figure 4.8 shows an optical layout for a coded aperture spectrometer. Because the mask is two-dimensional, unlike a slit, the figure shows the propagation of the (x, y, λ) datacube incident on the front of the mask, as the light propagates through the system. Because the aperture code contains equal numbers of transparent and opaque cells, the overall throughput of the system drops by half, but it increases by N over that of a conventional slit spectrometer, where N is the number of columns in the aperture code. Using modern detector arrays, it is not difficult to achieve throughput increases of a factor of 20 (Gehm and Brady 2006).

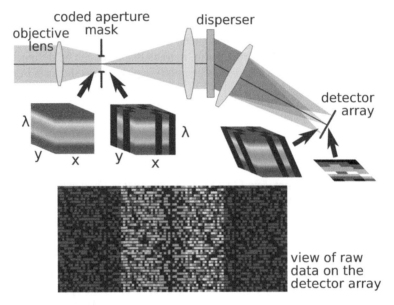

Figure 4.8. *An $f/1.8$ coded aperture spectrometer. The objective lens does not image the scene onto the coded aperture mask, since the mask is placed at the lens' pupil plane in order to illuminate the mask as uniformly as possible. The raw image on the detector array shows a view of what the data would look like when viewing a uniform scene containing three sharp spectral lines (shown here as blue, green and red) (figure adapted from Hagen and Kudenov (2013)). For a color version of this figure, see www. iste.co.uk/funatomi/computational.zip*

4.2.4. Echelle spectrometers

Echelle spectrometers take advantage of the increase in pixels in modern 2D detector arrays to disperse the spectrum in both x and y axes simultaneously (Nagaoka and Mishima 1923). A typical system layout is shown in Figure 4.9. As a result, the high-resolution spectrum is sliced into many different pieces which can then be re-assembled together by calibration and post-processing. With this approach, resolving powers of 10,000 to 70,000 are achievable with off-the-shelf parts (Eversberg 2016; Jones et al. 2021). Since typical commercial spectrometers deliver resolving powers of up to 1,000, echelle systems achieve a factor of 10 to 70 improvement in resolution. The tradeoffs include a larger optical system, a larger detector array and a lower optical efficiency.

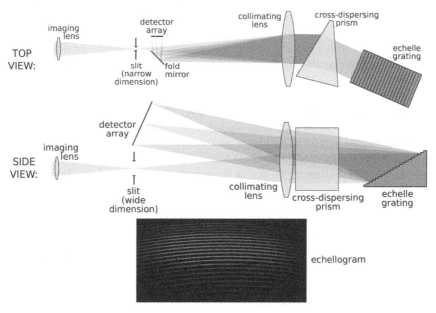

Figure 4.9. *An $f/6$ echelle spectrometer. An example raw image on the detector array shows the echellogram. The slit is kept short in order to maintain separation between the different diffraction orders. For a color version of this figure, see www.iste.co.uk/funatomi/computational.zip*

4.3. 1D scanning architectures

While point spectrometers can be made simple and compact, their use in imaging spectrometry is limited due to their low frame rate or, if a pinhole aperture is used, their poor optical throughput. As a result, most imaging spectrometers use an architecture that scans in one dimension of the datacube – either the x direction ("whiskbroom scanner"), the y direction ("pushbroom scanner") or the λ direction ("tunable filter").

One of the advantages that line-imaging spectrometers have for applications is that they can be placed into a pushbroom scanning architecture when the slit is oriented perpendicularly to the scan direction. If imaging objects moving along a conveyor belt, or if the spectrometer itself is moved at a constant velocity (such as when looking down from an airborne or spaceborne platform), then the scene itself does the scanning without any need for scanning hardware in the spectrometer.

4.3.1. *Dispersive spectrometers*

With the exception of the echelle spectrometers, the same architectures used in point spectrometers can be used for linear imaging spectrometers as well, though with some modifications. In the CZT design, for example, the main hindrance to imaging spectrometry is the large amount of astigmatism in the system (Simon et al. 1986). As shown in Figure 4.5(a), however, this can be corrected by either replacing one of the spherical mirrors with a toroidal mirror or equivalently by inserting a cylindrical lens in front of the detector array (Lee et al. 2010).

Another way to correct astigmatism is to use a spherically curved grating in place of the plane grating used in CZT designs, giving the Offner spectrometer (Figure 4.10). The resulting design is monocentric: the centers of curvature of the grating surface, collimating mirror and focusing mirror all coincide, and both the slit and image are co-planar with the center of curvature (Offner 1987). However, imaging performance can be improved by adjusting the design away from ideal concentricity (Prieto-Blanco et al. 2006).

The benefit of the Offner design over the CZT layout is improved imaging performance, allowing us the use of longer slits and astigmatism correction. However, for many years, a stumbling block to the widespread use of the Offner design was the difficulty with manufacturing high-quality gratings on spherical surfaces. Modern manufacturing methods have eased these difficulties, so that spectrometers based on these designs are now commercially available.

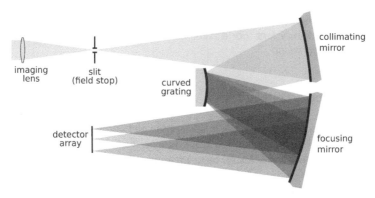

Figure 4.10. *(a) An $f/3.8$ Offner spectrometer. For a color version of this figure, see www.iste.co.uk/funatomi/computational.zip*

The coded aperture spectrometer (Figure 4.8 from section 4.2.3) can be adapted to line-imaging spectrometry, so that it can operate in the same way as a conventional slit spectrometer, but with a much higher throughput. One of the requirements for the technique was for the illumination of the mask to be uniform, but in fact, the illumination needs to be uniform only in the direction orthogonal to the dispersion axis (McCain et al. 2006). Thus, we can modify the objective lens for anamorphic imaging, so that the coded aperture is at an image plane for the vertical axis, but at the Fourier plane for the horizontal axis. One way to achieve this is to add a cylindrical lens to a standard objective lens so that its focal length on one axis is half that for the other axis.

Although the coded aperture spectrometer has demonstrated impressive improvements in light throughput, it has yet not seen widespread use. One of the limiting factors in the design is that the imaging quality must be significantly higher than for an equivalent slit spectrometer. However, there is a compromise available: we can replace the single slit with a multi-slit aperture, so that the system with four slits can effectively operate as three slit spectrometers all operating in parallel (Martin 1974; Shaw 1993). The result is a factor of four improvement in throughput, but the design must then either allow the detector and optical design to stretch to four times their previous length or must sacrifice a factor of four in resolving power. While these tradeoffs were problematic for a long time, the steady improvement in detector array sizes has eased the difficulty so that this approach is now attractive for some applications.

4.3.2. Interferometric methods

Fourier transform spectrometers (FTSs) use the Fourier domain properties of an interferogram to measure the optical spectrum as a function of optical path difference (OPD). Figure 4.11 shows a typical optical layout for a Michelson-type FTS. Detecting the light intensity while moving one of the interferometer mirrors gives the spectrum as a function of the OPD between the two arms of the interferometer. Taking the Fourier transform of the OPD spectrum produces the wavenumber spectrum.

Although the older literature contains many references to the throughput advantages of Fourier transform spectrometers over grating spectrometers, the classical advantages have eroded with improvements in technology. As a result, for modern detectors, Fourier transform and slit spectrometers generally provide comparable SNR. Non-imaging Fourier transform spectrometers are also available to take advantage of the two-dimensional extent of modern detector arrays to distribute the optical path difference between the two beams as a function of location on the array, so that scanning is unnecessary (Padgett and Harvey 1995; Waldron et al. 2021).

4.3.3. Interferometric filter methods

Filter-based approaches to spectral imaging are conceptually simple but often have surprising complexity underneath. The tunable Fabry–Perot interferometer was an early solution to the problem of passing only a narrow wavelength band while reflecting the remaining unwanted light. The equipment and expertise for depositing precise many-layer thin films were not yet developed in the early 20th century, but the expertise for depositing single-layer thin metal films was available, and by placing two of these partial mirrors together we can make a large-aperture filter whose bandpass can be tuned by adjusting the spacing between the two mirrors (Perot and Fabry 1899). In modern usage, Fabry–Perot filters are usually used to describe interference filters employing only one cavity, in contrast to interference filters that use multilayer thin films.

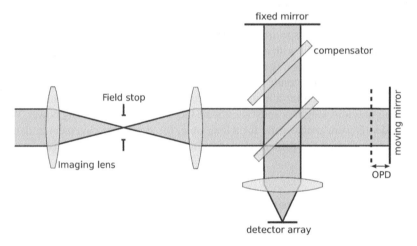

Figure 4.11. *An $f/2$ Michelson-type Fourier transform spectrometer. In practice, the flat mirrors are often replaced with corner mirrors to reduce system sensitivity to changes in mirror tilt angle during scanning. For a color version of this figure, see www.iste.co.uk/funatomi/computational.zip*

Figure 4.12 shows the image produced at the focal plane of the Fabry–Perot-filtered imager, where we see the familiar pattern of thin rings distributed across the face of the image. Each ring indicates a high transmission for the field angles that map to that location, for a given wavelength. If white light is incident on the system, then at each pixel in the image, there will be a series of peaks at different wavelengths, and these will be mixed together if the incident light has too wide a spectral band – when the incident light's spectral width exceeds the Fabry–Perot filter's free spectral range.

The Fabry–Perot transmission is given approximately by the Airy function

$$T = \left[\frac{(1-R)^2}{1 - 2R * \cos(\delta) + R^2} \right]^2$$

for $\delta = 4\pi n d \cos(\theta)/\lambda$, where R is the mirror reflectivity (Cooper 1971). The dependence on the field of view angle θ is incorporated into δ and scales with the mirror spacing d and the refractive index n of the medium between the mirrors. In order to scan over a wavelength for collecting a datacube, we can either adjust the Fabry–Perot cavity thickness d or the cavity refractive index n to tune the filter passband.

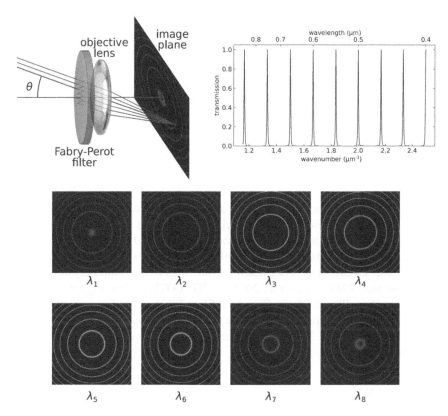

Figure 4.12. *Fabry–Perot filter. The transmission function shown here uses $\theta = 0$, $R = 0.85$, d=3 µm. While the poor reflectivity used here is unusual, it helps to visualize the shape of the function by broadening the peaks. For a color version of this figure, see www.iste.co.uk/funatomi/computational.zip*

Thin film interference filters are now much more common than Fabry–Perot filters, but are based on the same principle of controlling wavelength-dependent interference to generate constructive interference in the desired bands and deconstructive interference in the unwanted bands. Whereas a Fabry–Perot filter may have a thickness d of tens of microns, thin film interference filter layers generally have thicknesses of much less than a micron: a quarter-wave film, for example, has a thickness of $\lambda/(4n)$, or 92 nm for 550 nm light. As a result, thin film filters generally have a weaker angular dependence (Goossens et al. 2018). There are, however, bandpass interference filters that have been specifically engineered to shift smoothly in wavelength with the angle of incidence, so that the transmission bandpass can be accurately tuned by rotating the filter (Anderson et al. 2011).

In order to employ thin film filters for spectral imaging, the most common approach is to prepare a set of filters in a filter wheel that can then rotate each filter into position during the scan sequence. Rotation-engineered filters can achieve a similar behavior by simply placing the filter on a rotational stage for scanning over a wavelength. Similarly, linear-variable (Emadi et al. 2012) and circular-variable (Cabib and Orr 2012) filters provide a continuously variable passband across the face of a large filter, so that the filter can be translated or rotated across the pupil of an imaging system for scanning across the wavelength dimension.

4.3.4. *Polarization-based filter methods*

Several varieties of tunable filters have been developed that make use of polarization to induce constructive and destructive interferences for modulating the filter passband. These include the classic Lyot- and Şolc-type filters (Lyot 1944; Şolc et al. 1954) (these are pronounced "lee-oh" and "sholtz", respectively), plus acousto-optic (Harris and Wallace 1969; Vila-Francés et al. 2010; Katrašnik et al. 2013), electro-optic (Lotspeich et al. 1981) and liquid-crystal tunable filters (Masterson 1989; Berns et al. 2015) (AOTF, EOTF, LCTF). Figure 4.13 shows a three-stage Lyot filter and a six-stage Şolc filter. Whereas the fast axes of the Lyot filter waveplates are all oriented at 45°, the Şolc waveplates are successively tilted in azimuth. For a system of N waveplates, the fan-type Şolc waveplates are oriented at azimuth angles of

$$\theta_i = \frac{\pi}{4N}(2i + 1)$$

for $0 \leq i \leq N - 1$. When the overall thickness of crystal in each system is held constant, the two filters have similar transmission profiles. However, the Lyot-type filter exhibits a slightly narrower transmission peak, and a slightly lower peak transmittance than the Şolc-type filter (Evans 1958). Since the Şolc filter has a larger free spectral range (FSR) but broader peak, the overall resolving powers of the two systems are about the same.

Classic Lyot and Şolc filters use thick crystal waveplates that are rotated about the system optical axis in order to adjust the passband. Electro-optic and liquid-crystal tunable filters are based on a similar principle of using a sequence of waveplates, but can electronically tune not only the orientation but also the retardance value of each waveplate in the sequence. As a result, the EOTF and LCTF tuning speeds are much faster, with switching times of 5~50 ms. Because LCTFs are compact and easy to use and relatively inexpensive to build, these filters are currently the most widely used of the electrically tunable types. While electro-optic tunable filter systems also exist, they are not as common as the LCTFs and AOTFs.

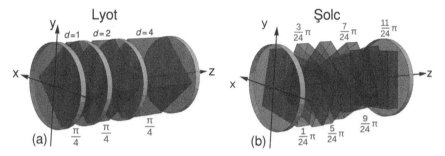

Figure 4.13. *(a) A three-stage Lyot filter and (b) a six-stage Şolc filter. In both diagrams, the polarizers (gray) are all oriented with their transmission axes oriented along the x-axis, while the retarders (blue) have their orientation angles shown (in blue). The Şolc filter waveplates are all of identical thickness. For a color version of this figure, see www.iste.co.uk/funatomi/computational.zip*

The acousto-optic tunable filter is based on a different principle than the previous types. In this case, an acoustic transducer creates a series of standing waves inside the crystal, such that the crystal's refractive index is modulated throughout its volume. In effect, the crystal becomes a volume grating, with the grating diffraction angle determined by matching the k-vectors of the acoustic wave, the ordinary wave and the extraordinary wave light polarizations propagating through the crystal (Chang 1981). Due to this Bragg matching condition, only a narrow range of wavelengths are k-matched such that their diffractions become cumulative over the length of the crystal. And since the Bragg deflection angle depends on the 2D angle of incidence, a line field becomes strongly curved at the detector array (Gorevoy et al. 2022).

Because the AOTF's acoustic transducer can create multiple frequencies, it is possible to create multiple standing pressure waves in the crystal that do not interact, so that it is possible to create multiple passbands. Some commercial AOTFs allow up to eight independent passbands, and these can be either spaced out as desired across the spectrum, or they can be placed side-by-side to produce a single wide passband.

Comparing the AOTF and LCTF tunable filter systems, we can say that the AOTF allows a finer spectral resolution but suffers from a narrower field of view.

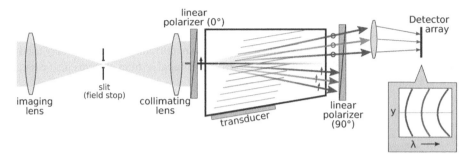

Figure 4.14. *An acousto-optic tunable filter (AOTF) used in a line-imaging configuration. The arrows show the rays diffracted by the acoustic waves in the medium. Since the diffraction angle varies strongly with position y, a line image becomes curved at the detector array. For a color version of this figure, see www.iste. co.uk/funatomi/computational.zip*

4.3.5. *Active illumination methods*

Each of the above methods uses *passive* measurement, in which no control is required over the illumination in order to resolve the incident light spectrum. Several of these techniques can be used in reverse to produce spectrally encoded illumination systems (Abramov et al. 2011). In this implementation, the object is illuminated with a well-defined spectral pattern, such that the imaging side no longer needs to have spectral resolution capability. For example, we can illuminate a scene with a broadband light source transmitted through a tunable filter, and time the image acquisition to coincide with steps in the filter wavelength to produce a datacube of the scene. One advantage of this active illumination setup is that it allows the user to vary the exposure time as a function of wavelength, thus optimizing SNR in situations where scene reflectivity and detection sensitivity vary over the spectral range. One widely implemented example of this approach is in swept-source optical coherence tomography (SS-OCT) (Zheng et al. 2008).

4.4. Snapshot architectures

Snapshot is best used as a synonym for *non-scanning* – i.e. systems in which the entire datacube is obtained during a single detector integration period. Thus, while snapshot systems can often offer much higher light collection efficiency than equivalent scanning instruments, "snapshot" by itself does not mean "high throughput" if the system architecture includes spatial and/or spectral filters. *Snapshot* also does not by itself imply *fast*. A dynamic scene can blur the image obtained using either a snapshot or a scanning device, but while this creates motion blur in a snapshot system, it induces measurement *artifacts* in a scanning system. Since motion blur is easy to understand and easy to correct, it is much less of a serious problem than artifact errors, which are more difficult to correct.

In recent years, some researchers and companies have started using the word "snapshot" to refer to spectral imaging systems that collect a single-wavelength image slice in one detector integration period. While this may be a snapshot from the standpoint of a camera, it is misleading nomenclature in the case of acquiring a datacube, since scanning along the wavelength dimension is still required.

Finally, when comparing scanning and snapshot devices, we can note that the division between the two is not as black and white as we might expect. For example, designers have produced sensor architectures that mix both snapshot and scanning techniques, in order to reduce the number of scans required. The earliest example of this of which we are aware (although it seems likely that astronomers tried this earlier) is a patent by Busch (1985), where the author illustrates a method for coupling multiple optical fibers such that each fiber is mapped to its entrance slit within a dispersive spectrometer's field of view. We can also find examples such as Chakrabarti et al., who use a grating spectrometer with four separate entrance slits simultaneously imaged by the system, with the respective slits spaced apart such that the dispersed spectra do not overlap at the image plane (Chakrabarti et al. 2012). This setup can be used to improve light collection by a factor of four, at the expense of either increasing the detector size, or reducing the spectral resolution, by a factor of four. Ocean Optics' SpectroCam is another example of a mixed approach, in which a spinning filter disk is combined with a pixel-level spectral filter array to improve the speed of multispectral image acquisition.

4.4.1. *Bowen–Walraven image slicer*

Bowen's original image slicer (Bowen 1938) used a series of tilted mirrors to slice the image into thin strips and re-organize them into a single long slit. Walraven later took this concept and created a design that was easier to manufacture, using only a thick glass plate (with a 45° angle cut into one end) cemented to a wedge-cut prism (Walraven and Walraven 1972; Avila and Guirao 2009; Tala et al. 2017). The prism-based approach made the device easier to align and assemble, but its use is primarily limited to "slow" beams with f-numbers above 30 (Cardona et al. 2010). The layout for the resulting device is shown in Figure 4.15. A beam of light entering the glass plate reflects multiple times due to the total internal reflection, except in those regions where the plate meets the prism edge, where the beam transmits. The succession of partial reflections and partial transmissions transforms the input beam into a long slit, which can then be used as the input to a slit spectrometer. One drawback is that although the individual slices of the original beam fall along a line, the slices do not share the same focal distance. Thus, this type of slicing works well for a high f-number system, but is problematic at low f-numbers.

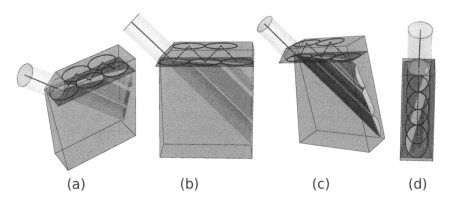

Figure 4.15. *Various views of a Bowen–Walraven image slicer, illustrating how the glass plate and wedge-cut prism combine to slice the optical beam into a long slit. Shapes shown in yellow indicate the light passing through the slicer; the beam reflecting inside the top plate is not shown for clarity (figure from Hagen and Kudenov (2013)). For a color version of this figure, see www.iste.co.uk/funatomi/ computational.zip*

4.4.2. *Image slicing and image mapping*

Figure 4.16 shows an example system layout for an image slicer: the slicer mirror is placed at the image plane and reflects each thin strip of the image in a different direction (Weitzel et al. 1996). In effect, each facet of the slicer mirror acts as the slit of a spectrometer. By placing a different pupil at each of the facet reflection directions, we can either collect all of the strips together into one long slit, or we can use independent spectrometer optics at each pupil to analyze the spectrum along each slice. By this process, if we slice an image into 50 strips, then we can increase the light collection capacity of the system by a factor of 50 without sacrificing spectral resolution. Instead, we trade off in system size and complexity by building an array of 50 line-imaging spectrometers for the back-end of the system.

Because the procedure for manufacturing image slicers requires precision alignment and high-quality optical surfaces, it is only in recent decades that advances in optical manufacturing have made this practical (Suematsu et al. 2017). The first slicer mirrors were manufactured as thin glass blades that were polished using conventional optical methods, coated with a reflecting layer, and then carefully aligned and fixed in place as a set (Laurent et al. 2004). Later groups attempted to use ultra-precision diamond machining to cut and polish all of the facets on a single monolithic metal substrate (Schmoll et al. 2006). Finally, some of the advanced image slicers use curved facets to improve the optical performance of the back-end array optics (Content 1998). Each of these options is shown in Figure 4.17.

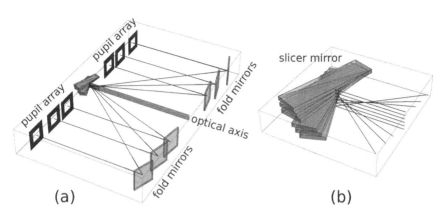

Figure 4.16. *The system layout (a) for an image slicer, and closeup (b) of the slicer mirror. For clarity, the layout only shows the chief rays corresponding to each mirror facet, and the spectrometer optics behind each pupil have been omitted. (These back-end optics are located behind each pupil in the array, and include a collimating lens, a disperser, a re-imaging lens and a detector array.) (figure from Hagen and Kudenov (2013))*

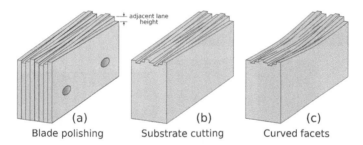

Figure 4.17. *A faceted mirror constructed with (a) a set of individually polished glass blades pressed together, (b) cutting a monolithic substrate with a diamond tool. The optically active surface is facing upwards. While all of the tilts shown here are along the long dimension of the facets, in practice the facets are tilted in 2D, so that each wide facet shown here would actually be composed of multiple thinner facets, each of which is tilted along the short dimension of the cuts (figure from Hagen (2020a))*

Conventional image slicing uses a separate pupil for each slice of the image, so that the number of pupils determines the lateral dimension of the datacube. Thus, if the short dimension of the mirror facets is defined to be the x-dimension of the datacube, then a system having 50 facets will have a datacube with $N_x = 50$. As a result, image slicing is good for systems in which the datacube $N_w \gg N_x$. This is problematic for applications such as microscopy, where it is more useful to have a large image and a small spectral resolving power. The "image mapping" technique uses a variation of image slicing that improves spatial sampling at the cost of making it more difficult to achieve high spectral resolution. To achieve this, we modify the slicer mirror to have a pattern of tilts that replicates periodically across the face of the mirror. Thus, multiple facets will share the same tilt, and thus will be directed into the same pupil at the back-end of the system (Gao et al. 2009, 2011). With this kind of setup, the number of pupils now defines not the N_x size of the datacube but rather N_w – a system with 50 pupils can measure spectral images with up to 50 spectral elements. Figure 4.18 shows an example image mapping spectrometer layout.

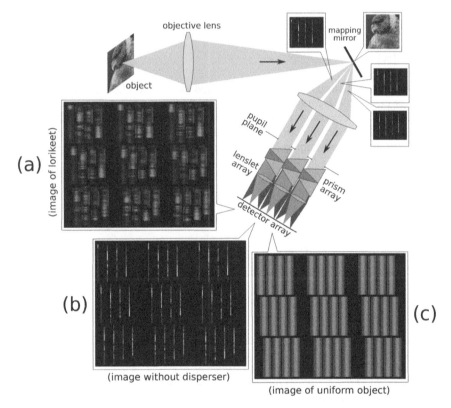

Figure 4.18. *The image mapping spectrometer (IMS) system layout. Three different raw images are shown, corresponding to a setup in which (a) the lorikeet is being imaged through the full system (as shown), (b) the lorikeet is being imaged in a system in which the prism array has been removed and (c) a spectrally and spatially uniform object is being imaged through the full system. (For clarity, in all three examples of raw detector data shown here, the data is shown in color as if imaged by a color detector array, even though a monochromatic array is used in existing instruments.) (figure from Hagen and Kudenov (2013)). For a color version of this figure, see www.iste.co.uk/funatomi/computational.zip*

Pawlowski et al. recently succeeded in adapting an image mapping spectrometer to produce a 16-bit $(N_x, N_y, N_w) = (210, 210, 46)$ datacube at 100 Hz frame rates (Pawlowski et al. 2019).

4.4.3. Integral field spectrometry with coherent fiber bundles (IFS-F)

In 1958, N. S. Kapany introduced the concept of placing a coherent fiber bundle at the image plane and then squeezing the output end of the bundle into a long thin line for easy adaptation to fit into the long entrance slit of a dispersive spectrometer (Kapany 2004). It appears that this idea had to wait until 1980 before it could be implemented (Vanderriest et al. 1980), but adaptations by other astronomers soon followed (Barden and Wade 1988; Arribas et al. 1991).

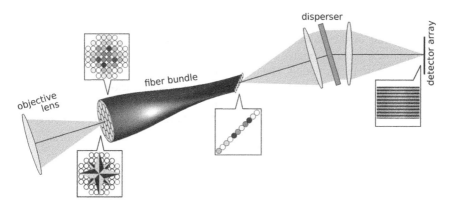

Figure 4.19. *The system layout for an integral field spectrometer with coherent fiber bundles (IFS-F): the object is imaged onto the face of a coherent fiber bundle. At the opposite end of the bundle, the fibers are splayed out ("reformatted") into a linear array, which is compatible with the input of a standard slit spectrometer. At the input and output faces of the fiber bundle, there may be lenslets coupled to each fiber in order to improve light throughput (figure from Hagen and Kudenov (2013)). For a color version of this figure, see www.iste.co.uk/funatomi/computational.zip*

Though the basic idea is simple, manufacturing these reformatted fiber bundles for high performance has proven to be surprisingly difficult. Squeezing the bundle into a thin line generally produces some broken fibers, and coupling light into the fibers also produces significant light loss. Finally, light exiting the fibers will have a larger numerical aperture – a phenomenon termed "focal ratio degradation" – so that the spectrometer optics must have sufficient performance to handle a low f-number input (Allington-Smith and Content 1998; Ren and Allington-Smith 2002). Following the work by astronomers, researchers in other fields began adapting similar approaches, the progress of which is summarized below:

Authors	Year	N_x	N_y	N_w	Frame rate
Matsuoka et al. 2002	2002	10	10	151	0.2 Hz
Fletcher-Holmes and Harvey 2005	2005	14	14	500	5 Hz
Kriesel 2012	2012	44	40	300	30 Hz
Wang et al. 2019	2019	188	170	61	3.6 Hz

4.4.4. *Integral field spectroscopy with lenslet arrays (IFS-L)*

Another method of mapping the 3D datacube onto a 2D detector array was developed by G. Courtes in 1960 (Courtès 1960, 1980). This approach uses a lenslet array to create an array of demagnified pupils that fill only a small portion of the available space in the image. This extra space can then be filled in with spectra by imaging through a slitless dispersive spectrometer. Figure 4.20 shows the optical layout for this approach.

As with all of the non-computational snapshot spectral imaging methods, the IFS-L approach takes advantage of unused étendue in the optical system in order to collect the spectral information. The front-end optics have a lower numerical aperture than the lenslets do, but a high-performance spectrometer back-end can be used to work at high numerical aperture to image the entire dataset. As with many other techniques, other fields began to adapt this method after the astronomers had started developing it, as summarized below:

Authors	Year	N_x	N_y	N_w	Frame rate
Courtès et al. 1988	1988	44	35	580	?
Bodkin et al. 2012	2012	180	180	28	30 Hz
Dwight and Tkaczyk 2017	2017	200	200	27	3 Hz
Zhang and Gross 2019	2019	21	29	40	?
Yu et al. 2021	2021	28	14	180	?

4.4.5. *Filter array camera (FAC)*

An FAC uses an array of filter elements to separate the different wavelength bands in the datacube. In a mosaic filter array, the individual filter elements are applied at the pixel level, so that the detector array itself becomes a spectral imager, as shown in Figure 4.21(a) (Eichenholz et al. 2010). This has the advantages of being extremely compact and allowing the user to change the objective lens to meet the application's needs. (Note, however, that accurate datacube measurements will require the system to be recalibrated for each lens used (Goossens et al. 2018; Hahn et al. 2020)). Mosaic filter array users also need to be careful with how they define the focus of the lens, since it is necessary to bandlimit the image to prevent aliasing. The typical way to do this is to defocus the image a little so that the PSF width is approximately matched to the tile size of the mosaic filter (i.e. 5 pixels across when using a 5×5 filter mosaic). Without proper bandlimiting, aliasing will produce spurious spectral features appearing on the edges of objects (Dubois 2005; Jaiswal et al. 2016).

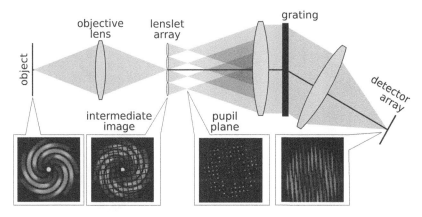

Figure 4.20. *The system layout for an integral field spectrometer with lenslet arrays (IFS-L) (figure from Hagen and Kudenov (2013)). For a color version of this figure, see www.iste.co.uk/funatomi/computational.zip*

In a tiled filter array, the individual filters are much bigger than in a mosaic array, so that each filter tile covers the pupil of a lens in an array of miniature cameras. This can take the form of using an array of individual cameras or of using a lenslet array and filter array matched to a monolithic detector array (Figure 4.21(a)) (Shogenji et al. 2004; Mathews 2008; Gupta et al. 2011; Mu et al. 2019). This approach has the advantage of making the filter array much easier to manufacture and customize, but requires careful registration between the different sub-imagers and increases problems that may arise due to parallax between the individual cameras. In addition, this approach also requires either an array of field baffles (as used by Shogenji et al. (2004)) or a field stop in the objective lens, in order to prevent images from the various sub-imagers from overlapping one another (Maione et al. 2019).

A drawback to any approach using filter arrays is that they have low light throughput in comparison to the unfiltered snapshot approaches, when comparing systems with the same pupil size. On the other hand, the FAC approach makes it easy to increase the system pupil size by adding additional cameras, and so we can compensate for the low throughput by increasing the system size. Additionally, low light throughput is not a problem for many applications, where either the light level may be sufficient in most circumstances (outdoor daylight measurements, for example) or where the user has control over the illumination and can increase the light level to compensate (such as in many industrial imaging applications). It is also possible to improve the light collection efficiency of filter-based systems by combining them with some kind of dispersive or beamsplitting approach that partially separates the bands before transmitting the light through the filters (Kobylinskiy et al. 2020).

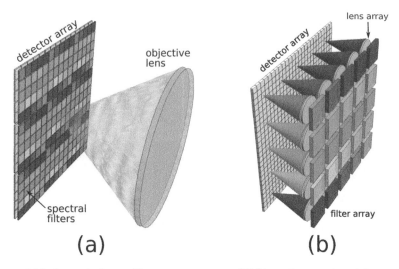

Figure 4.21. *Layouts for a filter array camera (FAC) system, using (a) a mosaic design (pixel-scale filters), or (b) a tiled design (with aperture-scale filters) (figure adapted from Hagen and Kudenov (2013)). For a color version of this figure, see www.iste.co.uk/funatomi/computational.zip*

4.4.6. *Computed tomography imaging spectrometry (CTIS)*

With the improvement in the performance of modern computers, researchers began developing computational sensing approaches to snapshot imaging spectroscopy. The first of these was the computed tomography imaging spectrometer (CTIS) (Okamoto and Yamaguchi 1991; Bulygin and Vishnyakov 1992; Descour 1994), which images a scene through a custom-designed 2D disperser (see Figure 4.22). Because this mixes the spatial and spectral elements on the detector array, it is necessary to reconstruct the datacube from the raw data using calibration data and reconstruction algorithms. Since the raw data on the detector array can be viewed as 2D tomographic projections of the 3D datacube taken at multiple view angles, the most commonly used reconstruction algorithms for CTIS have been MART and EM Kudenov (2012).

While the CTIS approach can be quite compact, it faces serious problems dealing with its computational complexity, calibration accuracy and measurement artifacts (Hagen and Dereniak 2008). This is a common theme among most computational sensors, and is not reflected well by the literature, where authors are discouraged from being forthright in acknowledging the weaknesses of the instrument that they are investigating. Thus, while CTIS was a promising concept, it has not yet shown the ability to achieve a performance level sufficient for use in scientific applications.

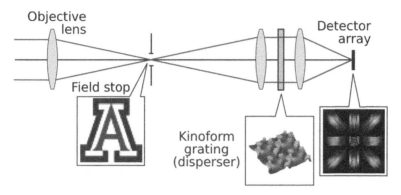

Figure 4.22. *The system layout for a computed tomography imaging spectrometry (CTIS) (figure from Hagen and Kudenov (2013)). For a color version of this figure, see www.iste.co.uk/funatomi/computational.zip*

4.4.7. *Coded aperture snapshot spectral imager (CASSI)*

The coded aperture snapshot spectral imager (CASSI) attempts to take advantage of compressive sensing theory to allow snapshot datacube measurement with simple hardware. Coded aperture spectrometers replace the entrance slit of a dispersive spectrometer with a field stop containing a wide binary coded mask. The encoded light passes into an imaging spectrometer back-end (i.e. collimating lens, disperser, re-imaging lens and detector array). As with the CTIS approach, the irradiance projected onto the detector array is a mix of spatial and spectral elements of the datacube (see Figure 4.23). If this aperture code satisfies the requirements of compressive sensing, then we can use compressive sensing reconstruction algorithms to estimate the object datacube (Gehm et al. 2007). The resulting system layout is not only extremely compact, but allows the use of a modest-size detector array, and so is capable of imaging at high frame rates. Although the CASSI looks almost identical to the coded aperture spectrometer (Figure 4.8), the differences are that CASSI images the object onto the aperture mask, uses a random binary code instead of an orthogonal code, and attempts to reconstruct a full datacube from the measurement rather than a single slice.

As with CTIS, however, CASSI systems have persistent difficulty in dealing with issues of calibration accuracy and reconstruction artifacts, and have not yet demonstrated science-quality data after more than a hundred research publications.

Moreover, most of the CASSI publications show systems that are not compressive, despite their name and their purported aims, since they use aperture codes that are 2×2 or 3×3 in size, and yet report reconstructed datacube sizes in terms of the raw pixel count and not the aperture code count.

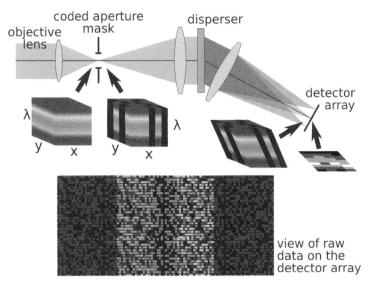

Figure 4.23. *(Top) The system layout for a coded aperture snapshot spectral imager (CASSI), showing only the "single-disperser" configuration. (Bottom) The pattern on the detector array due to imaging a coded aperture mask through a disperser, for an object that emits only three wavelengths (the wavelengths used in the example image here are the shortest, middle and longest wavelengths detected by the system) (figure adapted from Hagen and Kudenov (2013)). For a color version of this figure, see www.iste.co.uk/funatomi/computational.zip*

4.5. Comparison of snapshot techniques

There are many ways to compare the various imaging spectrometer implementations, such as compactness, speed, manufacturability, ease of use, light efficiency and cost. And while these are all important, different system designers and different applications have different opinions about each of these factors. However, one of the primary limitations to snapshot imaging spectrometer performance remains the size and speed of detector arrays. Thus, efficiently making use of detector array pixels leads to direct improvements in compactness, speed and cost. Allington-Smith (2006) has previously termed this metric the *specific information density*, but it can be given the more familiar name *instrument efficiency*, as it is the product of the optical efficiency η (i.e. average optical transmission times the detector quantum efficiency) with the efficiency μ of using the detector pixels:

$$Q = \eta\mu = \eta\left(\frac{N_x N_y N_w}{M_x M_y}\right)$$

In this formula, N is the number of Nyquist-resolved elements in the imaging spectrometer datacube divided, and M is the number of detector elements (pixels) M required to Nyquist-sample those voxels. Thus, the datacube dimensions are (N_x, N_y, N_w), while the detector array dimensions are (M_x, M_y). The reason that μ differs among snapshot techniques is that each technique requires some space to be left between different slices of the datacube to prevent crosstalk (Bacon and Monnet 2017).

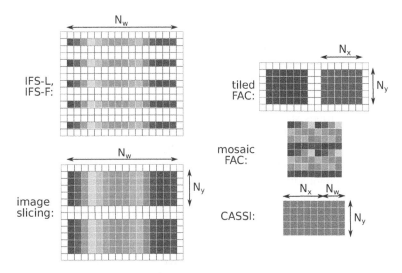

Figure 4.24. *Diagrams showing how the detector utilization efficiency μ is calculated for each technique, given the basic layout of how the datacube is projected onto the 2D detector array. Each square shown here represents a single pixel on the detector array, and each subfigure assumes a datacube of dimensions $(N_x, N_y, N_w) = (5, 5, 16)$, and spacing of $s = 1$ between slices. In general, a spacing of $s = 2$ or 3 is more practical (figure adapted from Hagen and Kudenov (2013)). For a color version of this figure, see www.iste.co.uk/funatomi/computational.zip*

Figure 4.24 gives a comparison of how each snapshot technique lays out its slices on the detector array, allowing us to calculate the utilization efficiency μ. For example, each spectrum in an IFS-L or IFS-F needs $3(N_w + 2s)$ pixels for the spectrum plus surrounding margins. Multiplying this by the total number of spectra in the datacube gives the required number of detector pixels:

$$[M_x M_y]_{\text{IFS-F}} = N_x N_y (N_w + s)(2s + 1) .$$

The resulting efficiency for each of the snapshot techniques is listed in Table 4.1. (CTIS and CASSI are not listed there because it is difficult to assess the number of resolved elements in their datacubes.) Aside from the mosaic filter array, we can see that the tiled FAC approach offers the highest μ for high spatial/low spectral resolution datacubes ("squat-shaped" cubes), whereas image slicing offers the highest μ for low spatial/high spectral resolution datacubes ("tall" cubes). The latter do especially well when the spatial dimensions of the datacube are rectangular $N_x > N_y$.

Instrument	$M_x M_y$	μ
IFS-F	$3N_x N_y (N_w + 2s)$	$3(N_w + 2s)/N_w$
IFS-L	$3N_x N_y (N_w + 2s)$	$3(N_w + 2s)/N_w$
Image slicing	$N_x(N_y + 2s)(N_w + 2s)$	$(N_y + 2s)(N_w + 2s)/N_y N_w$
Tiled FAC	$(N_x + 2s)(N_y + 2s)N_w$	$(N_x + 2s)(N_y + 2s)/N_x N_y$
Mosaic FAC	$N_x N_y N_w$	1

Table 4.1. *The required number of detector pixels $M_x M_y$ and the utilization efficiency μ for each snapshot technique*

Fellgett and Jacquinot were the first researchers to investigate the fundamental limits to μ for various spectrometer technologies (Fellgett 1951; Jacquinot 1954), leading to the Fellgett (multiplex) and Jacquinot (throughput) advantages, now widely associated with Fourier transform and Fabry–Perot spectroscopy (Griffiths et al. 1977). The modern availability of large 2D detector arrays, however, has eliminated the classical Jacquinot advantage, while the improvement in noise performance such that most measurements are now shot-noise-limited has eliminated the classical Fellgett advantage (Schumann and Lomheim 2002). Thus, for most modern filterless snapshot imaging spectrometers, both of the traditional advantages are automatically satisfied: no exit slit is used, and the detector dwell time on every voxel is equal to the full measurement period. These two characteristics can even be considered as a definition of the *snapshot advantage* (Hagen et al. 2012).

4.5.1. *The disadvantages of snapshot*

Snapshot approaches are of course not without their tradeoffs. The system design is generally more complex than for scanning systems, and makes use of recent technology such as large focal plane arrays, high speed data transmission, advanced manufacturing methods and high-precision optics. In different ways, all of the various snapshot techniques encode the spectral information by expanding the system étendue. For example, if all things are held constant in an instrument design except for the number of wavelength resolution elements in the datacube, then, every technique requires increasing étendue[1] in the optical system in order to achieve the improvement.

---

1. The light collection capacity of a given instrument is its throughput (or étendue).

This quickly runs into difficult design constraints – increasing étendue generally means using larger and more expensive optics.

We can also recognize that in many cases the snapshot advantage in light collection is often fully realized only by tailoring the instrument design to its intended application, rather than by employing a general-purpose design. To give an example, we can use a given snapshot imaging spectrometer to take measurements outdoors in daylight, giving a $(N_x, N_y, N_w) = (490, 320, 32)$ datacube. However, if this system has a frame rate of 7 Hz but an exposure time of only 6 ms in that environment, then the system's light collection advantage will be thrown away. An application for which these are much better matched would require a much dimmer scene or a much faster readout rate.

There are measurement configurations for which snapshot spectral imaging is impossible to realize, such as confocal microscopy. In this case, light is confined by a small aperture to reject light scatter from outside the focal volume (Pawley 2006). Thus, this method of rejecting unwanted light means that only one spatial point is in view at any given time, and we must scan the optics about the sample in a raster fashion in order to generate a complete (x, y, λ) spectral image. By its nature, this prevents a snapshot implementation, although we can of course resort to using multiple apertures at the same time, such as in spinning disk confocal microscopy. For volumetric imaging microscopy, the use of a snapshot imaging spectrometer requires an alternate approach to measurement, such as structured illumination microscopy, which is compatible with widefield imaging (Gao et al. 2011).

An additional difficulty with using snapshot systems is the sheer volume of data that must be dealt with in order to take full advantage of them. Only recently have commercial data transmission formats (Full Cameralink, USB3 and multi-lane CoaXPress, for example) become fast enough to fully use a large-format snapshot imaging spectrometer for daylight scenes.

4.6. Conclusion

Imaging spectrometry has now been around for over four decades, and is now used in applications throughout science and engineering. As we see continued improvements in optical manufacturing, detector arrays, data transmission and computation speed, we slowly push instrumentation closer to the fundamental limits of optical measurement. There is still quite a long way to go. In this survey, we have spent a lot of space on snapshot techniques for the reason that these instruments are at the forefront of pushing the boundaries of what we can measure, in terms of efficiency, speed and interactivity. Whereas improvements in scanning approaches have been large over the last four decades, snapshot approaches can provide orders of magnitude improvements for applications that are light-starved. Some applications are currently crippled due to lack of sufficient light collection, and snapshot imaging spectrometers can therefore make applications possible that were previously unattainable. To give an example, snapshot spectral imaging was developed for the longwave infrared for use in gas imaging, which had the benefits of improving measurement accuracy and reducing system cost so that a truly *autonomous* remote sensing of gas became possible for the first time Hagen (2020b). Other applications also await a similar cultivation effort.

4.7. References

Abramov, A., Minai, L., Yelin, D. (2011). Spectrally encoded spectral imaging. *Optics Express*, 19, 6913–6922.

Allington-Smith, J. (2006). Basic principles of integral field spectroscopy. *New Astronomy Reviews*, 50, 244–251.

Allington-Smith, J. and Content, R. (1998). Sampling and background subtraction in fiber-lenslet integral field spectrographs. *Publications of the Astronomical Society of the Pacific*, 110, 1216–1234.

Anderson, N., Beeson, R., Erdogan, T. (2011). Angle-tuned thin-film interference filters for spectral imaging. *Opt. & Phot. News*, 22, 12–13.

Arribas, S., Mediavilla, E., Rasilla, J.L. (1991). An optical fiber system to perform bidimensional spectroscopy. *Astrophysics Journal*, 369, 260–272.

Avila, G. and Guirao, C. (2009). A Bowen-Walraven image slicer with mirrors [Online]. Available at: https://spectroscopy.wordpress.com/2009/05/12/inexpensive-image-slicer-with-mirrors/ [Accessed January 2013].

Bacon, R. and Monnet, G. (2017). *Optical 3D-Spectroscopy for Astronomy*. Wiley.

Barden, S.C. and Wade, R.A. (1988). DensePak and spectral imaging with fiber optics. In *Fiber Optics in Astronomy*, Barden, S.C. (ed.). Astronomical Society of the Pacific Conference Series 3.

Berns, R.S., Cox, B.D., Abed, F.M. (2015). Wavelength-dependent spatial correction and spectral calibration of a liquid crystal tunable filter imaging system. *Applied Optics*, 54, 3687–3693.

Bodkin, A., Sheinis, A., Norton, A., Daly, J., Roberts, C., Beaven, S., Weinheimer, J. (2012). Video-rate chemical identification and visualization with snapshot hyperspectral imaging. *Proceedings of SPIE*, 8374, 83740C.

Bowen, I.S. (1938). The image slicer, a device for reducing loss of light at slit of stellar spectrograph. *Astrophysics Journal*, 88, 113–124.

Bulygin, F.V. and Vishnyakov, G.N. (1992). Spectrotomography – A new method of obtaining spectrograms of two-dimensional objects. *Proceedings of SPIE*, 1843, 315.

Busch, K.W. (1985). Multiple entrance aperture optical spectrometer. USA patent 4494872.

Cabib, D. and Orr, H. (2012). Circular variable filters (CVF) at CI, progress and new performance. *Proceedings of SPIE*, 8542, 85420U.

Cardona, O., Cornejo-Rodríguez, A., García-Flores, P.C. (2010). Star image shape transformer for astronomical slit spectroscopy. *Revista Mexicana de Astronomía y Astrofísica*, 46, 431–438.

Chakrabarti, S., Jokiaho, O.-P., Baumgardner, J., Cook, T., Martel, J., Galand, M. (2012). High-throughput and multislit imaging spectrograph for extended sources. *Optics Engineering*, 51, 013003.

Chang, I.C. (1981). Acousto-optic tunable filters. *Optics Engineering*, 20, 824–829.

Content, R. (1998). Image slicer for integral field spectroscopy with NGST. *Proceedings of SPIE*, 3356, 122–133.

Cooper, V.G. (1971). Analysis of Fabry-Perot interferograms by means of their Fourier transforms. *Applied Optics*, 10, 525–530.

Courtès, G. (1960). Méthodes d'observation et étude de l'hydrogène interstellaire en émission. *Annales d'Astrophysique*, 23, 115–244.

Courtès, G. (1980). Le télescope spatial et les grands télescopes au sol. In *Application de la photométrie bidimensionelle à l'astronomie*, Roy, J.-R. (ed.). Astropresse.

Courtès, G., Georgelin, Y., Monnet, R.B.G., Boulesteix, J. (1988). A new device for faint objects high resolution imagery and bidimensional spectrography – First observational results with TIGER at CFHT 3.6-meter telescope. In *Instrumentation for Ground-Based Optical Astronomy*, Robinson, L.B. (ed.). Springer.

Czerny, M. and Turner, A.F. (1930). Über den Astigmatismus bei Spiegelspektrometern [On astigmatism in mirror spectrometers]. *Zeitschrift für Physik A*, 61, 792–797.

Descour, M.R. (1994). Non-scanning imaging spectrometry. PhD Dissertation, University of Arizona.

Dubois, E. (2005). Frequency-domain methods for demosaicking of Bayer-sampled color images. *IEEE Signal Processing Letters*, 12, 847–850.

Dwight, J.G. and Tkaczyk, T.S. (2017). Lenslet array tunable snapshot imaging spectrometer (LATIS) for hyperspectral fluorescence microscopy. *Biomedical Optics Express*, 8, 1950–1964.

Eichenholz, J.M., Barnett, N., Juang, Y., Fish, D., Spano, S., Lindsley, E., Farkas, D.L. (2010). Real time megapixel multispectral bioimaging. *Proceedings of SPIE*, 7568, 75681L.

Emadi, A., Wu, H., de Graaf, G., Wolffenbuttel, R. (2012). Design and implementation of a sub-nm resolution microspectrometer based on a linear-variable optical filter. *Optics Express*, 20, 489–507.

Evans, J.W. (1958). Solc birefringent filter. *Journal of the Optical Society of America A*, 48, 142–145.

Eversberg, T. (2016). Off-the-shelf Echelle spectroscopy: Two devices on the test block. *Publications of the Astronomical Society of the Pacific*, 128, 115001–115006.

Fellgett, P.B. (1951). The theory of infra-red sensitivities and its application to investigations of stellar radiation in the near infra-red. PhD Dissertation, University of Cambridge.

Fletcher-Holmes, D.W. and Harvey, A.R. (2005). Real-time imaging with a hyperspectral fovea. *Journal of Optics A*, 7, S298–S302.

Gao, L., Kester, R.T., Tkaczyk, T.S. (2009). Compact image slicing spectrometer (ISS) for hyperspectral fluorescence microscopy. *Optics Express*, 17, 12293–12308.

Gao, L., Bedard, N., Hagen, N., Kester, R.T., Tkaczyk, T.S. (2011). Depth-resolved image mapping spectrometer (IMS) with structured illumination. *Optics Express*, 19, 17439–17452.

Gehm, M.E. and Brady, D.J. (2006). High-throughput hyperspectral microscopy. *Proceedings of SPIE*, 6090, 609007.

Gehm, M.E., John, R., Brady, D.J., Willett, R.M., Schulz, T.J. (2007). Single-shot compressive spectral imaging with a dual-disperser architecture. *Optics Express*, 15, 14013–14027.

Goossens, T., Geelen, B., Pichette, J., Lambrechts, A., Hoof, C.V. (2018). Finite aperture correction for spectral cameras with integrated thin-film Fabry-Perot filters. *Applied Optics*, 57, 7539–7549.

Gorevoy, A., Machikhin, P.A., Martynov, G., Pozhar, V. (2022). Computational technique for field-of-view expansion in AOTF-based imagers. *Optics Letters*, 47, 585–588.

Griffiths, P., Sloane, H.J., Hannah, R.W. (1977). Interferometers vs monochromators: Separating the optical and digital advantages. *Applied Spectroscopy*, 31, 485–495.

Gupta, N., Ashe, P.R., Tan, S. (2011). Miniature snapshot multispectral imager. *Optical Engineering*, 50, 033203.

Hagen, N. (2020a). Manufacturing image slicing and image mapping mirrors. *Transactions of the 13rd MIRAI Conference on Microfabrication and Green Technology*, 8.

Hagen, N. (2020b). Survey of autonomous gas leak detection and quantification with snapshot infrared spectral imaging. *Journal of Optics*, 22, 103001.

Hagen, N. and Dereniak, E.L. (2008). Analysis of computed tomographic imaging spectrometers I. Spatial and spectral resolution. *Applied Optics*, 47, F85–F95.

Hagen, N. and Kudenov, M.W. (2013). Review of snapshot spectral imaging technologies. *Optical Engineering*, 52, 090901.

Hagen, N. and Tkaczyk, T.S. (2011). Compound prism design principles I. *Applied Optics*, 50, 4998–5011.

Hagen, N., Kester, R.T., Gao, L., Tkaczyk, T.S. (2012). Snapshot advantage: A review of the light collection improvement for parallel high-dimensional measurement systems. *Optics Engineering*, 51, 111702.

Hahn, R., Hämmerling, F.-E., Haist, T., Fleischle, D., Schwanke, O., Hauler, O., Rebner, K., Brecht, M., Osten, W. (2020). Detailed characterization of a mosaic based hyperspectral snapshot imager. *Optics Engineering*, 59, 125102.

Harris, S.E. and Wallace, R.W. (1969). Acousto-optic tunable filter. *Journal of the Optical Society of America A*, 59, 744–747.

Harwit, M. and Sloane, N.J.A. (1979). *Hadamard Transform Optics*. Academic Press.

Ibsen Photonics (2017). Why are transmission gratings less angle sensitive than reflection gratings? [Online]. Available at: https://ibsen.com/resources/grating-resources/transmission-grating-angle-sensitivity/.

Jacquinot, P. (1954). The luminosity of spectrometers with prisms, gratings, or Fabry-Perot etalons. *Journal of the Optical Society of America A*, 44, 761–765.

Jaiswal, S.P., Fang, L., Jakhetiya, V., Pang, J., Mueller, K., Au, O.C. (2016). Adaptive multispectral demosaicking based on frequency-domain analysis of spectral correlation. *IEEE Transactions on Image Processing*, 26, 953–968.

Jones, H.R.A., Martin, W.E., Anglada-Escudé, G., Errmann, R., Campbell, D.A., Baker, C., Boonsri, C., Choochalerm, P. (2021). A small actively controlled high-resolution spectrograph based on off-the-shelf components. *Publications of the Astronomical Society of the Pacific*, 133, 25001–25013.

Kapany, N.S. (2004). Fiber optics. In *Concepts of Classical Optics*, Strong, J. (ed.). Appendix N, pp. 553–579. Dover.

Katrašnik, J., Pernuš, F., Likar, B. (2013). Radiometric calibration and noise estimation of acousto-optic tunable filter hyperspectral imaging systems. *Applied Optics*, 52, 3526–3537.

Kobylinskiy, A., Laue, B., Förster, E., Höfer, B., Shen, Y., Hillmer, H., Brunner, R. (2020). Substantial increase in detection efficiency for filter array-based spectral sensors. *Applied Optics*, 59, 2443–2451.

Kriesel, J., Scriven, G., Gat, N., Nagaraj, S., Willson, P., Swaminathan, V. (2012). Snapshot hyperspectral fovea vision system (HyperVideo). *Proceedings of SPIE*, 8390, 83900T.

Kudenov, M.W., Craven-Jones, J., Aumiller, R., Vandervlugt, C., Dereniak, E.L. (2012). Faceted grating prism for a computed tomographic imaging spectrometer. *Optics Engineering*, 51, 044002.

Laurent, F., Macaire, C., Blanc, P.-E., Prieto, E., Moreaux, G., Robert, D., Ferruit, P., Bonneville, C., Hénault, F., Robertson, D., Schmoll, J. (2004). Designing, manufacturing and testing of an advanced image slicer prototype for the James Webb Space Telescope. *Proceedings of SPIE*, 5494, 188–199.

Lee, K.-S., Thompson, K.P., Rolland, J.P. (2010). Broadband astigmatism-corrected Czerny–Turner spectrometer. *Optics Express*, 18, 23378–23384.

Lotspeich, J.F., Stephens, R.R., Henderson, D.M. (1981). Electro-optic tunable filter. *Optical Engineering*, 20, 830–836.

Lyot, B. (1944). The birefringent filter and its applications in solar physics. Reprinted in: Title, A. and Rosenberg, W. (1979). Research on Spectroscopic Imaging. NASA Tech. Report NSAA-CR-158702.

Maione, B.D., Baldridge, C., Kudenov, M.W. (2019). Microbolometer with a multi-aperture polymer thin-film array for neural-network-based target identification. *Applied Optics*, 58, 7285–7297.

Martin, S.F., Ramsey, H.E., Carroll, G.A., Martin, D.C. (1974). A multi-slit spectrograph and Hα Doppler system. *Solar Physics*, 37, 343–350.

Masterson, H.J., Sharp, G.D., Johnson, K.M. (1989). Ferroelectric liquid-crystal tunable filter. *Optics Letters*, 14, 1249–1251.

Mathews, S.A. (2008). Design and fabrication of a low-cost, multispectral imaging system. *Applied Optics*, 47, F71–F76.

Matsuoka, H., Kosai, Y., Saito, M., Takeyama, N., Suto, H. (2002). Single-cell viability assessment with a novel spectro-imaging system. *Journal of Biotechnology*, 94, 299–308.

McCain, S.T., Gehm, M.E., Wang, Y., Pitsianis, N.P., Brady, D.J. (2006). Coded aperture Raman spectroscopy for quantitative measurements of ethanol in a tissue phantom. *Applied Spectroscopy*, 60, 663–671.

Mu, T., Han, F., Bao, D., Zhang, C., Liang, R. (2019). Compact snapshot optically replicating and remapping imaging spectrometer (ORRIS) using a focal plane continuous variable filter. *Optics Letters*, 44, 1281–1284.

Nagaoka, H. and Mishima, T. (1923). A combination of a concave grating with a Lummer-Gehrcke plate or an echelon grating for examining fine structure of spectral lines. *Astrophysical Journal*, 57, 92–97.

Offner, A. (1987). Annular field systems and the future of optical microlithography. *Optical Engineering*, 26, 294–299.

Okamoto, T. and Yamaguchi, I. (1991). Simultaneous acquisition of spectral image information. *Optics Letters*, 16, 1277–1279.

Padgett, M.J. and Harvey, A.R. (1995). A static Fourier-transform spectrometer based on Wollaston prisms. *Review of Scientific Instruments*, 66, 2807–2811.

Pawley, J.B. (2006). Points, pixels, and gray levels. In *Handbook of Biological Confocal Microscopy*, 3rd edition, Pawley, J.B. (ed.). Springer.

Pawlowski, M.E., Dwight, J.G., Nguyen, T.-U., Tkaczyk, T.S. (2019). High performance image mapping spectrometer (IMS) for snapshot hyperspectral imaging applications. *Optics Express*, 27, 1597–1612.

Perot, A. and Fabry, C. (1899). Sur l'application de phénomènes d'interférence à la solution de divers problèmes de spectroscopie et de métrologie [On the application of interference phenomena to the solution of various problems in spectroscopy and metrology]. *Bulletin Astronomique, Serie I*, 16, 5–32.

Pribram, J.K. and Penchina, C.M. (1968). Stray light in Czerny–Turner and Ebert spectrometers. *Applied Optics*, 7, 2005–2014.

Prieto-Blanco, X., Montero-Orille, C., Couce, B., la de Fuente, R. (2006). Analytical design of an Offner imaging spectrometer. *Optics Express*, 14, 9156–9168.

Ren, D. and Allington-Smith, J. (2002). On the application of integral field unit design theory for imaging spectroscopy. *Publications of the Astronomical Society of the Pacific*, 114, 866–878.

Schmoll, J., Dubbeldam, C.M., Robertson, D.J., Yao, J. (2006). The Durham micro-optics programme. *New Astronomy Reviews*, 50, 337–341.

Schumann, L.W. and Lomheim, T.S. (2002). Infrared hyperspectral imaging Fourier transform and dispersive spectrometers: Comparison of signal-to-noise based performance. *Proceedings of SPIE*, 4480, 1–14.

Shaw, G.D. (1993). New techniques in astronomical multi-slit spectroscopy. PhD Dissertation, University of Durham.

Shogenji, R., Kitamura, Y., Yamada, K., Miyatake, S., Tanida, J. (2004). Multispectral imaging using compact compound optics. *Optics Express*, 12, 1643–1655.

Simon, J.M., Gil, M.A., Fantino, A.N. (1986). Czerny–Turner monochromator: Astigmatism in the classical and in the crossed beam dispositions. *Applied Optics*, 25, 3715–3720.

Şolc, I. (1954). A new type of birefringent filter. *Czech. Journal of Physics*, 4, 53–66. Reprinted in: Title, A. and Rosenberg, W. (1979). Research on Spectroscopic Imaging. NASA Tech. Report NSAA-CR-158702.

Suematsu, Y., Saito, K., Koyama, M., Enokida, Y., Okura, Y., Nakayasu, T., Sukegawa, T. (2017). Development of micro-mirror slicer integral field unit for space-borne solar spectrographs. *CEAS Space Journal*, 9, 421–431.

Tala, M., Vanzi, L., Avila, G., Guirao, C., Pecchioli, E., Zapata, A., Pieralli, F. (2017). Two simple image slicers for high resolution spectroscopy. *Experimental Astrophysics*, 43, 167–176.

Vanderriest, C. (1980). A fiber-optics dissector for spectroscopy of nebulosities around quasars and similar objects. *Publications of the Astronomical Society of the Pacific*, 92, 858–862.

Vila-Francés, J., Calpe-Maravilla, J., Gómez-Chova, L., Amorós-López, J. (2010). Improving the performance of acousto-optic tunable filters in imaging applications. *Journal of Electronic Imaging*, 19, 043022.

Waldron, A., Allen, A., Colón, A., Chance Carter, J., Angel, S.M. (2021). A monolithic spatial heterodyne Raman spectrometer: Initial tests. *Applied Spectroscopy*, 75, 57–69.

Walraven, T. and Walraven, J.H. (1972). Some features of the Leiden radial velocity instrument. In *Proceedings of ESO/CERN Conference on Auxiliary Instrumentation for Large Telescopes*, Laustsen, S. and Reiz, A. (eds), pp. 175–183.

Wang, Y., Pawlowski, M.E., Cheng, S., Dwight, J.G., Stoian, R.I., Lu, J., Alexander, D., Tkaczyk, T.S. (2019). Light-guide snapshot imaging spectrometer for remote sensing applications. *Optics Express*, 27, 15701–15725.

Weitzel, L., Krabbe, A., Kroker, H., Thatte, N., Tacconi-Garman, L.E. (1996). 3D: The next generation near-infrared imaging spectrometer. *Astronomy and Astrophysics, Supplement Series*, 119, 531–546.

Yu, C., Feng, S., Yang, J., Song, N., Sun, C., Wang, M. (2021). Microlens array snapshot hyperspectral microscopy system for the biomedical domain. *Applied Optics*, 60, 1896–1902.

Zhang, Y. and Gross, H. (2019). Systematic design of microscope objectives. Part III: Miscellaneous design principles and system synthesis. *Advanced Optical Technologies*, 8, 385–402.

Zheng, K., Liu, B., Huang, C., Brezinski, M.E. (2008). Experimental confirmation of potential swept source optical coherence tomography performance limitations. *Applied Optics*, 47, 6151–6158.

5

Spectral Modeling and Separation of Reflective-Fluorescent Scenes

Ying Fu[1], Antony Lam[2], Imari Sato[3], Takahiro Okabe[4], and Yoichi Sato[5]

[1]*Beijing Institute of Technology, China*
[2]*Mercari, Inc., Tokyo, Japan*
[3]*Computational Imaging and Vision Lab,*
National Institute of Informatics, Tokyo, Japan
[4]*Department of Artificial Intelligence, Kyushu Institute*
of Technology, Fukuoka, Japan
[5]*Institute of Industrial Science, The University of Tokyo, Japan*

5.1. Introduction

Hyperspectral reflectance data are beneficial to many applications including but not limited to archiving for cultural e-heritage (Balas et al. 2003), medical imaging (Styles et al. 2006) and also color relighting of scenes (Johnson and Fairchild 1999). As a result, many methods for acquiring the spectral reflectance of scenes have been proposed (Maloney and Wandell 1986; Tominaga 1996; Gat 2000; DiCarlo et al. 2001; Park et al. 2007; Chi et al. 2010). Despite the success of these methods, they have all made the assumption that fluorescence is absent from the scene. However,

Computational Imaging for Scene Understanding,
coordinated by Takuya FUNATOMI and Takahiro OKABE.
© ISTE Ltd 2024.

fluorescence does frequently occur in many objects, such as natural gems and corals, fluorescent dyes used for clothing and plant containing chlorophyll to name a few. In fact, Barnard shows that fluorescent surfaces are present in 20% of randomly constructed scenes (Barnard 1999). This is a significant proportion of scenes that have not been considered by most of the past methods.

(a) White light	(b) Reflection	(c) Fluorescence

Figure 5.1. *(a) The scene captured under white light. (b) The recovered reflective component. (c) The recovered fluorescent component. For a color version of this figure, see www.iste.co.uk/funatomi/computational.zip*

Another important point is that reflective and fluorescent components behave very differently under different illuminants (Johnson and Fairchild 1999; Zhang and Sato 2011). Thus, to accurately predict the color of objects, separate modeling of all spectral properties of both reflective and fluorescent components is essential. Specifically, when a reflective surface is illuminated by incident light, it reflects back light of the same wavelength. Fluorescent surfaces, on the other hand, first absorb incident light and then emit at longer wavelengths. This wavelength shifting property is known as the Stokes shift (Rost 1992; Skoog et al. 2007), and the question of which wavelengths of light are absorbed and which wavelengths are emitted is defined by the fluorescent surface's absorption and emission spectra. As the properties of fluorescence are very different from ordinary reflection, neglecting fluorescence can result in completely incorrect color estimation. This in turn negatively affects many methods that rely on accurate color estimation. For example, algorithms for relighting and color constancy would be affected.

The goal of this work is to accurately recover the full spectral reflective and fluorescent components of an entire scene. Typical fluorescent objects exhibit both reflection and fluorescence (Figure 5.1). So, the question of how these components can be accurately separated also needs to be addressed. In this work, we show that the reflectance and fluorescence spectra of a scene can be efficiently separated and measured through the use of high-frequency illumination in the spectral domain. Our approach only assumes that the absorption spectrum of the fluorescent material is a smooth function with respect to the frequency of the lighting in the spectral domain. With this assumption, it is possible to separate reflective and fluorescent components using just two hyperspectral images taken under a high-frequency illumination pattern and its shifted version in the spectral domain. We show that the reflectance and fluorescence emission spectra can then be fully recovered by our separation method.

In addition to recovering reflectance and fluorescence emission spectra, we also make the observation that materials with similar emission spectra tend to have similar absorption spectra as well. Using this observation, we devise a method to estimate the absorption spectra by taking the corresponding recovered emission spectra from high-frequency lighting.

In summary, the contributions in this study are that an efficient method is presented for the separation and recovery of full reflectance and fluorescent emission spectra using high-frequency illumination in the spectral domain. In addition, we present a method for estimating the absorption spectrum of a material given its emission. Since the reflective and fluorescent emission and absorption spectra of the scene are completely recovered, the ability to accurately predict the relighting of scenes under novel lighting is also shown. Furthermore, we provide a more in-depth analysis of our method and also show that filters can be used in conjunction with standard light sources to generate the required high-frequency illuminants. Thus, bypassing the need for a programmable light source, we extend our method to work under ambient light.

5.2. Related Work

As noted earlier, there have been a number of papers on recovering the spectral reflectance of scenes (Maloney and Wandell 1986; Tominaga 1996; Gat 2000; DiCarlo et al. 2001; Park et al. 2007; Chi et al. 2010). Despite the effectiveness of these methods for spectral reflectance capture, they do not take the effects of fluorescence into account.

Unfortunately, not accounting for fluorescence can have a detrimental effect on color accuracy. For example, Johnson and Fairchild (1999) showed that considering fluorescence can dramatically improve color renderings. Later, Wilkie et al. (2006) showed accurate results by rendering fluorescence emissions using diffuse surfaces that can reflect light at a wavelength different from its incident illuminant wavelength. Hullin et al. (2010) showed the importance of modeling and rendering of reflective-fluorescent materials using their bidirectional reflectance and reradiation distribution functions (BRRDF). Besides color rendering, the observation of fluorescence emissions on an object's surface has also been applied to photometric stereo for shape reconstruction (Sato et al. 2012; Treibitz et al. 2012). As mentioned earlier, Barnard concluded that fluorescent surfaces are present in 20% of randomly constructed scenes (Barnard 1999). Thus, the presence of fluorescence is significant and warrants attention.

In practice, fluorescent objects typically exhibit both reflection and fluorescence so the joint occurrence of these phenomena in scenes needs to be considered. Some methods in the literature have given this issue attention. Lee et al. (2001) provided a mathematical description for fluorescence processes and recovered the additive spectra of reflective and fluorescent components but did not separate them. Alterman et al. (2010) separated the appearance of fluorescent dyes from a mixture by unmixing multiplexed images. Zhang and Sato (2011) derived an independent component analysis-based method to estimate the RGB colors of reflectance and fluorescence emission but not their spectral distributions. They also did not estimate the absorption spectrum of the fluorescent component and so cannot predict intensity changes in fluorescence emission due to different illumination spectra. Tominaga et al. (2011) estimated fluorescence emission spectra using multispectral images taken under two ordinary light sources. A limitation is that they assumed fluorescence emissions to be constant for all absorption wavelengths and thus cannot accurately predict the brightness of fluorescent components under varying illumination. Finally, none of these methods fully recover all reflectance and fluorescence spectral components of scenes.

In recent work, methods for hyperspectral imaging of reflective-fluorescent scenes have been proposed. Lam and Sato (2013) provided a method for recovering the full spectral reflectance and fluorescence absorption and emission spectra of scenes but they needed to capture the scene about 30 times using a multiband camera under multiple narrowband illuminants. Suo et al. (2014) presented a bispectral coding scheme which was rooted in the classical bispectral measurement method (Springsteen 1999) where dozens of images also had to be captured under shifting narrowband illuminations. Zheng et al. (2014) also recovered all the different types of fluorescence and reflectance spectra using off-the-shelf lights and three hyperspectral images. Fu et al. (2014) recovered all these spectra by using an RGB camera and capturing them under different illuminants. Both methods have advantages in that conventional light sources or cameras can be used but at the expense of accuracy. Our method only uses two hyperspectral images to recover all these spectra, and achieves highly accurate results with fewer illuminations.

As mentioned earlier, one of the key challenges in our problem is the separation of reflective and fluorescent components from composite objects exhibiting both phenomena. There have been a number of methods in the literature on separating components in images. For example, Farid and Adelson (1999) used independent component analysis to separate reflections on glass and a painting on the side of the glass opposite the observer. Nayar et al. (1993) separated specular reflections from diffuse reflections. It is interesting that an analogy can be made between our spectral domain work and the spatial domain work of Nayar et al. (2006). Whereas previous work (Nayar et al. 2006) used high-frequency spatial light patterns to separate lighting components in the spatial domain, we use high-frequency light spectra to separate lighting components in the spectral domain.

5.3. Separation of reflection and fluorescence

In this section, we describe the reflection and fluorescence models used in our method, present the separation method for reflective and fluorescent components by using high-frequency illumination, discuss the conditions required for the illumination frequencies and analyze the errors of our method.

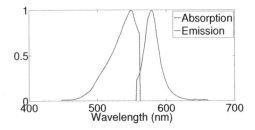

Figure 5.2. *An example of absorption and emission spectra from the McNamara and Boswell fluorescence spectral dataset (McNamara et al. 2006). For a color version of this figure, see www.iste.co.uk/funatomi/computational.zip*

5.3.1. *Reflection and fluorescence models*

We begin with a brief review of how reflective-fluorescent materials are modeled (Zhang and Sato 2013). Since reflection and fluorescence have different physical behaviors, they need to be described by different models.

The radiance of a reflective surface depends on incident light and its reflectance. The observed radiance of an ordinary reflective surface at wavelength λ is computed as

$$p_r(\lambda) = l(\lambda)r(\lambda), \tag{5.1}$$

where $l(\lambda)$ is the spectrum of the incident light at wavelength λ and $r(\lambda)$ is the spectral reflectance of the surface at wavelength λ.

The observed radiance of a pure fluorescent surface depends on the incident light, the material's absorption spectrum and its emission spectrum. Fluorescence typically absorbs light at some wavelengths and emits them at longer wavelengths. The way this works is that when incident light hits a fluorescent surface, the surface's absorption spectrum will determine how much of the light is absorbed. Some of the absorbed energy is then released in the form of an emission spectrum at longer wavelengths than the incident light. The remainder of the absorbed energy is released as heat. The reason for this phenomenon is that fluorescence emission occurs after an orbital electron of a molecule, atom or nanostructure absorbs light and is excited, the electron relaxes to its ground state by emitting a photon of light and sends out heat after several nanoseconds. The shorter the light's wavelength is, the more energy the light carries.

Since some of the absorbed energy is lost as heat, the fluorescence emission is at a longer wavelength. Figure 5.2 illustrates an example of the absorption and emission spectra for a fluorescent material over the visible spectrum.

Let $l(\lambda')$ represent the intensity of the incident light at wavelength λ'; the observed spectrum of a pure fluorescent surface (Zhang and Sato 2013) at wavelength λ is described as

$$p_f(\lambda) = \left(\int l(\lambda')a(\lambda')d\lambda' \right) e(\lambda), \qquad [5.2]$$

where $a(\lambda')$ and $e(\lambda)$ represent the absorption and emission spectrum, respectively. $\left(\int l(\lambda')a(\lambda')d\lambda' \right)$ is determined by the absorption spectrum and the spectrum of the incoming light, and is independent of the emission spectrum. Replacing this part by scale factor k, equation [5.2] can be rewritten as $p_f(\lambda) = ke(\lambda)$, which means that the shape or the distribution of the emitted spectrum is constant, but the scale k of the emitted spectrum changes under different illuminations. In other words, the radiance of the fluorescence emission changes under different illuminations, but its color (specifically, chromaticity) stays the same regardless of illumination color.

The radiance of a reflective-fluorescent surface point can be expressed as a linear combination of the reflective component p_r and fluorescent component p_f, i.e. $p = p_r + p_f$. Thus:

$$p(\lambda) = l(\lambda)r(\lambda) + \left(\int l(\lambda')a(\lambda')d\lambda' \right) e(\lambda). \qquad [5.3]$$

5.3.2. Separation using high-frequency illumination

In our method, we use high-frequency illumination defined in the spectral domain for separating reflective and fluorescent components. Let us start with simple binary illuminants to describe the key idea of our method. We denote a high-frequency illumination pattern shown in Figure 5.3(b) by $l_1(\lambda)$ and its complement shown in Figure 5.3(c) by $l_2(\lambda)$. The illuminants are defined such that when $l_1(\lambda)$ has intensity, $l_2(\lambda)$ has no intensity and vice versa. Let us consider a certain wavelength λ_1, where the wavelength λ_1 is lit directly under the illuminant $l_1(\lambda)$, so that $l_1(\lambda_1) = 1$ and then it is not lit under the illuminant l_2, so $l_2(\lambda_1) = 0$. Since reflection occurs at the same wavelength with the illumination, we obtain

$$p_1(\lambda_1) = r(\lambda_1) + k'e(\lambda_1),$$
$$p_2(\lambda_1) = k'e(\lambda_1). \qquad [5.4]$$

Here, we assume that $\int l_1(\lambda')a(\lambda')d\lambda' = \int l_2(\lambda')a(\lambda')d\lambda' = k'$. That is, the absorptions due to our high-frequency illumination patterns are the same. We show in section 5.3.3 that this is true when the absorption $a(\lambda')$ is smooth with respect to the frequency of the illumination patterns in the spectral domain. With the same absorptions under the two illuminants, we obtain the reflectance and emission spectra at λ_1 as

$$r(\lambda_1) = p_1(\lambda_1) - p_2(\lambda_1),$$
$$k'e(\lambda_1) = p_2(\lambda_1). \qquad [5.5]$$

The reflectance and emission spectra at λ_2 where $l_1(\lambda_2) = 0$ and $l_2(\lambda_2) = 1$ are obtained in a similar manner.

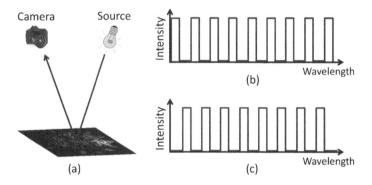

Figure 5.3. *An example of a captured scene (a). When a reflective-fluorescent point in the scene is lit by the illuminant (b), which is a high-frequency binary illumination pattern in the wavelength domain, each lit wavelength includes both reflective and fluorescent components while the unlit wavelengths have only the fluorescent component. Its complement is shown in (c). For a color version of this figure, see www.iste.co.uk/funatomi/computational.zip*

In our work, we use high-frequency sinusoidal illuminants (Figure 5.4) in the spectral domain to achieve the same effect as the binary lighting patterns because they are more practical and also fit into the theory of our framework. The illuminants can be represented as

$$l_1(\lambda) = \alpha + \beta \cos(2\pi f_l \lambda),$$
$$l_2(\lambda) = \alpha + \beta \cos(2\pi f_l \lambda + \phi), \qquad [5.6]$$

where f_l is the frequency of illumination. The radiance of a surface under these two illuminants can be described as

$$p_1(\lambda) = l_1(\lambda)r(\lambda) + k_1 e(\lambda),$$
$$p_2(\lambda) = l_2(\lambda)r(\lambda) + k_2 e(\lambda),$$
$$k_n = \int l_n(\lambda')a(\lambda')\mathrm{d}\lambda'. \qquad [5.7]$$

Here, assuming that k_n is constant for l_1 and l_2, i.e. $k_1 = k_2 = k$, the reflectance $r(\lambda)$ and fluorescence emission $ke(\lambda)$ can be recovered as

$$r(\lambda) = \frac{p_1(\lambda) - p_2(\lambda)}{l_1(\lambda) - l_2(\lambda)},$$

[5.8]

$$ke(\lambda) = p_1(\lambda) - \frac{p_1(\lambda) - p_2(\lambda)}{l_1(\lambda) - l_2(\lambda)} l_1(\lambda).$$

Thus, to recover the reflectance $r(\lambda)$ and fluorescence emission $ke(\lambda)$ completely, we first need to make $k_1 = k_2 = k$.

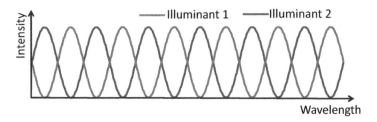

Figure 5.4. *Sinusoidal illuminant patterns. The blue and pink solid lines denote two illumination patterns. There is a phase shift between them. For a color version of this figure, see www.iste.co.uk/funatomi/computational.zip*

5.3.3. *Discussion on the illumination frequency*

Here, we discuss how to satisfy the condition $k_1 = k_2 = k$. In the following, we consider the requirements for our illuminants based on the Nyquist sampling theorem (Oppenheim et al. 1996) and on an analysis of the McNamara and Boswell fluorescence spectral dataset (McNamara et al. 2006).

Let $a_n(\lambda) = l_n(\lambda)a(\lambda)$ $\{n = 1, 2\}$, where $l_n(\lambda)$ can be considered as a sampling or modulating function of $a(\lambda)$. The sampling theorem, which is most easily explained in terms of impulse-train sampling, establishes the fact that a band-limited signal is uniquely represented by its samples. In practice, however, narrow, large-amplitude pulses, which approximate impulses, are relatively difficult to generate and transmit. Instead, we use sinusoidal illuminant patterns in the spectral domain as shown in Figure 5.4. These patterns are similar to amplitude modulation functions in communication systems.

The spectrum of sinusoidal illumination $l_1(\lambda)$ in the frequency domain (Oppenheim et al. 1996) is

$$L_1(f) = \frac{1}{2}[\beta\delta(f - f_l) + 2\alpha\delta(f) + \beta\delta(f + f_l)],$$

[5.9]

where $\delta(f)$ is the Dirac delta function. Let $A(f)$ and $A_n(f)$ denote the Fourier transform of $a(\lambda)$ and $a_n(\lambda)$, respectively. Since the product $l_n(\lambda)a(\lambda)$ in the spectral domain corresponds to a convolution in its Fourier domain, i.e.

$$A_n(f) = L_n(f) * A(f), \qquad [5.10]$$

the Fourier transform of $a_1(\lambda)$ is

$$A_1(f) = \frac{1}{2} [\beta A(f - f_l) + 2\alpha A(f) + \beta A(f + f_l)]. \qquad [5.11]$$

That is, a replication of the Fourier transform of the original signal $A(f)$ is centered around $+f_l$ and 0 and $-f_l$.

The Fourier transforms of $l_1(\lambda)$ and $l_2(\lambda)$ with the phase offset ϕ are related as $L_2(f) = e^{i\phi} L_1(f)$, and thus the frequency spectrum of $a_2(\lambda)$ is

$$A_2(f) = \frac{1}{2} [\beta e^{i\phi} A(f - f_l) + 2\alpha A(f)$$
$$+ \beta e^{-i\phi} A(f + f_l)]. \qquad [5.12]$$

From the definition of the Fourier transform $A_n(f) = \int_{-\infty}^{+\infty} a_n(\lambda)e^{-i2\pi f\lambda}\mathrm{d}\lambda$, substituting $f = 0$ into this definition, we obtain

$$A_n(0) = \int_{-\infty}^{+\infty} a_n(\lambda)\mathrm{d}\lambda = \int_{-\infty}^{+\infty} l_n(\lambda)a(\lambda)\mathrm{d}\lambda = k_n. \qquad [5.13]$$

Therefore, k_n corresponds to $A_n(f)$'s zero-frequency component. This tells us that we need to satisfy the condition $A_1(0) = A_2(0)$ so that $k_1 = k_2 = k$. In equations [5.11] and [5.12], substituting $f = 0$, we obtain

$$A_1(0) = \frac{1}{2} [\beta A(-f_l) + 2\alpha A(0) + \beta A(f_l)],$$
$$A_2(0) = \frac{1}{2} [\beta e^{i\phi} A(-f_l) + 2\alpha A(0) + \beta e^{-i\phi} A(f_l)]. \qquad [5.14]$$

Let us define f_a as $a(\lambda)$'s maximum frequency. When $f_l > f_a$, $A(-f_l)$ and $A(f_l)$ become zero. This means that we obtain $A_1(0) = A_2(0) = 2\alpha A(0)$ for $f_l > f_a$ to achieve $k_1 = k_2 = k$. Thus, the frequency of the illuminants in the spectral domain f_l needs to be greater than $a(\lambda)$'s maximum frequency or bandwidth f_a.

We now discuss the maximum frequency of $a(\lambda)$ on the McNamara and Boswell fluorescence spectral dataset. We examine the maximum frequency of all 509 materials in the dataset, and obtain the maximum frequency of each absorption spectrum while retaining 99% of the energy[1]. The mean of the maximum frequency for all absorption spectra in the dataset is $1/45.9[nm^{-1}]$ and its standard deviation is $1/24.1[nm^{-1}]$. As mentioned previously, the illumination frequency f_l needs to be greater than $a(\lambda)$'s maximum frequency f_a. As the period is the reciprocal of the frequency, the period of the illumination – called the "sampling interval" – in the spectral domain needs to be less than the minimum sampling interval of all absorption spectra of fluorescent materials in the scene. Figure 5.5 shows the percentage of absorption spectra in the McNamara and Boswell fluorescence spectral dataset that satisfy the condition $k_1 = k_2$ under different periods of the illumination. We set the period of the illumination to 40 nm in our experiments due to limitations of our light source. This is still less than the mean minimum sampling interval of all absorption spectra (45.9 nm) found in the dataset and works well in practice.

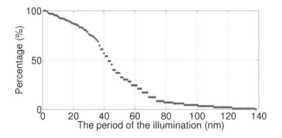

Figure 5.5. *The percentage of absorption spectra in the McNamara and Boswell fluorescence spectral dataset where $k_1 = k_2$ given different the period of the illumination. The smaller the period of the illumination, the more absorption spectra satisfy our requirement that $k_1 = k_2$. For a color version of this figure, see www.iste. co.uk/funatomi/computational.zip*

5.3.4. *Error analysis*

Due to the limitations of the light source, we cannot produce ideal and arbitrary high-frequency illuminations. It is thus unlikely for k to be the exact constant for all kinds of fluorescent materials in the scene under realistic conditions. Therefore, to reduce errors, we substitute the recovered $r(\lambda)$ into both $p_1(\lambda) = l_1(\lambda)r(\lambda) + k_1 e(\lambda)$ and $p_2(\lambda) = l_2(\lambda)r(\lambda) + k_2 e(\lambda)$, and average the recovered $ke(\lambda)$ from these two equations. Thus, the fluorescence emission $ke(\lambda)$ is recovered by

$$ke(\lambda) = \frac{1}{2}\left[k_1 e(\lambda) + k_2 e(\lambda)\right]$$

$$= \frac{1}{2}\left[p_1(\lambda) - \frac{p_1(\lambda) - p_2(\lambda)}{l_1(\lambda) - l_2(\lambda)}l_1(\lambda) + p_2(\lambda) - \frac{p_1(\lambda) - p_2(\lambda)}{l_1(\lambda) - l_2(\lambda)}l_2(\lambda)\right].$$

[5.15]

---

1. Since there exists some noise in the original spectra, ignoring some high-frequency components is reasonable.

If $k_1 \neq k_2$:

$$r(\lambda) = \frac{(p_1(\lambda) - p_2(\lambda)) - (k_1 - k_2)e(\lambda)}{l_1(\lambda) - l_2(\lambda)}. \qquad [5.16]$$

Let $r_{error}(\lambda)$ and $e_{error}(\lambda)$ denote the errors of $r(\lambda)$ and $ke(\lambda)$ (where $k = (k_1 + k_2)/2$ when $k_1 \neq k_2$). These errors can be expressed as

$$r_{error}(\lambda) = abs\left[\frac{(k_1 - k_2)e(\lambda)}{l_1(\lambda) - l_2(\lambda)}\right],$$

$$e_{error}(\lambda) = abs\left[(k_1 - k_2)e(\lambda)\frac{l_1(\lambda) + l_2(\lambda)}{2\,[l_1(\lambda) - l_2(\lambda)]}\right]. \qquad [5.17]$$

For the sinusoidal illuminant $l_n(\lambda)$ in equation [5.6], the maximum and minimum intensities over all wavelengths λ are $\alpha + \beta$ and $\alpha - \beta$. Each value in an illuminant's spectrum has to be positive, so $\alpha/\beta \geq 1$.

In equation [5.17], the errors for the reflective and fluorescence emission are directly proportional to $k_1 - k_2$. This means that the less difference between k_1 and k_2 there is, the smaller the errors. As α/β becomes larger, $A_1(0)$ and $A_2(0)$ in equation [5.14] are less affected by the $\beta A(-f_l)$ and $\beta A(f_l)$ terms. As a result, the difference between k_1 and k_2 can be decreased under the same illumination frequency. Thus, the k term is more robust under different illumination conditions, when α/β is large.

Nevertheless, when the scene is captured by the camera, the noise from the camera cannot be totally avoided. As $l_1(\lambda) - l_2(\lambda)$ shrinks, $r(\lambda)$ and $e(\lambda)$ are increasingly affected by noise, as can be seen in equation [5.8]. In order to make the proposed method more robust to noise, we need to make the difference $l_1(\lambda) - l_2(\lambda)$ greater. In practice, we set the phase shift of illuminations $l_1(\lambda)$ and $l_2(\lambda)$ to π (Figure 5.4) and capture the scene at the illumination's peaks or crests to maximize the observed difference in $l_1(\lambda) - l_2(\lambda)$.

It is also interesting to note that the need to maximize $l_1(\lambda) - l_2(\lambda)$ also means that α/β should be closer to 1, which is at odds with the need to make α/β large to allow for a more robust k as discussed above. We discuss the influence and tradeoffs of the value of α/β in real data in sections 5.5.4 and 5.5.5.

5.4. Estimating the absorption spectra

In this section, we explain how we estimate the absorption spectrum of a material from its emission spectrum that was obtained using our method in section 5.3.2.

The basic observation behind our method is that fluorescent materials with similar emission spectra tend to have similar absorption spectra (Figure 5.6). From this observation, we derive a method that uses a dictionary of known emission and absorption spectrum pairs to estimate an absorption spectrum from a given novel emission.

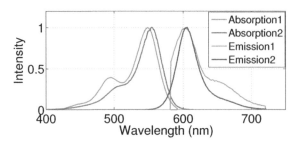

Figure 5.6. *Absorption and emission spectra of two fluorescent materials. For a color version of this figure, see www.iste.co.uk/funatomi/computational.zip*

Specifically, let $\hat{e}$ be a known emission spectrum whose absorption spectrum $\hat{a}$ is unknown. Let $\{e_j\}$ be a dictionary of emission spectra and $\{a_j\}$ be the known corresponding absorption spectra. Representing all these spectra as vectors, we first determine the linear combination of $\{e_j\}$ to reconstruct $\hat{e}$ by solving

$$\hat{e} = \sum_j w_j e_j. \tag{5.18}$$

The weights $\{w_j\}$ are then used to calculate the corresponding absorption spectrum $\hat{a}$ by

$$\hat{a} = \sum_j w_j a_j. \tag{5.19}$$

Let $\{e_j'\}$ and $\{a_j'\}$ denote the subsets of $\{e_j\}$ and $\{a_j\}$ with corresponding weights $\{w_j \neq 0\}$. Note that using the same $\{w_j\}$ in equations [5.18] and [5.19] requires that the linear combination be kept between the subspaces spanned by $\{e_j'\}$ and $\{a_j'\}$. We assert that an emission spectrum can typically be well represented by a sparse basis. To show this, we perform leave-one-out cross-validation where for each emission spectrum in the McNamara and Boswell fluorescence spectral dataset, we set $\hat{e}$ as the testing sample and use the remaining emission spectra in $\{e_j\}$ as the dictionary. We find that any given emission $\hat{e}$ can on average be well represented by 10 emission spectra from the dictionary, which is very sparse compared to the size of the whole dictionary. Thus, $\hat{e}$ can be considered to live in a low-dimensional sub-space spanned by $\{e_j'\}$. Therefore, to minimize the number of basis vectors used from $\{e_j\}$, we seek to reconstruct $\hat{e}$ by sparse weights w through l_1-norm minimization (Efron et al. 2004; Candes and Tao 2006; Donoho et al. 2012), according to

$$\min \|w\|_1 \quad s.t. \quad w_j \geq 0 \text{ and } \left\| \hat{e} - \sum_j w_j e_j \right\|_2^2 \leq \epsilon. \tag{5.20}$$

To test the accuracy of our method, we chose a subset of materials from the McNamara and Boswell fluorescence spectral dataset where both the emission and absorption spectra are present in the visible range (400–720 nm). This results in a collection of 183 materials. We then perform leave-one-out cross-validation using our method and the 183 emission and absorption spectra. The estimated absorption spectrum is then compared against the ground truth using the mean root square error $\sqrt{\left(\sum_\lambda \left(a^{gt}(\lambda) - a^{re}(\lambda)\right)^2 d\lambda\right)/N}$, where $a^{gt}(\lambda)$ and $a^{re}(\lambda)$ are the ground truth and recovered spectra at wavelength λ, respectively. N is the discrete number of wavelengths representing the spectrum in the visible range. The ground truth and estimation are also normalized for scale by setting them to be unit length vectors.

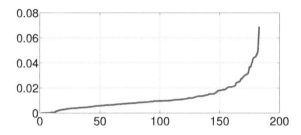

Figure 5.7. *All test errors sorted in ascending order. Sixty-seven percent of cases were below the average error of 0.012. For a color version of this figure, see www.iste.co.uk/funatomi/computational.zip*

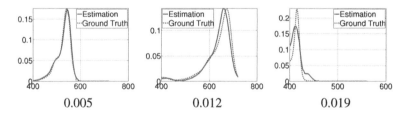

Figure 5.8. *Examples of estimated absorption spectra and their root mean square errors. For a color version of this figure, see www.iste.co.uk/funatomi/computational.zip*

In our results, we obtain an average error of 0.012. See Figure 5.7 for a plot of all the errors for the 183 estimated absorption spectra. We do find a minority of cases with high errors that violate our assumption that similar emission spectra map to the same absorption spectra. Despite this, the majority of materials fit our assumption and absorption spectra are accurately estimated, as shown in Figure 5.8. We also note that absorption only determines the scale of the emission and not the color of the material.

Thus, some minor loss in accuracy for estimated absorption does not have a dramatic effect on the predicted color of scenes.

5.5. Experiment results and analysis

In our experiments, we first demonstrate the importance of high-frequency illumination using quantitative results on the recovery of reflectance and fluorescence spectra from real scenes. We then present visual examples of separated reflective and fluorescent components using images captured under high-frequency illuminations produced by a programmable light source, and use our recovered spectra to accurately relight fluorescent scenes. We then bypass the need for an expensive programmable light source by using filters in conjunction with standard light sources to generate the required high-frequency illuminants. Finally, we show that our method also works under ambient light.

5.5.1. *Experimental setup*

With the exception of near-UV light and ambient lights used in section 5.5.5, for all other illuminants, we use a Nikon equalized light source (ELS). The ELS is a programmable light source that can produce light with arbitrary spectral patterns from 400 nm to 720 nm. We use a PR-670 SpectraScan Spectroradiometer to collect ground truth spectra. For our proposed method, we use a hyperspectral camera (EBA Japan NH-7) to capture whole scenes.

Figure 5.9(a) shows two high-frequency illuminants produced by the ELS. Under these illuminants, we use the hyperspectral camera to capture the scene at wavelengths where either one of these illuminants have peaks so that the difference between l_1 and l_2 would be large and allow for reliable separation.

5.5.2. *Quantitative evaluation of recovered spectra*

Here, we first compare quantitative results on recovering the reflectance and fluorescence spectral components using high- and low-frequency lights on fluorescent colored sheets. To make the quantitative evaluation, we measure the root mean square error (RMSE) of the estimated spectra, with respect to their corresponding ground truth. Figure 5.9(a) and (b) shows the spectral distributions of the high- and low-frequency illuminants used in our experiments. These illuminants are then used to recover spectra that are compared against the ground truth spectra.

The ground truth reflectance and fluorescent absorption and emission spectra of the fluorescent material are captured by bispectral measurements (Springsteen 1999). In this procedure, narrowband illuminants are employed across the visible spectrum. The reflectance spectra are measured at the same wavelength as the narrowband illuminant, fluorescence emission spectra are measured at longer wavelengths than the illuminations and fluorescence absorption spectra are measured by observing the emission at a certain wavelength λ while varying the illuminant wavelength λ' for $\lambda' < \lambda$.

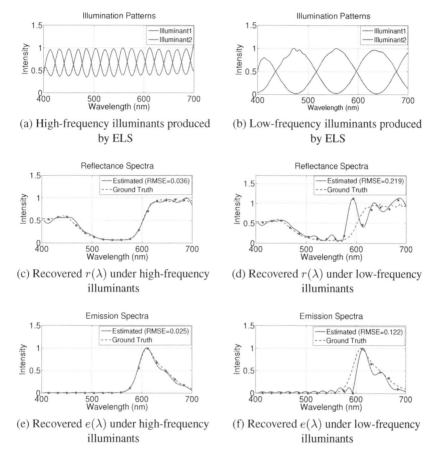

(a) High-frequency illuminants produced by ELS

(b) Low-frequency illuminants produced by ELS

(c) Recovered $r(\lambda)$ under high-frequency illuminants

(d) Recovered $r(\lambda)$ under low-frequency illuminants

(e) Recovered $e(\lambda)$ under high-frequency illuminants

(f) Recovered $e(\lambda)$ under low-frequency illuminants

Figure 5.9. *Evaluation of our separation method on a pink sheet. (a) Two high-frequency illuminations. (c) and (e) The recovered reflectance and fluorescence emission spectra under these high-frequency illuminations are shown, respectively. (b) Two low-frequency illuminations. (d) and (f) The recovered reflectance and fluorescence emission spectra under these low-frequency illuminations are shown, respectively. The red lines show the ground truths, and the blue lines show the estimated results. For a color version of this figure, see www.iste.co.uk/funatomi/ computational.zip*

In Figure 5.9(c)–(f), we see the recovered reflectance and fluorescence emission spectra of a pink fluorescent sheet under different frequency illuminants. The recovered reflectance (Figure 5.9(c)) and fluorescence emission spectra (Figure 5.9(e)) under the high-frequency illuminants approximate the ground truth well. When the object is captured under the low-frequency illuminants, the recovered reflectance (Figure 5.9(d)) and fluorescence emission (Figure 5.9(f)) have obvious errors. Figure 5.10(a)–(d) shows the recovered reflectance and fluorescence emission spectra of the red and yellow fluorescent sheets under the high-frequency illuminants. All these results demonstrate that our method is able to recover reflectance and fluorescence emission spectra efficiently under high-frequency illuminants.

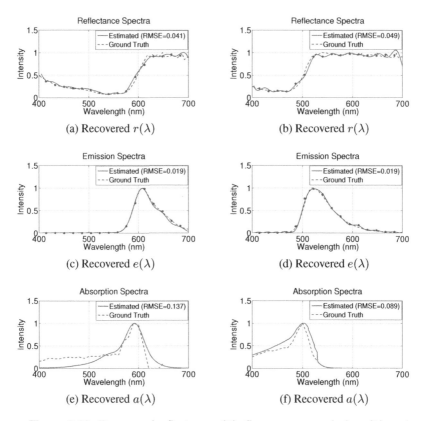

Figure 5.10. *Recovered reflectance $r(\lambda)$, fluorescence emission $e(\lambda)$ and absorption $a(\lambda)$ spectra of the red and yellow sheets. For a color version of this figure, see www.iste.co.uk/funatomi/computational.zip*

In Figure 5.10(e) and (f), the recovered fluorescence absorption spectra of the red and yellow fluorescent sheets are shown. Due to limitations of our capture equipment, the ground truth could not be accurately measured in the short wavelength region in cases where absorption was relatively weak. This issue can be seen in the shorter wavelengths for the red sheet (Figure 5.10(e)). However, we can see that the recovered absorption spectra and the ground truth measurements still agree quite well.

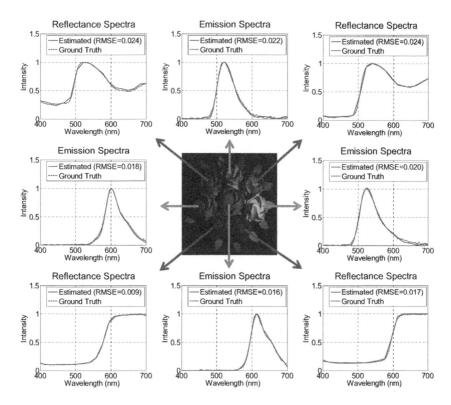

Figure 5.11. *Recovered reflectance spectra for the ordinary reflective materials (red arrows) and fluorescence emission spectra for the fluorescent materials (green arrows). For a color version of this figure, see www.iste.co.uk/funatomi/computational. zip*

We now show that our method works well for both ordinary reflective materials and fluorescent materials. For ordinary reflective materials, the reflectance spectrum can be easily recovered by capturing the scene under white light across the visible spectrum, while the emission spectrum for a fluorescent material can be easily captured at longer wavelengths under near-UV light. Generally, a scene consists of both ordinary reflective materials and fluorescent materials. Here, we first evaluate the recovered reflectance spectra for ordinary reflective materials and the emission spectra for fluorescent materials. Their ground truth data are captured under white light and near-UV light, respectively. In our method, the reflectance spectrum $r(\lambda)$ and the emission spectrum $e(\lambda)$ are estimated by equations [5.8] at the same time. Figure 5.11 shows recovered reflectance spectra for ordinary materials (red arrows) and fluorescence emission spectra for the fluorescent materials (green arrows) by using high-frequency illuminations. We can see that all recovered spectra (blue line) approximate the ground truth (red line) well. This demonstrates that our method can effectively separate fluorescent emission spectra $e(\lambda)$ from the fluorescent material, and also works for ordinary reflective materials, in which $e(\lambda) = 0$, $r(\lambda)$ can be well recovered by the first equation in [5.8].

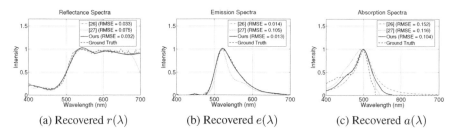

| (a) Recovered $r(\lambda)$ | (b) Recovered $e(\lambda)$ | (c) Recovered $a(\lambda)$ |

Figure 5.12. *Comparison results on the fluorescent yellow sheet. For a color version of this figure, see www.iste.co.uk/funatomi/computational.zip*

We also compared our method against state-of-art works (Fu et al. 2014; Zheng et al. 2014). To allow for the fairest comparison under ideal conditions with all these methods, we performed synthetic tests. As shown in Figure 5.12, we can see that our method achieves similar accuracy to Zheng et al. (2014), which uses three hyperspectral images and needs to capture enough hyperspectral bands to separate reflective and fluorescent components, while our method uses two hyperspectral images and can also separate reflective and fluorescent components for any number of narrowband channels under these two high-frequency light spectra. Compared against Fu et al. (2014), which uses multiple color RGB images, we achieve high accuracy results.

5.5.3. *Visual separation and relighting results*

Here, we show the results for the separation of reflection and fluorescence as well as accurate relighting performance on visual images. Our original results are in the form of hyperspectral images. Figure 5.13 shows four channels of the hyperspectral images and separated results by using high-frequency illuminations for a fluorescent scene. From the left to the right columns, Figure 5.13(a) shows the scene captured at the peak, crest, peak and crest wavelengths of illuminant l_1. Correspondingly, Figure 5.13(b) shows the scene captured at the crest, peak, crest and peak of illuminant l_2 at the same wavelengths. The wavelengths get longer from the left to right columns. Their separated reflective and fluorescent components are shown in Figure 5.13(c) and (d), respectively. The separated fluorescent components (Figure 5.13(d)) only contain the fluorescent material, and demonstrate that our method can effectively separate the reflective and fluorescent components. We also see that the green and yellow fluorescent roses are clearly visible in the shorter wavelengths (the first and second columns in Figure 5.13(d)) and the orange and red fluorescent roses are clearly visible in the longer wavelengths (the third and fourth columns in Figure 5.13(d)).

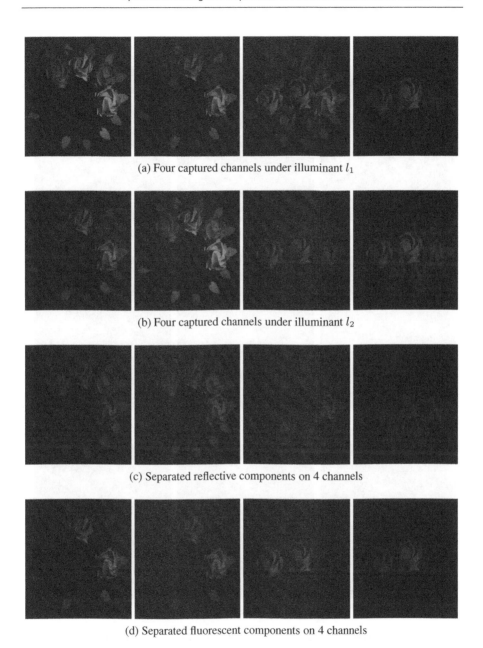

(a) Four captured channels under illuminant l_1

(b) Four captured channels under illuminant l_2

(c) Separated reflective components on 4 channels

(d) Separated fluorescent components on 4 channels

Figure 5.13. *The separation results on four channels of the hyperspectral images for a scene with fluorescent and non-fluorescent roses. These four channels, from left to right, are at 520, 540, 600 and 620 nm. For a color version of this figure, see www.iste.co.uk/funatomi/computational.zip*

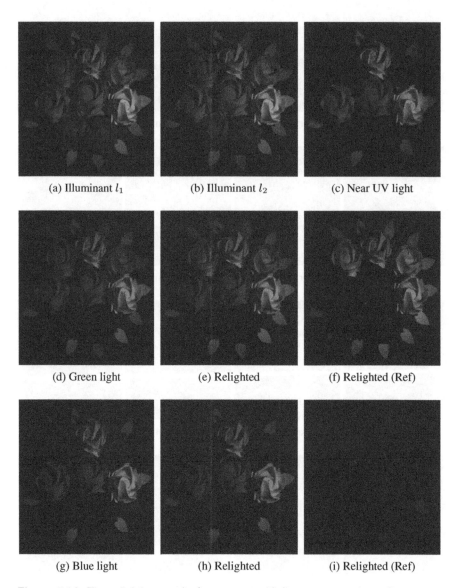

(a) Illuminant l_1 (b) Illuminant l_2 (c) Near UV light

(d) Green light   (e) Relighted   (f) Relighted (Ref)

(g) Blue light   (h) Relighted   (i) Relighted (Ref)

Figure 5.14. *The relighting results for a scene with fluorescent and non-fluorescent roses. "Ref" denotes relighting with only the reflective component. For a color version of this figure, see www.iste.co.uk/funatomi/computational.zip*

To easily visualize hyperspectral images, we have converted them all to RGB images in the following. The first scene is an image consisting of fluorescent and non-fluorescent roses and is taken under two high-frequency illuminants (Figure 5.14(a) and (b)). Figure 5.1(b) and (c) is the corresponding separated reflective and fluorescent components. The roses in four corners (the red arrows in Figure 5.11) only have ordinary reflection so their colors in the recovered reflective component (Figure 5.1(b)) are the same as those seen under white light (Figure 5.1(a)). Looking at the center rose, which is made from the red sheet in Figure 5.1(a), we see that the recovered fluorescent component appears to be red. The measured emission spectrum of the red sheet (Figure 5.10(c)) indicates that the color of the fluorescent component is indeed red. In addition, the scene captured under near-UV light (Figure 5.14(c)) shows nearly pure fluorescent emission colors that also agree with our results in Figure 5.1(c). We note that since each fluorescent material has its own absorption spectrum, the value for $\left(\int l(\lambda')a(\lambda')d\lambda' \right)$ is different between fluorescent materials captured under near-UV light and high-frequency light. As a result, under different lightings, fluorescent objects can exhibit different scales of emission, but the chromaticities match well as can be seen by comparing the images under near-UV light (Figure 5.14(c)) and for the recovered fluorescent component (Figure 5.1(c)).

Since our method is able to recover the full reflectance, fluorescence emission and fluorescence absorption spectra for an entire scene, we are also able to relight scenes. Figure 5.14 shows that real scenes can be accurately relighted using our method. The scenes are captured under green (Figure 5.14(d)) and blue (Figure 5.14(g)) illuminants. The corresponding relighting results are shown in Figure 5.14(e) and (h). We can see that the relighting results are very similar to the ground truths (Figure 5.14(d) and (g)), and demonstrate the effectiveness of our method in recovering the reflectance and fluorescence emission and absorption spectra. When the scene is relighted using the reflective component only (Figure 5.14(f) and (i)), this leads to many fluorescent materials appearing as black, especially under blue light (Figure 5.14(i)).

Figures 5.15 and 5.16 show additional separation results on two other fluorescent scenes and their relighting results. They are a fluorescent color checker with a Macbeth color chart, and fluorescent and non-fluorescent notebooks. The separated reflective component (Figures 5.15 and 5.16(b)) for the ordinary reflective material is the same as those seen under white light (Figures 5.15 and 5.16(c)), and the separated fluorescent component (Figures 5.15 and 5.16(h)) also approximates the scene captured under near-UV light (Figures 5.15 and 5.16(i)) which shows nearly pure fluorescence emission colors. The relighting results (Figures 5.15 and 5.16(e)(k)) were all close to the ground truth (Figures 5.15 and 5.16(d)(j)). These additional results on real scenes show that our method is effective for different scenes.

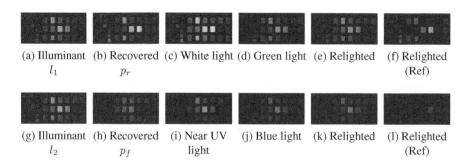

(a) Illuminant (b) Recovered (c) White light (d) Green light (e) Relighted (f) Relighted
l_1 p_r (Ref)

(g) Illuminant (h) Recovered (i) Near UV (j) Blue light (k) Relighted (l) Relighted
l_2 p_f light   (Ref)

Figure 5.15. *Separation and relighting results for a fluorescent and a non-fluorescent color chart. For a color version of this figure, see www.iste.co.uk/funatomi/computational.zip*

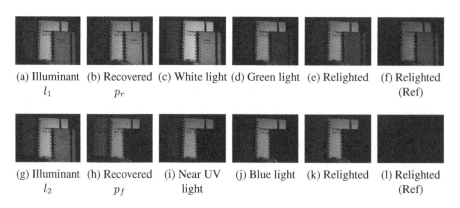

(a) Illuminant (b) Recovered (c) White light (d) Green light (e) Relighted (f) Relighted
l_1 p_r (Ref)

(g) Illuminant (h) Recovered (i) Near UV (j) Blue light (k) Relighted (l) Relighted
l_2 p_f light   (Ref)

Figure 5.16. *Separation and relighting results for a scene with fluorescent and non-fluorescent objects. For a color version of this figure, see www.iste.co.uk/funatomi/computational.zip*

5.5.4. *Separation by using high-frequency filters*

In the previous parts, we employ a programmable light source known as the ELS to produce complementary high-frequency illuminants, by which the excellent experimental results can be obtained. However, programmable light sources such as the ELS are prohibitively expensive for many laboratories and consumers. They are also heavy and thus not portable. Due to these limitations, we designed two complementary high-frequency filters, which are portable, as shown in Figure 5.17. These two filters are put in front of a light source to modulate an illuminant into high-frequency illuminations that are the same as the lights produced by the programmable light source. They are designed as two complementary sinusoidal patterns with periods of 20 nm^2.

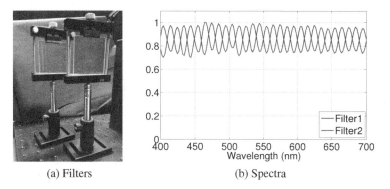

(a) Filters   (b) Spectra

Figure 5.17. *The two high-frequency filters (a) and their spectra (b). For a color version of this figure, see www.iste.co.uk/funatomi/computational.zip*

Let us denote one of the filters as F_1 and its complement as F_2. The illumination $l(\lambda)$ after going through the filters can be described as

$$l_1^f(\lambda) = F_1(\lambda)l(\lambda),$$
$$l_2^f(\lambda) = F_2(\lambda)l(\lambda).$$

[5.21]

When the spectrum of light $l(\lambda)$ is flat (constant for all wavelengths λ), we can see that l_n^f ($n = 1, 2$) are equivalent to the lights l_n ($n = 1, 2$) discussed in section 5.3.

However, common light sources such as daylight and off-the-shelf lights are not exactly flat, so we need to evaluate how an arbitrary light source affects the resulting high-frequency illuminations. We first replace $l_n(\lambda)$ with $l_n^f(\lambda)$ in equation [5.7] and obtain

$$p_1(\lambda) = F_1(\lambda)l(\lambda)r(\lambda) + k_1 e(\lambda),$$
$$p_2(\lambda) = F_2(\lambda)l(\lambda)r(\lambda) + k_2 e(\lambda),$$
$$k_n = \int F_n(\lambda')l(\lambda')a(\lambda')d\lambda'.$$

[5.22]

As discussed in section 5.3.4, the errors for the recovered reflectance and fluorescence emission spectra are directly proportional to $k_1 - k_2$. In this case, the

---

2. Due to the limitations in manufacturing of the filters, we cannot produce the exact sinusoidal spectra in the filters.

difference between k_1 and k_2 is decided by the spectra of the filters $F_n(\lambda)$, illuminant $l(\lambda)$ and fluorescent absorption $a(\lambda)$. Table 5.1 shows the mean percent differences between k_1 and k_2 for 183 absorption spectra, where each row corresponds to a CIE standard illuminant. To explore the influences due to the frequency and the appearance of the filters, we calculate the mean percentage on three kinds of filters: "Ideal (40 nm)", "Ideal (20 nm)" and "Real filters".

Illuminant	Ideal (40 nm)	Ideal (20 nm)	Real filters
E	2.56	0.34	0.44
A	2.79	0.33	0.37
B	2.64	0.29	0.36
C	2.59	0.28	0.36
D50	2.70	0.46	0.42
D55	2.66	0.45	0.42
D65	2.61	0.44	0.43
D75	2.57	0.43	0.44
F1	8.06	1.41	0.44
F2	9.57	1.74	0.59
F3	11.25	2.13	0.77
F4	13.00	2.56	0.97
F5	7.92	1.36	0.41
F6	9.92	1.80	0.62
F7	7.82	1.33	0.43
F8	8.74	1.50	0.54
F9	9.86	1.74	0.65
F10	14.56	3.12	1.02
F11	16.58	3.73	1.21
F12	19.20	4.52	1.47

Table 5.1. *The mean percent difference between k_1 and k_2 for 183 absorption spectra on CIE standard illuminants (Hunt and Pointer 2011) with the ideal sinusoidal pattern filters and real filters. "Ideal (20 nm)" and "Ideal (40 nm)" denote filters with ideal sinusoidal patterns and periods of 20 nm and 40 nm. "Real Filters" are the filters used in our experiments and their spectra are shown in Figure 5.17*

"Ideal (40 nm)" and "Ideal (20 nm)" denote ideal sinusoidal patterned filters with 40 and 20 nm periods and ratio $\alpha/\beta = 1$. According to the discussion in section 5.3.3, when the frequency of the filters is higher, the difference between k_1 and k_2 will be lower. In Table 5.1, we can see that the differences under "Ideal (20 nm)" are less than those under "Ideal (40 nm)" under the same illuminant $l(\lambda)$.

The real filters shown in Figure 5.17 approximate the ideal sinusoidal patterns well and have the same period as the "Ideal (20 nm)" filters, but the value α/β of the real filters is larger. Recall that in section 5.3.4, we discussed that the difference between k_1 and k_2 relies on the high-frequency component $\beta \cos(2\pi f_l \lambda)$ in the illuminant and is not related to the direct current (DC) component α. Thus, when the ratio α/β becomes larger, the difference between k_1 and k_2 is reduced. As shown in Table 5.1, under most standard illuminants, the difference between k_1 and k_2 under "Real Filters" is lower than for the "Ideal (20 nm)" filters for the same standard illuminant. Nevertheless, in some cases, the difference between k_1 and k_2 under "Real Filters" is a little larger than those in the "Ideal (20 nm)" filters. This is because the real filters are not the exact sinusoidal spectra and exhibit some distortions. Figure 5.18 shows the recovered reflectance $r(\lambda)$ and fluorescence emission $e(\lambda)$ of a pink fluorescent sheet under high-frequency illuminants produced by the ELS, C light through high-frequency filters and D55 light through high-frequency filters. All these recovered spectra approximate the ground truth well. This indicates that the errors for recovered reflectance and emission spectra are acceptable and our method is effective using high-frequency filters on real data.

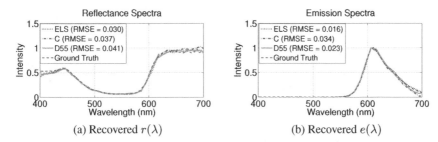

(a) Recovered $r(\lambda)$ (b) Recovered $e(\lambda)$

Figure 5.18. *Recovered reflectance $r(\lambda)$ and fluorescence emission $e(\lambda)$ of a pink fluorescent sheet under high-frequency illuminants produced by the ELS, C light through high-frequency filters and D55 light through high-frequency filters. For a color version of this figure, see www.iste.co.uk/funatomi/computational.zip*

Figure 5.19 shows the separated reflective and fluorescent components under different illuminants through the high-frequency filters. The first column shows the high-frequency illuminations' spectra, resulting from using different light sources. Taking a channel at 550 nm as an example, the second and third columns show the captured images under two high-frequency illuminations, and their separation results are shown in the third and fourth columns. The fifth and sixth columns show separation results over all captured spectra in RGB images. Figure 5.19(b)–(d) shows the separation results under C, D55 and F5 lights with the high-frequency filters, respectively. Compared with the separation results under the high-frequency illuminations produced by the ELS (Figure 5.19(a)), we can see that the separation results using the high-frequency filters are competitive but there is more noise, for example, in the two fluorescent flowers. The decreased amount of light going through the filters likely caused the camera to exhibit more noise. The ratio α/β for the high-frequency filters is also much larger than for the spectra produced by the ELS, which makes the separation results more sensitive to the noise from the camera and also contributes to the noise.

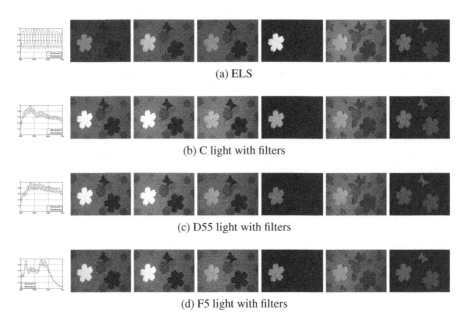

(a) ELS

(b) C light with filters

(c) D55 light with filters

(d) F5 light with filters

Figure 5.19. *The separation results with the high-frequency filters. The spectra of the illuminants are shown in the first column. Taking a channel at 550 nm as an example, the second and third columns show the captured images under two high-frequency illuminations, and their separated reflective and fluorescent components are shown in the third and fourth columns. The fifth and sixth columns show reflective and fluorescent components over all captured spectra in RGB images. To make a comparison, the first row shows the separation results under high-frequency illuminations produced by ELS. From the second to fourth rows, the separation results under C, D55 and F5 lights with the high-frequency filters are shown, respectively. For a color version of this figure, see www.iste.co.uk/funatomi/computational.zip*

5.5.5. *Ambient illumination*

So far, we have extended our method to more general light sources by using high-frequency filters instead of the ELS, but we did not consider ambient light. In the following, we discuss the effect that ambient light has on our approach. Let us denote an ambient illuminant as $l_a(\lambda)$. Without loss of generality, the two illuminants produced by either the ELS or flat light source with filters are defined as $l_1(\lambda)$ and $l_2(\lambda)$. So, the illuminations with ambient light can be described as

$$l_1^a(\lambda) = l_1(\lambda) + l_a(\lambda),$$
$$l_2^a(\lambda) = l_2(\lambda) + l_a(\lambda).$$

[5.23]

Replacing the $l_n(\lambda)$ by $l_n^a(\lambda)$ in equation [5.7], we obtain

$$p_1(\lambda) = [l_1(\lambda) + l_a(\lambda)]r(\lambda) + k_1 e(\lambda),$$
$$p_2(\lambda) = [l_2(\lambda) + l_a(\lambda)]r(\lambda) + k_2 e(\lambda),$$

$$k_n = \int [l_n(\lambda) + l_a(\lambda)]a(\lambda')\mathrm{d}\lambda'.$$

[5.24]

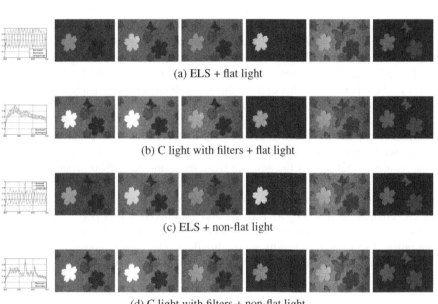

(a) ELS + flat light

(b) C light with filters + flat light

(c) ELS + non-flat light

(d) C light with filters + non-flat light

Figure 5.20. *The separation results with ambient light. The spectra of illuminants are shown in the first column, where the green curves in the first and third rows are the spectra of the flat and non-flat ambient lights. Taking a channel at 550 nm as an example, the second and third columns show the captured images under two high-frequency illuminations, and their separated reflective and fluorescent components are shown in the third and fourth columns. The fifth and sixth columns show reflective and fluorescent components over all captured spectra in RGB images. From the first to fourth rows, we see separation results under high-frequency illuminants produced by the ELS with flat ambient light (D50), C light through high-frequency filters with flat ambient light, high-frequency illuminants produced by the ELS with non-flat ambient light (typical of a fluorescent lamp) and C light through high-frequency filters with non-flat ambient light, respectively. For a color version of this figure, see www.iste.co.uk/funatomi/computational.zip*

Since the ambient illuminant l_a is the same under the two different high-frequency illuminants $l_1^a(\lambda)$ and $l_2^a(\lambda)$, the difference between k_1 and k_2 is only related to the high-frequency illuminants $l_1(\lambda)$ and $l_2(\lambda)$. Intuitively, the intensity of the ambient illuminant can be considered as a part of the DC component α, which is also the same under two complementary high-frequency illuminations. Thus, our method can be directly used on scenes with ambient light, and the reflectance and fluorescence emission spectra can be recovered by equation [5.8], in which the illuminations $l_n(\lambda)$ are replaced by $l_n^a(\lambda)$.

Figure 5.20 shows the separation results under different high-frequency illuminants and ambient light. The spectra of flat and non-flat ambient lights are shown as green lines in the first column of Figure 5.20(a) and (c). We can see that all separation results under flat ambient light (Figure 5.20(a) and (b)) and non-flat ambient light (Figure 5.20(c) and (d)) are clear. These results demonstrate that our method works well under both flat and non-flat ambient light sources. Compared with the results under the high-frequency illuminations produced by the ELS (Figure 5.20(a) and (c)), the separation results under lights through the high-frequency filters (Figure 5.20(b) and (d)) contain noise like in Figure 5.19, but are also acceptable.

We also capture the scene under a strong ambient illuminant which is typical of a fluorescent lamp, and show the separation results in Figure 5.21. The results on the crest located at 520 nm of the ambient light spectrum show clear separation of the components (Figure 5.21(b) and (c)), while the results on the peak at 540 nm of the ambient light spectrum are clearly wrong (Figure 5.21(d) and (e)). This is because the DC component in the $l_n^a(\lambda)$ is larger and the observation of two high-frequency illuminations will be almost the same under the strong ambient illuminant spectrum in that range. The separation results are easily affected by camera noise. Therefore, we need to choose a higher intensity light source when the ambient illuminant is strong in practice.

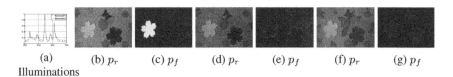

(a)   (b) p_r (c) p_f (d) p_r (e) p_f (f) p_r (g) p_f
Illuminations

Figure 5.21. *The separation results under strong ambient light, which is typical of fluorescent lamps. (a) The two illuminations used to illuminate the scene are shown. (b) and (c) Separation results at crests of the ambient light spectrum at 520 nm are shown, (d) and (e) results at peaks of the ambient light spectrum 540 nm are shown, and (f) and (g) results over the visible spectrum as RGB images are shown. For a color version of this figure, see www.iste.co.uk/funatomi/computational.zip*

5.6. Limitations and conclusion

In this work, we presented a method to simultaneously recover the reflectance and fluorescence emission spectra of an entire scene by using high-frequency illumination in the spectral domain. Afterward, we presented our method for estimating the fluorescence absorption spectrum of a material given its emission spectrum. Through our method, we also showed that similar emission spectra tend to map to similar absorption spectra. The effectiveness of the proposed method was successfully demonstrated with experiments using real data taken by a spectroradiometer and camera, both in conjunction with a programmable light source. To extend our method to much more general light sources, we designed two high-frequency filters and employed them under CIE standard illuminants. We demonstrated that when the light source is flat enough compared with the frequency of the filters, our method also works well. We also implemented our method under high-frequency illuminations with flat/non-flat ambient light, and the results show that our method works well under different types of ambient light.

There are still a few limitations in our research that are worth attention and further investigation. First, the two high-frequency filters used in our experiments affect the separation results, due to their distortions of the sinusoidal patterns and the differences between the peaks and crests of the light spectra. In the future, we will design/employ much better high-frequency filters, which have much smaller distortions, and especially smaller values of α/β, to make them more robust to noise. Second, we did not consider shading from the light source and specularity from the materials in our work. More importantly, shading and specularity can provide more information about a scene. Therefore, it is worth investigating a more comprehensive model for reflective and fluorescent separation to make it applicable to more real cases.

5.7. References

Alterman, M., Schechner, Y., Weiss, A. (2010). Multiplexed fluorescence unmixing. *IEEE International Conference on Computational Photography (ICCP)*, 1–8.

Balas, C., Papadakis, V., Papadakis, N., Papadakis, A., Vazgiouraki, E., Themelis, G. (2003). A novel hyper-spectral imaging apparatus for the non-destructive analysis of objects of artistic and historic value. *Journal of Cultural Heritage*, 4(1), 330–337.

Barnard, K. (1999). Color constancy with fluorescent surfaces. *CIC, Proceedings of the IS&T/SID Color Imaging Conference*, 1–5.

Candes, E. and Tao, T. (2006). Near-optimal signal recovery from random projections: Universal encoding strategies? *IEEE Transactions on Information Theory*, 52(12), 5406–5425.

Chi, C., Yoo, H., Ben-Ezra, M. (2010). Multi-spectral imaging by optimized wide band illumination. *International Journal of Computer Vision*, 86(2–3), 140–151.

DiCarlo, J.M., Xiao, F., Wandell, B.A. (2001). Illuminating illumination. *CIC, IS&T/SID*, 27–34.

Donoho, D., Tsaig, Y., Drori, I., Starck, J.-L. (2012). Sparse solution of underdetermined systems of linear equations by stagewise orthogonal matching pursuit. *IEEE Transactions on Information Theory*, 58(2), 1094–1121.

Efron, B., Hastie, T., Johnstone, I., Tibshirani, R. (2004). Least angle regression. *The Annals of Statistics*, 32(2), 407–451.

Farid, H. and Adelson, E. (1999). Separating reflections and lighting using independent components analysis. *Computer Vision and Pattern Recognition (CVPR)*, Fort Collins, CO, 262–267.

Fu, Y., Lam, A., Kobashi, Y., Sato, I., Okabe, T., Sato, Y. (2014). Reflectance and fluorescent spectra recovery based on fluorescent chromaticity invariance under varying illumination. *Proceedings of IEEE Conference on Computer Vision and Pattern Recognition*, pp. 2171–2178.

Gat, N. (2000). Imaging spectroscopy using tunable filters: A review. *Wavelet Applications VII*, vol. 4056, SPIE, 1–15.

Hullin, M.B., Hanika, J., Ajdin, B., Seidel, H.-P., Kautz, J., Lensch, H.P.A. (2010). Acquisition and analysis of bispectral bidirectional reflectance and reradiation distribution functions. *ACM Transactions on Graphics*, 29(4), 1–7.

Hunt, R.W.G. and Pointer, M.R. (2011). *Measuring Colour*, 4th edition. John Wiley & Sons, Ltd.

Johnson, G. and Fairchild, M. (1999). Full-spectral color calculations in realistic image synthesis. *IEEE Computer Graphics and Applications*, 19(4), 47–53.

Lam, A. and Sato, I. (2013). Spectral modeling and relighting of reflective-fluorescent scenes. *IEEE Conference on Computer Vision and Pattern Recognition*, 1–8.

Lee, B.-K., Shen, F.-C., Chen, C.-Y. (2001). Spectral estimation and color appearance prediction of fluorescent materials. *Optical Engineering*, 40, 1–8.

Maloney, L.T. and Wandell, B.A. (1986). Color constancy: A method for recovering surface spectral reflectance. *Journal of the Optical Society of America A*, 3(1), 92–33.

McNamara, G., Gupta, A., Reynaert, J., Coates, T.D., Boswell, C. (2006). Spectral imaging microscopy web sites and data. *Cytometry. Part A: The Journal of the International Society for Analytical Cytology*, 69(8), 863–871.

Nayar, S., Fang, X., Boult, T. (1993). Removal of specularities using color and polarization. *IEEE Conference on Computer Vision and Pattern Recognition*, 583–590.

Nayar, S.K., Krishnan, G., Grossberg, M.D., Raskar, R. (2006). Fast separation of direct and global components of a scene using high frequency illumination. *ACM Transactions on Graphics*, 25(3), 935–944.

Oppenheim, A.V., Willsky, A.S., Hamid, W.S. (1996). *Signals and Systems*, 2nd edition. Prentice Hall.

Park, J., Lee, M., Grossberg, M.D., Nayar, S.K. (2007). Multispectral imaging using multiplexed illumination. *ICCV*, IEEE, 1–8.

Rost, F.W.D. (1992). *Fluorescence Microscopy*. Cambridge University Press.

Sato, I., Okabe, T., Sato, Y. (2012). Bispectral photometric stereo based on fluorescence. *Color Imaging Conference: Color Science, Systems, and Applications*, 270–277.

Skoog, D.A., Holler, F.J., Crouch, S.R. (2007). *Principles of Instrumental Analysis*. Thomson Publishers.

Springsteen, A. (1999). Introduction to measurement of color of fluorescent materials. *Analytica Chimica Acta*, 380(2–3), 183–192.

Styles, I.B., Calcagni, A., Claridge, E., Orihuela-Espina, F., Gibson, J.M. (2006). Quantitative analysis of multi-spectral fundus images. *Medical Image Analysis*, 10(4), 578–597.

Suo, J., Bian, L., Chen, F., Dai, Q. (2014). Bispectral coding: Compressive and high-quality acquisition of fluorescence and reflectance. *Optics Express*, 22(2), 1697–1712.

Tominaga, S. (1996). Multichannel vision system for estimating surface and illumination functions. *Journal of the Optical Society of America A*, 13(11), 2163–2173.

Tominaga, S., Horiuchi, T., Kamiyama, T. (2011). Spectral estimation of fluorescent objects using visible lights and an imaging device. *Proceedings of the IS&T/SID Color Imaging Conference*, 352–356.

Treibitz, T., Murez, Z., Mitchell, B.G., Kriegman, D. (2012). Shape from fluorescence. *European Conference on Computer Vision*, 292–306.

Wilkie, A., Weidlich, A., Larboulette, C., Purgathofer, W. (2006). A reflectance model for diffuse fluorescent surfaces. *International Conference on Computer Graphics and Interactive Techniques*, 321–499.

Zhang, C. and Sato, I. (2011). Separating reflective and fluorescent components of an image. *IEEE Conference on Computer Vision and Pattern Recognition*, 185–192.

Zhang, C. and Sato, I. (2013). Image-based separation of reflective and fluorescent components using illumination variant and invariant color. *IEEE Transactions on Pattern Analysis and Machine Intelligence*, 35(12), 2866–2877.

Zheng, Y., Sato, I., Sato, Y. (2014). Spectra estimation of fluorescent and reflective scenes by using ordinary illuminants. *Proceedings of the 9th European Conference on Computer Vision*, pp. 188–202.

6

Shape from Water

Yuta ASANO[1], Yinqiang ZHANG[2],
Ko NISHINO[3], and Imari SATO[4]

[1]*Digital Content and Media Sciences Research Division,
National Institute of Informatics, Tokyo, Japan*
[2]*Next Generation Artificial Intelligence Research Center,
The University of Tokyo, Japan*
[3]*Graduate School of Informatics, Kyoto University, Japan*
[4]*Computational Imaging and Vision Lab,
National Institute of Informatics, Tokyo, Japan*

6.1. Introduction

Recovering 3D scene geometry has been one of the most important tasks in computer vision, and its methodologies have progressed significantly during the last several decades. Numerous 3D reconstruction methods have been proposed, including triangulation, time of flight and shape-from-X, where X can be shading, texture, focus or other surface or image formation properties. The fundamental but often neglected assumption of these different approaches is that the light, either actively or passively shed on the object surface including environmental illumination, can be measured unaltered between the object's surface and the camera. Although there have been studies on shape recovery of objects in a non-air medium where this assumption does not hold (e.g. participating medium like dilute milk), their focus is on undoing adversarial optical perturbations such as scattering in order to apply the same recovery principles that were designed for objects in clear air. In other words, the medium is treated as an unwanted nuisance that violates the assumed geometry recovery principle.

Computational Imaging for Scene Understanding,
coordinated by Takuya FUNATOMI and Takahiro OKABE.
© ISTE Ltd 2024.

In the past, scattering has been modeled to restore the appearance of a clear-day scene from images taken in bad weather (e.g. fog (Narasimhan and Nayar 2002)), and in the process, the scene depth can also be recovered (e.g. Nishino et al. 2012). However, this is limited to accidental imaging in bad weather conditions and cannot be used as a general shape recovery method. In this chapter, we focus on light absorption in the infrared spectrum as a light propagation characteristic that encodes depth. When light travels through a homogeneous isotropic medium, it usually gets absorbed at certain wavelengths. Light absorption is dictated by the Beer–Lambert law, which says that the absorption at a certain wavelength is proportional to the length of the light travel path and to the absorption coefficient of the medium (Reinhard et al. 2008). This suggests that we may be able to recover the distance of a surface point to the camera by measuring the amount of light absorbed between the surface and the camera. In other words, we may be able to recover the depth of an object by measuring the light path distance in the medium.

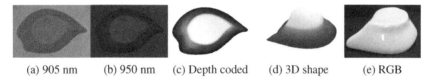

| (a) 905 nm | (b) 950 nm | (c) Depth coded | (d) 3D shape | (e) RGB |

Figure 6.1. *(a) and (b) The scene at 905 nm and 950 nm after normalizing the illumination and camera and filter sensitivity functions is shown. The intensity difference between (a) and (b) is due to the difference in water absorption at these two wavelengths. The color-coded depth is shown in (c), and the recovered 3D shape is given in (d). (e) The object for this example is shown: a textureless ceramic object with strong specularity. For a color version of this figure, see www.iste.co.uk/ funatomi/computational.zip*

We focused on water as a medium for a few important reasons. In addition to the fact that water is a familiar liquid that is almost everywhere, geometry recovery in water in itself finds applications in many areas of science, such as oceanography, geography and biology, as well as in engineering for purposes including underwater surveillance and navigation. Furthermore, multi-spectral light propagation in water is mostly dominated by absorption and scattering has little effect as long as the water is sufficiently clear, which would otherwise compound the optical length computation. Few of the previous methods have directly applied depth-recovery principles in air to underwater scenarios and have found light absorption to adversely affect the results (Dancu et al. 2014). We instead take advantage of light absorption in water and introduce shape from water as a shape recovery approach.

In what follows, we introduce a shape recovery method based on monochromatic images captured at two different infrared wavelengths, which we refer to as *the bispectral principle of depth recovery*. The key idea is to exploit the difference in the amount of light absorption that takes place at two distinct wavelengths and cancel out light interaction effects, including those due to surface texture and reflectance, other than those proportional to the optical length to the object surface. Figure 6.1 shows an example of recovering a textureless, specular object which would be a challenging object for conventional depth recovery methods.

6.2. Related works

Most of the popular shape recovery methods can be categorized based on the underlying shape or depth probing principle: triangulation, time of flight and shape-from-X where shading is most prominent for X but may include other surface and image formation properties like texture or focus. The literature on each method is vast, and readers may find suitable survey articles elsewhere.

The triangulation method is one of the major techniques of 3D shape recovery, which estimates the depth from the corresponding locations in binocular or multiview stereo, and structure from motion (Hartley and Zisserman 2003; Schonberger and Frahm 2016; Qian et al. 2018). If reliable correspondences can be established, sparse or even dense 3D shapes can be recovered. The fundamental limitation of the triangulation method is that the object's surface must have sufficient discriminative texture to establish correspondences. Structured active light can mitigate this limitation (Batlle et al. 1998) by actively projecting light patterns and essentially putting a texture on the object's surface. However, it is still difficult to recover the 3D shape of textureless objects passively by using triangulation methods.

Time of flight, i.e. the travel time of a light pulse to hit a surface and return to the source, directly encodes the distance of the surface from the source (Hansard et al. 2013). Coherent light (e.g. laser) can be used for long-distance depth sensing, while infrared light has recently been used for short-distance measurements (e.g. Kinect 2). Accurate measurement of time of flight is challenging due to the very high speed of light, and poor accuracy can limit the resolution of the depth image.

Shape-from-X refers to shape recovery methods that exploit specific surface or image formation properties. Among the many radiometric cues, shading so far has been one of the most popular. Shape-from-shading (Zhang et al. 1999) and photometric stereo (Woodham 1980) model the surface brightness change to infer its gradients (i.e. surface normals) from which the shape can be recovered. In contrast to triangulation-based methods, texture as well as complex reflectance become nuisances that hinder the applicability of these methods.

In this chapter, we introduce a bispectral depth imaging principle based on light absorption. It clearly differs from triangulation and other shape-from-X, especially shading, methods that require neither feature correspondence nor a known or simplistic surface reflectance. Unlike time-of-flight methods, it recovers depth by measuring pixel intensity differences, in contrast to light travel time, which obviates the need for often expensive hardware for making an accurate temporal measurement.

Shape recovery in non-air media has been studied in the past. Narasimhan et al. (2005) applied light stripe range scanning and photometric stereo to objects in participating media. They modeled sub-surface scattering and accounted for it to recover the geometry of objects in murky liquid (e.g. dilute milk). Light scattering has also been studied in computer vision for other participating media such as fog (Schechner et al. 2001; Narasimhan and Nayar 2002; Fattal 2008; Tan 2008; He et al. 2009; Kratz and Nishino 2009; Nishino et al. 2012) in which depth can be recovered in the process of removing light propagation effects on the appearance (i.e. defogging). Our depth recovery principle is similar to such approaches in that it actively exploits the light propagation characteristics in the medium to decode the optical length and thus depth. Our focus is, however, light absorption, not scattering.

A number of depth recovery methods for underwater scenes have been developed. For example, Queiroz-Neto et al. (2004) proposed a stereo matching method that tries to remove scattering and absorption effects. It focuses on undoing adversarial optical perturbations. Yano et al. (2013) proposed a reconstruction method for underwater objects using multiview stereo that accounts for the refractive effect of water and the shape of the interfacing layer. Photometric stereo was applied to underwater objects (Tsiotsios et al. 2014; Murez et al. 2015; Fujimura et al. 2018) in an attempt to deal with the scattering problem due to water turbidity. Structured active stereo was also applied to underwater objects (Kawahara et al. 2016; Kawasaki et al. 2017) in an attempt to consider light refraction. The performance of various time-of-flight sensors was evaluated in recovering underwater scenes, and it was found that time-of-flight methods do not work well for slightly deep water because infrared light is strongly absorbed by water (Dancu et al. 2014). A structure-from-motion method for underwater images was presented (Saito et al. 1995; Jordt-Sedlazeck and Koch 2013) that considers light refraction at the water surface.

In addition to shape recovery, wavelength-dependent color formation and restoration of underwater objects have recently attracted much attention. For example, Akkaynak et al. (2017) measured, and later modeled, light attenuation in the ocean due to absorption by water and forward scattering of light. A more accurate model with wavelength-dependent backscattering was presented by Akkaynak and Treibitz (2018). Chiang and Chen (2012) proposed a wavelength-dependent compensation algorithm to eliminate the bluish tone of underwater images and the effects of artificial light. Berman et al. (2017) transformed the underwater color restoration task into single-image dehazing by estimating two global attenuation ratios. Different from 2D image restoration, they took advantage of the light absorption of water in the visible range to estimate the camera relative pose by using three point correspondences, rather than five in the standard setting.

6.3. Light absorption in water

First, we review the basics of light absorption in pure water. Light gets absorbed at certain wavelengths when it travels in clear water. The absorption curve in Figure 6.2(a) shows how light is attenuated as it travels through water (6 mm depth here) in the wavelength range from 400 nm to 1,400 nm. As can be seen from the figure, water rarely absorbs visible light, which explains why water appears transparent to the human eye. In contrast, it absorbs infrared light from 900 nm to 1,400 nm.

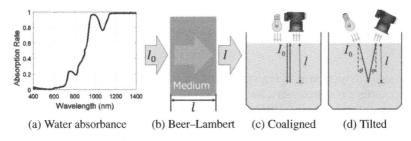

(a) Water absorbance   (b) Beer–Lambert   (c) Coaligned   (d) Tilted

Figure 6.2. *(a) The water absorption curve in the range from 400 nm to 1400 nm is shown. (b) The setup of the Beer–Lambert law is shown. (c) and (d) Our depth imaging in the coaligned and tilted configurations is shown, respectively. For a color version of this figure, see www.iste.co.uk/funatomi/computational.zip*

As illustrated in Figure 6.2(b), the Beer–Lambert law (Reinhard et al. 2008) accurately expresses light absorption at a given wavelength λ as the relation between incident light intensity I_0 and outgoing attenuated intensity I

$$I = I_0 e^{-\alpha(\lambda)l}, \tag{6.1}$$

in which l represents the light pathlength in millimeters (mm), $\alpha(\lambda)$ denotes the wavelength-dependent absorption coefficient in mm^{-1}, and $e^{-\alpha(\lambda)l}$ is the natural exponential of $-\alpha(\lambda)l$. Figure 6.3 shows 950 nm images of water pouring into a cup. As the water depth becomes deeper, more infrared light is absorbed. In other words, depth can be encoded as the intensities of the outgoing light.

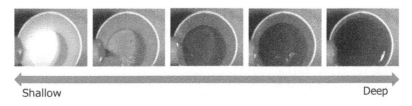

Shallow   Deep

Figure 6.3. *Images of 950 nm of water pouring into a cup*

6.4. Bispectral light absorption for depth recovery

We will use the wavelength dependence of light absorption in water for depth recovery of objects. We assume that the camera is orthographic and the incident light rays to the object surface are parallel, coming from an infinitely distant point source. We also assume that the reflectances of the same point at two wavelengths are similar, which will be described in section 6.4.1. However, we do not make the assumption that the reflective characteristic of an object is Lambertian or diffuse plus specular, except that the geometric and spectral characteristics of the reflectance are separable. This is a very mild assumption wherein the reflectance function $f(\omega, \lambda) = r(\omega)s(\lambda)$ can be factorized into its geometric properties (e.g. incident and reflected light angles), $r(\omega)$, and spectral characteristics (i.e. color), $s(\lambda)$, which applies to most real-world surfaces.

We also envision an imaging system placed in the water, in which consideration of water surface is unnecessary, but for practical necessity, we set up the camera and light source outside the water. We could safely ignore the effects of light refraction at the water surface by assuming directional light and orthographic cameras.

6.4.1. Bispectral depth imaging

As illustrated in Figure 6.2(c), we first consider an ideally coaligned light-camera configuration, where both the optical axis of the camera and the directional light are perpendicular to the planar water surface. Monochromatic light of wavelength λ_1 and intensity I_0 reaches an opaque scene point at a water depth l. After being reflected back from the scene point at water depth l, the light intensity received by the camera is

$$I(\lambda_1) = r(\omega)s(\lambda_1)I_0 e^{-2\alpha(\lambda_1)l}, \qquad\qquad\qquad [6.2]$$

in which $2l$ is the underwater optical pathlength, which is twice as long as the water depth l.

The geometric and spectral characteristics of light reflection at the object surface, $r(\omega)$ and $s(\lambda_1)$, respectively, are related to the underlying surface material composition, which is, of course, unknown. To cancel out these unknown components, we use a second monochromatic observation at wavelength λ_2 with a corresponding light source of the same intensity I_0. The light intensity received by the camera for the second light beam is

$$I(\lambda_2) = r(\omega)s(\lambda_2)I_0 e^{-2\alpha(\lambda_2)l}. \qquad\qquad\qquad [6.3]$$

By dividing equation [6.2] by equation [6.3], the depth l can be calculated as

$$l = \frac{1}{2(\alpha(\lambda_2) - \alpha(\lambda_1))} \ln\left(\frac{I(\lambda_1)}{I(\lambda_2)} \frac{s(\lambda_2)}{s(\lambda_1)} \right). \qquad\qquad [6.4]$$

It is worth noting that the geometric factor of the reflectance function $r(\omega)$ has been eliminated, no matter how complex it is. Provided that we can find two wavelengths such that the reflectance spectrum values at these two wavelengths are almost identical, i.e. $s(\lambda_1) \simeq s(\lambda_2)$, the approximate depth can be estimated as

$$l \simeq \frac{1}{2\left(\alpha(\lambda_2) - \alpha(\lambda_1)\right)} \ln \frac{I(\lambda_1)}{I(\lambda_2)}. \tag{6.5}$$

Equation [6.5] allows us to estimate depth simply by measuring the difference in the light intensity of two properly chosen wavelengths without knowing any information on the arbitrarily general reflectance function of the scene material. This stays at the core of our bispectral depth recovery principle. Regarding the mild assumption that the reflectance spectrum values at two neighboring wavelengths are almost identical, we can make a correspondence to the difference in the reflectance spectrum values by using trispectral light absorption. This is discussed in section 6.7 as a way to address some challenging reflectances.

6.4.2. Depth accuracy and surface reflectance

Next, we analyze the relative depth error $\triangle l$ with respect to the relative difference $\triangle s$ between $s(\lambda_1)$ and $s(\lambda_2)$, which is defined as $\triangle s = s(\lambda_1)/s(\lambda_2)\text{-}1$. From equation [6.4] and equation [6.5], the relative depth error $\triangle l$ can be calculated as

$$\triangle l = \frac{\ln\left(\frac{I(\lambda_1)}{I(\lambda_2)}\frac{s(\lambda_2)}{s(\lambda_1)}\right) - \ln\frac{I(\lambda_1)}{I(\lambda_2)}}{\ln\left(\frac{I(\lambda_1)}{I(\lambda_2)}\frac{s(\lambda_2)}{s(\lambda_1)}\right)} = \frac{\ln(1 + \triangle s)}{\ln(1 + \triangle s) - \ln\frac{I(\lambda_1)}{I(\lambda_2)}}. \tag{6.6}$$

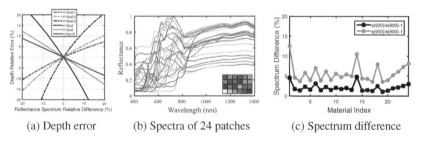

(a) Depth error	(b) Spectra of 24 patches	(c) Spectrum difference

Figure 6.4. *(a) The relative depth error with respect to the reflectance spectrum difference under varying intensity ratios is shown. (b) The spectra of the 24 patches on the color checker in the range from 400 nm to 1,400 nm are shown. The reflectance spectrum difference for spectral pairs of 900 nm and 920 nm, as well as of 900 nm and 950 nm, for each patch spectrum is shown in (c). For a color version of this figure, see www.iste.co.uk/funatomi/computational.zip*

The relative depth error is plotted against the relative reflectance difference for varying intensity ratios $I(\lambda_1)/I(\lambda_2)$ in Figure 6.4(a). From this figure, we can see that the estimated depth becomes less sensitive to differences in the reflectance spectrum, as the intensity ratio increases away from one (i.e. the difference between the two wavelengths becomes larger). This suggests a criterion for choosing the two wavelengths for bispectral depth recovery. Specifically, we should choose two wavelengths whose water absorption coefficients' difference is maximized, while the corresponding reflectance spectrum difference is minimized.

As shown in Figure 6.2(a), the amount of light absorption in water changes quickly in the range between 900 nm and 1,000 nm. Surprisingly, we empirically find that the reflectance spectra of a great variety of materials tend to be flat in this range. We start our investigation by examining the spectra of the standard color checker board, as shown in Figure 6.4(b). From this figure, we can clearly observe that the spectral variance for all patches drastically reduces in the range longer than 900 nm. From Figure 6.4(c), the average relative spectrum difference for 900 nm and 950 nm is 5.7%, and the difference for 900 nm and 920 nm is further reduced to 2.1%, although there are a few patches with larger differences.

We also examined several other classes of common materials, including wood, cloth, leather and metal, as shown in Figure 6.5(a). There are 24 different materials in each class, except for metal, which has only 18. We measured their reflectance spectra and evaluated the difference for wavelength pairs of 900 nm and 920 nm, as well as 900 nm and 950 nm. The average relative spectrum differences of these four classes for the bispectral pair 900 nm and 950 nm are 3.8, 2.1, 6.0 and 11.1%, respectively. For the bispectral pair 900 nm and 920 nm, the corresponding average differences reduce to 1.4, 1.1, 1.9 and 5.0%. Although the scale of our database is limited, the evaluation result suggests that the reflectance spectrum difference is usually very small for two close near-infrared wavelengths.

6.5. Practical shape from water

Here, we derive algorithms for shape from water with practical setups based on the bispectral depth recovery principle. In particular, we devise two algorithms that correct distorted depth estimates resulting from non-idealities in the imaging setup.

6.5.1. *Non-collinear/perpendicular light-camera configuration*

So far, we have considered the collinear light-camera configuration, in which both the optical axis of the orthographic camera and the directional light are perpendicular to the water level. From now on, we consider a practical situation, in which the light rays and/or the camera might be slightly tilted from the water surface, due to the practical requirements of the system setup. We will show that the depth distortion can be corrected if the depth of a single point is given.

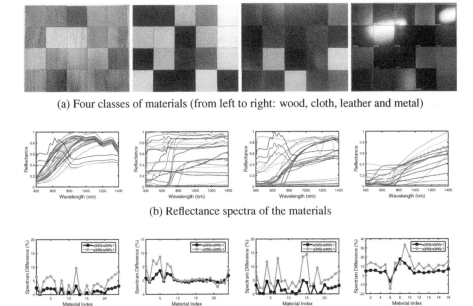

(a) Four classes of materials (from left to right: wood, cloth, leather and metal)

(b) Reflectance spectra of the materials

(c) Relative reflectance spectrum difference

Figure 6.5. *Reflectance spectra database in vis–NIR range from 400 to 1,400 nm. We empirically find that the spectral reflectance difference between two close near-infrared wavelengths is usually negligible. For a color version of this figure, see www.iste.co.uk/funatomi/computational.zip*

As illustrated in Figure 6.2(d), the tilt angles in water for the illuminant and the camera are denoted by θ and ψ, respectively. We can assume that these two angles do not change at the two working wavelengths because the refractive ratio of water is almost constant in the near-infrared range. The light pathlength in water is stretched to $l(\frac{1}{\cos\theta} + \frac{1}{\cos\psi})$, rather than $2l$. Similar to equation [6.5], the depth can be estimated as

$$l(\frac{1}{\cos\theta} + \frac{1}{\cos\psi}) \simeq \frac{1}{\alpha(\lambda_2) - \alpha(\lambda_1)} \ln \frac{I(\lambda_1)}{I(\lambda_2)}, \qquad [6.7]$$

from which we can see that the distortion factor $(\frac{1}{\cos\theta} + \frac{1}{\cos\psi})$ can be easily estimated when the depth of a single point is provided.

6.5.2. *Perspective camera with a point source*

Now, we try to relax the assumption of the orthographic projection and parallel illumination. In a practical situation, the incident light rays coming from a nearby point source to the object surface are not parallel, and the perspective effect becomes larger when the object is relatively large. With a perspective camera under a point light source with diverging rays, the tilt angles of the light rays differ depending on the spatial position. Therefore, we need to calculate the tilt angles of the light rays and the light pathlength for each scene point. The depth distortion can be corrected given the position of the camera and point light source, view angle, and the position of the reflection at the water surface.

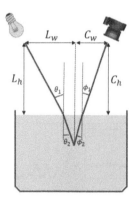

Figure 6.6. *Configuration of perspective light and camera. For a color version of this figure, see www.iste.co.uk/funatomi/computational.zip*

As illustrated in Figure 6.6, the tilt angles at the water surface for an incident light ray are denoted by θ_1 and θ_2, respectively, while those for the reflected ray are denoted by ϕ_1 and ϕ_2, respectively. Accordingly, we have four equations about light refraction:

$$\frac{\sin\theta_1}{\sin\theta_2} = \frac{\sin\phi_1}{\sin\phi_2} = n,$$

$$l_d = l_1\cos\theta_2 = l_2\cos\phi_2, \tag{6.8}$$

$$L_h\tan\theta_1 = L_w - l_1\sin\theta_2,$$

$$C_h\tan\phi_1 = C_w - l_2\sin\phi_2,$$

where n is the refractive index of water, l_1 represents the incident light pathlength from the water surface to the reflection point on the object, l_2 represents the reflection light pathlength from the reflection point of the object to the water surface, L_h and C_h are the heights from the water surface to the point light source and the camera, L_w and C_w are the widths from the reflection point to the point light source and the camera, and l_d denotes depth. We can calculate the depth easily from equation [6.8].

6.5.3. *Non-ideal narrow-band filters*

When constructing the bispectral imaging system for shape from water, it would be ideal to use a wide-band illuminant and narrowband filters in front of the camera. Until now, we have implicitly assumed that the response function of the band-pass filters is a delta function, which is hard to achieve in practice.

Here, we denote the spectral response functions of two non-ideal narrowband filters which are each centered at λ_1 and λ_2 as $\beta_1(\lambda)$ and $\beta_2(\lambda)$, respectively. If the band-pass filters are sufficiently narrow, we can assume that the reflectance spectrum of the scene point is flat between the two wavelengths. The imaging equation [6.2] becomes

$$I(\lambda_1) = r(\omega)s(\lambda_1)I_0 \int_0^\infty \beta_1(\lambda)e^{-2\alpha(\lambda)l}d\lambda. \tag{6.9}$$

A similar equation can be established for equation [6.3]. The depth l can be corrected by solving the following equation:

$$I(\lambda_1)\int_0^\infty \beta_2(\lambda)e^{-2\alpha(\lambda)l}d\lambda = I(\lambda_2)\int_0^\infty \beta_1(\lambda)e^{-2\alpha(\lambda)l}d\lambda, \tag{6.10}$$

using standard one-dimensional zero-finding techniques. Note that we do not explicitly consider the illumination spectrum or the camera spectral sensitivity function, since they can be merged into the spectral response function of the filters.

6.6. Co-axial bispectral imaging system and experiment results

We built a co-axial bispectral imaging system. The system uses co-axial cameras to simultaneously capture the scene in two wavelengths, recording bispectral image pairs at video rate. From the image sequence, we can recover the geometry of complex and dynamic objects immersed in water.

6.6.1. *System configuration and calibration*

As shown in Figure 6.7(a), the co-axial bispectral imaging system consists of a beam splitter and two grayscale cameras (POINTGREY GS3-U3-41C6NIR), which can sense NIR light with limited spectral sensitivity. We use two narrow band-pass filters centered at 905 nm and 950 nm, whose spectral response curves are shown in Figure 6.7(b). For illumination, we use an incandescent lamp with sufficient irradiance in the NIR range, as shown in Figure 6.7(c). We synchronize the two cameras and adjust the position of the beam splitter (half mirror) to capture spatially aligned bispectral image pairs of the same scene.

The water absorption coefficient needs to be known in advance for shape from water to work; it can easily be measured beforehand. We use a spectrophotometer and a standard white target for calibration. By immersing the white target in water at a known depth and reflected light intensity, we can calculate the water absorption coefficient from the Beer–Lambert law. Figure 6.7(d) shows the calibrated absorption coefficient for different wavelengths.

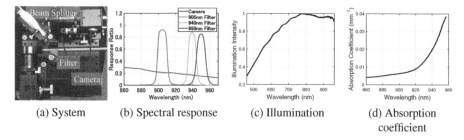

(a) System   (b) Spectral response   (c) Illumination   (d) Absorption coefficient

Figure 6.7. *(a) Our co-axial bispectral imaging system and (b) the spectral response functions of the camera and the three filters are shown. (c) The spectrum of the incandescent illuminant is shown. (d) The calibrated water absorption coefficient is shown. For a color version of this figure, see www.iste.co.uk/funatomi/computational. zip*

6.6.2. *Depth and shape accuracy*

We used planar plates made from different materials for a depth accuracy evaluation. We put the plates in water and measured the water depth with a ruler for the ground truth. We varied the water depth from 0 mm to 50 mm. We captured bispectral images with our co-axial bispectral system and estimated the depth using equation [6.5]. To evaluate the effectiveness of our algorithms in section 6.5, we also corrected the depth by using equation [6.7] and equation [6.10].

As shown in Figure 6.8, we used three plates in the experiments, including a piece of a white target, an orange plastic board and a piece of white marble. On each plate, we randomly chose 100 points (pixels of images) and calculated the average depth for these points. From Figure 6.8, it can be seen that when the water depth is deeper, the estimated value is smaller than the ground truth. This is because the camera cannot acquire a sufficient amount of light. Since the camera acquires a higher value than the correct intensity, the estimated depth is lower than the ground truth in the case of a low luminance value.

To evaluate the effectiveness of our algorithms in section 6.5.2, we set up the co-axial bispectral imaging system and perspective point light source in a tilted configuration and estimated the depth in the same way as above. Figure 6.9 shows the uncorrected estimate and the corrected estimate made by the above method. From Figure 6.9, we can see that with the correction, the average depth estimates are very close to the ground truth, and the correction algorithms play a critical role in improving the estimation accuracy when using a near-field point light source.

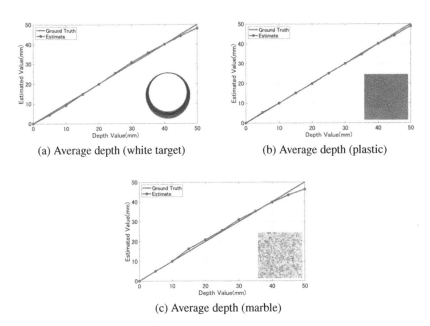

(a) Average depth (white target)   (b) Average depth (plastic)

(c) Average depth (marble)

Figure 6.8. *Depth estimation error by the bispectral method for three planar plates, including a white target, an orange plastic board and a white marble. For a color version of this figure, see www.iste.co.uk/funatomi/computational.zip*

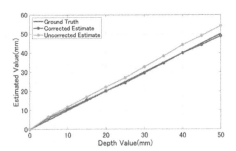

Figure 6.9. *Depth estimation error by the bispectral method for white target in a perspective light source. For a color version of this figure, see www.iste.co.uk/funatomi/computational.zip*

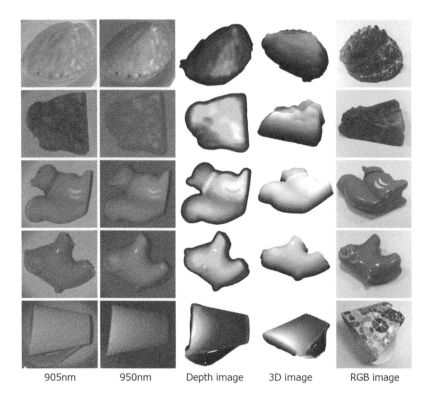

905nm   950nm   Depth image   3D image   RGB image

Figure 6.10. *Shape recovery of objects by the bispectral method with complex geometry, texture, and reflection properties. On each row are shown, from left to right, the input images at 905 nm and 950 nm, the depth-coded 3D shape, the virtually shaded shape and the RGB appearance of the object. For a color version of this figure, see www.iste.co.uk/funatomi/computational.zip*

6.6.3. *Complex static and dynamic objects*

We applied shape from water to objects with complex reflectance and dynamically moving objects whose shape deforms. Figure 6.10 shows the recovery results for several opaque objects. We can see that our system and method worked well for textureless objects with strong specularities. The surface reflectance and geometry of the seashell and rock in the first and second rows of Figure 6.10 are particularly complicated and would pose significant challenges to other shape recovery methods. These results clearly show, as in the theory, that shape from water is insensitive to such intricacies. This property is verified again by the compelling results for the colorful cups in the last row of Figure 6.10. Note that artifacts due to specularities sometimes occur (fourth row), which are attributed to camera saturation, rather than the method itself. Figure 6.11 shows the recovery results of some more challenging objects with translucence. The recovered shape looks reasonable when compared with its RGB appearance. Our co-axial shape from the water system is suited to capturing dynamic scenes. As shown in Figure 6.12, we demonstrate this by recovering the geometry of a hand moving in water.

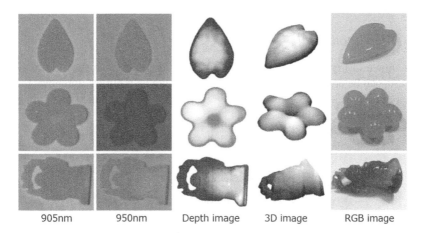

905nm   950nm   Depth image   3D image   RGB image

Figure 6.11. *Shape recovery of translucent objects by the bispectral method. On each row are shown, from left to right, the input images at 905 nm and 950 nm, the depth-coded 3D shape, the virtually shaded shape and the RGB appearance of the object. For a color version of this figure, see www.iste.co.uk/funatomi/computational.zip*

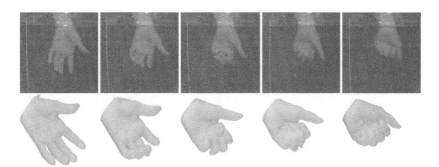

Figure 6.12. *Shape recovery of a moving hand captured at video rate by the bispectral method. The input images at 905 nm are shown in the upper row, and the 3D shape images are shown in the bottom row. For a color version of this figure, see www.iste.co.uk/funatomi/computational.zip*

6.7. Trispectral light absorption for depth recovery

The bispectral depth imaging method assumes that the reflectance spectrum values at the two capturing wavelengths λ_1 and λ_2 are almost identical, i.e. $s(\lambda_1) \simeq s(\lambda_2)$.

This assumption holds true for most materials in the database that we prepared, as shown in section 6.4.2. However, there are still some extreme cases, especially for metal materials, which violate this assumption to a nontrivial extent. Here, we show that, by introducing a third capturing wavelength λ_3 between λ_1 and λ_2, this drawback can be largely resolved.

6.7.1. Trispectral depth imaging

As shown in Figure 6.13, we now assume that the reflectance spectrum between λ_1 and λ_2 can be approximated by a line segment with an arbitrary (unknown) slope. Given a third wavelength λ_3 such that $\lambda_1 < \lambda_3 < \lambda_2$, the reflectance spectrum value $s(\lambda_3)$ at λ_3 can be calculated as

$$s(\lambda_3) = \frac{\lambda_2 - \lambda_3}{\lambda_2 - \lambda_1} s(\lambda_1) + \frac{\lambda_3 - \lambda_1}{\lambda_2 - \lambda_1} s(\lambda_2)$$

$$= \rho_1 s(\lambda_1) + \rho_2 s(\lambda_2),$$

[6.11]

where $\rho_1 = \frac{\lambda_2 - \lambda_3}{\lambda_2 - \lambda_1}$ and $\rho_2 = \frac{\lambda_3 - \lambda_1}{\lambda_2 - \lambda_1}$. In particular, when $\lambda_3 = \frac{\lambda_1 + \lambda_2}{2}$, $\rho_1 = \rho_2 = \frac{1}{2}$.

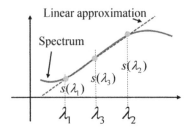

Figure 6.13. *Linear approximation of the reflectance spectrum in the wavelength range $[\lambda_1, \lambda_2]$. For a color version of this figure, see www.iste.co.uk/funatomi/computational.zip*

Similar to equation [6.2] and equation [6.3], we have three equations as follows:

$$I(\lambda_1) = r(\omega)s(\lambda_1)I_0 e^{-2\alpha(\lambda_1)l},$$

$$I(\lambda_2) = r(\omega)s(\lambda_2)I_0 e^{-2\alpha(\lambda_2)l},$$

$$I(\lambda_3) = r(\omega)s(\lambda_3)I_0 e^{-2\alpha(\lambda_3)l} = r(\omega)\left[\rho_1 s(\lambda_1) + \rho_2 s(\lambda_2)\right] I_0 e^{-2\alpha(\lambda_3)l}.$$

[6.12]

By dividing the first equation by the second equation in equation [6.12], the reflectance spectrum value ratio $\frac{s(\lambda_1)}{s(\lambda_2)}$ can be expressed as

$$\frac{s(\lambda_1)}{s(\lambda_2)} = \frac{I(\lambda_1)}{I(\lambda_2)} e^{2(\alpha(\lambda_1) - \alpha(\lambda_2))l}.$$

[6.13]

This ratio can be plugged into the quotient between the second equation and the third equation as follows:

$$\frac{I(\lambda_3)}{I(\lambda_2)} = \left[\rho_1 \frac{s(\lambda_1)}{s(\lambda_2)} + \rho_2\right] e^{2(\alpha(\lambda_2) - \alpha(\lambda_3))l}$$

$$= \rho_1 \frac{I(\lambda_1)}{I(\lambda_2)} e^{2(\alpha(\lambda_1) - \alpha(\lambda_3))l} + \rho_2 e^{2(\alpha(\lambda_2) - \alpha(\lambda_3))l},$$

[6.14]

from which the depth l can be easily estimated.

6.7.2. *Evaluation on the reflectance spectra database*

Here, we show that our assumption is valid for all materials in the database, including the challenging metal materials. We chose $\lambda_1, \lambda_2, \lambda_3$ to be 900, 950 and 925 nm, respectively. Given a reflectance spectrum $s(\lambda)$, we first drew a line $e(\lambda)$ passing through the two endpoints $[\lambda_1, s(\lambda_1)]$ and $[\lambda_2, s(\lambda_2)]$. Then, we calculated the relative difference between $s(\lambda_3)$ and $e(\lambda_3)$, which is $\frac{e(\lambda_3)}{s(\lambda_3)} - 1$.

Figure 6.14 shows the results of the evaluation, from which we can see that the linear approximation is very powerful. The average differences for wood, cloth, leather and metal are only 0.1, 0.4, 0.4 and 0.6%, respectively, in contrast to 3.8, 2.1, 6.0 and 11.1% in the bispectral case.

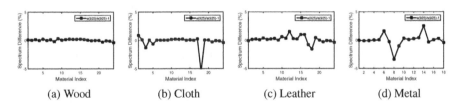

(a) Wood   (b) Cloth   (c) Leather   (d) Metal

Figure 6.14. *Relative reflectance spectrum difference for wood (a), cloth (b), leather (c) and metal (d) in the trispectral case. For a color version of this figure, see www.iste.co.uk/funatomi/computational.zip*

6.8. Discussions

Bispectral or trispectral depth recovery in its current form is not able to directly handle environment illumination. In practice, it can be eliminated by taking another image under the environment illumination only and subtracting them away from the input images under mixed illumination. Similar to most of the existing depth imaging principles and techniques, our principle is vulnerable to interreflection, which is often not negligible for concave surfaces and tends to smooth out shape details, as can be observed for the statue in the third row of Figure 6.11. This is due to the fact that interreflection causes the optical length in water to be longer than the depth.

Shape reconstruction of transparent objects is challenging for contact-free depth imaging. Fortunately, shape from water can be used to recover the geometry of transparent objects as long as they do not absorb light, or the absorption coefficients for capturing wavelengths have almost the same values, which is the case for most glasses and plastic, or have an opaque back side. Such objects are invisible to laser scanners too, and their geometry cannot be recovered without altering their surface reflectances (e.g. by painting them). In contrast, shape from water achieves geometry recovery while leaving the object untouched.

The analysis in section 6.4.2 implies that the larger the difference in absorption coefficients of the two near-infrared lights we use (e.g. 905 nm and 950 nm), the higher the depth recovery accuracy. Unfortunately, due to stronger absorption, for instance at 950 nm, the image will be darker, leading to poorer SNR. One idea to resolve the SNR issue is to capture multiple images with different exposure times and use only bright pixels to extend the dynamic range. For an easier way, we can also use shorter wavelengths to get a slightly brighter image. In so doing, it would become possible to widen the depth range that can be estimated with high accuracy.

6.9. Conclusion

In this chapter, we introduced shape from water, a depth recovery method based on light absorption in water. Shape from water builds on the derived bispectral or trispectral depth sensing principle based on the idea of leveraging the light absorption difference between two or three near-infrared wavelengths to estimate depth regardless of the surface reflectance. We constructed a co-axial bispectral depth imaging system using low-cost off-the-shelf hardware to capture bispectral image pairs for shape from water. Experimental results show that shape from water can accurately recover the geometry of objects with complex reflectance and dynamically deforming shapes. We also showed that the trispectral variant is capable of approximating reflectance spectra at higher accuracy.

6.10. References

Akkaynak, D. and Treibitz, T. (2018). A revised underwater image formation model. *CVPR*, pp. 776–792.

Akkaynak, D., Treibitz, T., Shlesinger, T., Tamir, R., Loya, Y., Iluz, D. (2017). What is the space of attenuation coefficients in underwater computer vision? *CVPR*, pp. 568–577.

Batlle, J., Mouaddib, E., Salvi, J. (1998). Recent progress in coded structured light as a technique to solve the correspondence problem: A survey. *Pattern Recognition*, 31(7), 963–982.

Berman, D., Treibitz, T., Avidan, S. (2017). Diving into hazelines: Color restoration of underwater images. *BMVC*, 1–12.

Chiang, J. and Chen, Y. (2012). Underwater image enhancement by wavelength compensation and dehazing. *IEEE Transactions on Image Processing*, 21(4), 1756–1769.

Dancu, A., Fourgeaud, M., Franjcic, Z., Avetisyan, R. (2014). Underwater reconstruction using depth sensors. *SIGGRAPH Asia Technical Briefs*, pp. 1–4.

Fattal, R. (2008). Single image dehazing. *Proceedings of ACM SIGGRAPH*, vol. 27, pp. 1–9.

Fujimura, Y., Iiyama, M., Hashimoto, A., Minoh, M. (2018). Photometric stereo in participating media considering shape-dependent forward scatter. *Proceedings of the IEEE Conf. on Computer Vision and Pattern Recognition*, pp. 7445–7453.

Hansard, M., Lee, S., Choi, O., Horaud, R. (2013). *Time-of-Flight Cameras: Principles, Methods and Applications*. Springer.

Hartley, R. and Zisserman, A. (2003). *Multiple View Geometry in Computer Vision*, 2nd edition. Cambridge University Press.

He, K., Sun, J., Tang, X. (2009). Single image haze removal using dark channel prior. *Proceedings of IEEE Int'l Conf. on Comp. Vision and Pattern Recognition*, vol. 1, 1956–1963.

Jordt-Sedlazeck, A. and Koch, R. (2013). Refractive structure-from-motion on underwater images. *Proceedings of the IEEE International Conference on Computer Vision*, pp. 57–64.

Kawahara, R., Nobuhara, S., Matsuyama, T. (2016). Dynamic 3D capture of swimming fish by underwater active stereo. *Methods in Oceanography*, 17, 118–137.

Kawasaki, H., Nakai, H., Baba, H., Sagawa, R., Furukawa, R. (2017). Calibration technique for underwater active oneshot scanning system with static pattern projector and multiple cameras. *2017 IEEE Winter Conference on Applications of Computer Vision (WACV)*, pp. 302–310.

Kratz, L. and Nishino, K. (2009). Factorizing scene albedo and depth from a single foggy image. *IEEE International Conference on Computer Vision*, pp. 1701–1708.

Murez, Z., Treibitz, T., Ramamoorthi, R., Kriegman, D. (2015). Photometric stereo in a scattering medium. *ICCV*, pp. 3415–3423.

Narasimhan, S.G. and Nayar, S.K. (2002). Vision and the atmosphere. *Int'l Journal on Computer Vision*, 48(3), 233–254.

Narasimhan, S.G., Nayar, S.K., Sun, B., Koppal, S. (2005). Structured light in scattering media. *IEEE International Conference on Computer Vision*, 1–8.

Nishino, K., Kratz, L., Lombardi, S. (2012). Bayesian defogging. *Int'l Journal of Computer Vision*, 98(2), 232–255.

Qian, Y., Zheng, Y., Gong, M., Yang, Y.-H. (2018). Simultaneous 3D reconstruction for water surface and underwater scene. *Proceedings of the European Conference on Computer Vision (ECCV)*, pp. 754–770.

Queiroz-Neto, J.P., Carceroni, R., Barros, W., Campos, M. (2004). Underwater stereo. *Proceedings of 17th Brazilian Symposium on Computer Graphics and Image Processing*. IEEE, pp. 170–177.

Reinhard, E., Khan, E.A., Akyuz, A.O., Johnson, G. (2008). *Color Imaging: Fundamentals and Applications*. CRC Press.

Saito, H., Kawamura, H., Nakajima, M. (1995). 3D shape measurement of underwater objects using motion stereo. *Proceedings of IECON'95-21st Annual Conference on IEEE Industrial Electronics*, vol. 2, pp. 1231–1235.

Schechner, Y.Y., Narasimhan, S.G., Nayar, S.K. (2001). Instant dehazing of images using polarization. *Proceedings of IEEE Int'l Conf. on Computer Vision and Pattern Recognition*, vol. 1, pp. 325–332.

Schonberger, J.L. and Frahm, J.-M. (2016). Structure-from-motion revisited. *Proceedings of the IEEE Conf. on Computer Vision and Pattern Recognition*, pp. 4104–4113.

Tan, R.T. (2008). Visibility in bad weather from a single image. *Proceedings of IEEE Int'l Conf. on Computer Vision and Pattern Recognition*, pp. 1–8.

Tsiotsios, C., Angelopoulou, M.E., Kim, T.-K., Davison, A.J. (2014). Backscatter compensated photometric stereo with 3 sources. *Proceedings of the IEEE Conf. on Computer Vision and Pattern Recognition*, pp. 2251–2258.

Woodham, R. (1980). Photometric method for determining surface orientation from multiple images. *Optical Engineering*, 19(1), 139–144.

Yano, T., Nobuhara, S., Matsuyama, T. (2013). 3D shape from silhouettes in water for online novel-view synthesis. *IPSJ Transactions on Computer Vision and Applications*, 5, 65–69.

Zhang, R., Tsai, P., Cryer, J., Shah, M. (1999). Shape-from-shading: A survey. *IEEE TPAMI*, 21(8), 690–706.

7

Far Infrared Light Transport Decomposition and Its Application for Thermal Photometric Stereo

Kenichiro TANAKA

Vision and Imaging Group, Ritsumeikan University, Shiga, Japan

7.1. Introduction

Light transport is a study of the complex interaction between light and matter. Decomposing light transport powers low-level computer vision tasks that range from shape recovery to reflectance estimation. Previous work has studied light transport at visible light wavelengths; here, we lay the foundation for light transport at long-wave infrared wavelengths. At these wavelengths, light transport is very unique due to the interplay between heat and long-wave infrared light.

Previously, color (Shafer 1985), polarization (Treibitz and Schechner 2009) and active illumination (Nayar et al. 2006) have been used for light transport decomposition. The transient behavior of optical components varies in the order of tens of picoseconds (Wu et al. 2014), thus paving the way for time-resolved approaches. So far, multiple time-resolved approaches have been proposed – with the use of a femto-pulsed laser, interferometer (Gkioulekas et al. 2015), time-of-flight camera modifications (Heide et al. 2013; Kadambi et al. 2013; Kitano et al. 2017) and single-photon sensor (O'Toole et al. 2017).

Computational Imaging for Scene Understanding,
coordinated by Takuya FUNATOMI and Takahiro OKABE.
© ISTE Ltd 2024.

Unlike visible light imaging, time-resolved light transport decomposition using thermal imaging is feasible at a video frame rate. This is because of the important observation that the speed of heat propagation is much slower than the speed of light propagation. Inspired by this, we develop a novel time-resolved decomposition technique for far infrared light transport (Tanaka et al. 2018, 2021).

Thermal imaging has been traditionally considered as being different from visible light imaging: while the thermal image is a representation of the temperature of the object, the visible light image is a description of the visual information. However, we show that when the imaging environment is appropriately controlled, similar patterns are observed in both the thermal and visible light imaging. This is because thermal imaging is a technique to observe light on far infrared wavelengths. Figure 7.1 shows an image captured by a color camera and a thermal image, where a ball is illuminated by a point light source. Both the color image and the thermal image exhibit the similar shading. This observation implies that computer vision techniques can also be applied for thermal images.

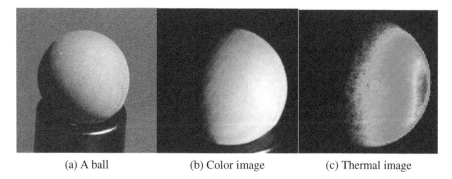

(a) A ball   (b) Color image   (c) Thermal image

Figure 7.1. *A ball captured by a conventional color camera and a thermal camera. (a) The target object. (b) Reflection image using a conventional camera. (c) Thermal image of the same object. When the object is carefully illuminated, shading of both images is the same, which implies that conventional computer vision techniques can be applied to thermal images. For a color version of this figure, see www.iste.co.uk/funatomi/computational.zip*

7.1.1. *Contributions*

In this study, a time-resolved approach for decomposition of the far infrared light transport is proposed for a photometric stereo application. With the help of the physical law of heat and thermodynamics, we show that far infrared light transport can be considered as a composition of multiple optical and thermal effects similar to the visible light transport. We propose definitions of the ambient, the reflection, the diffuse radiation and the global radiation components in the thermal framework and describe the transient properties of each. We also provide an application of our approach. We show that the surface normal can be estimated based on the Lambertian photometric stereo, because the diffuse component of the far infrared light, which follows the cosine law, is separated.

The chief contributions of this study are threefold. First, we extend the visible light transport model to the far infrared light transport. We show that the thermal image is a composition of ambient, reflection, diffuse and global components, which is similar to the visible light transport. Second, a novel approach for time-resolved light transport decomposition is provided based on the difference of the transient property of the far infrared light transport. Finally, we show that ordinary computer vision techniques can be straightforwardly applied to thermal images. As a proof of the concept, we propose a method to recover the surface normal using photometric stereo after decomposing the far infrared light transport. The surface normal of challenging objects that have complicated optical effects can be recovered. The proposed thermal photometric stereo (TPS) can be applied for any objects that absorb light and convert it into heat, including black body, transparent and translucent objects. It has a wide applicability compared with the photometric stereo using visible light.

7.2. Related work

This study lies at the intersection of thermal imaging, light transport decomposition and photometric stereo. We summarize prior work in these three areas.

7.2.1. *Light transport decomposition*

Light transport decomposition represents a principled study of how light propagates from an illumination source to a camera sensor. This area of research is organized into subfields based on the physics of capture setups.

Color-based approaches were among the first to be employed for light transport decomposition. The dichromatic reflectance model by Shafer (1985) proposes that specular reflection depends on the color of the light source, while diffuse reflection depends on the color of the object. Several methods (Sato and Ikeuchi 1994; Nguyen et al. 2014; Ren et al. 2017) use this model for the separation of the diffuse and the specular reflections.

Polarization-based approaches use polarizers to separate light transport. Separation of diffuse and specular reflections using linear polarization was demonstrated by Wolff and Boult (1991). Treibitz and Schechner (2009) showed the separation of diffuse reflection and volumetric scattering using circular polarization.

Active illumination has also proved to be a successful tool to decompose light transport. Nayar et al. (2006) used high-frequency projection patterns to separate direct and global reflection components of the visible imaging light transport. There have been several extensions proposed to this – environmental illumination for the separation of diffuse and specular reflections by Lamond et al. (2007), transmission and scattering (Tanaka et al. 2013), and single scattering and multiple scattering (Mukaigawa et al. 2010). O'Toole et al. (2012, 2014b, 2015) decompose light transport using an epipolar constraint. Although we use active illumination, to the best of our knowledge, this is the first study in the light transport decomposition of thermal images in the far infrared wavelength.

Time-resolved decomposition is one of the latest technologies in the domain of light transport decomposition. The temporal response of a femtosecond-pulsed laser was exploited by Wu et al. (2014) to develop a method to decompose diffuse reflection, inter-reflection and subsurface scattering. Resolving the multi-path interference problem in the time-of-flight camera is an active research topic and has been studied by assuming the two-bounce or simplified reflection models (Fuchs 2010; Dorrington et al. 2011; Godbaz et al. 2012; Jimenez et al. 2012), K-sparsity (Bhandari et al. 2014; Freedman et al. 2014; Qiao et al. 2015), parametric model (Kirmani et al. 2013; Heide et al. 2014b), consistency between ToF and stereo (Lee and Shim 2015), simplified indirect reflections (Naik et al. 2015) and large-scale multi-path (Kadambi et al. 2016). It can be used to measure a slice of BRDF (Naik et al. 2011), perform non-line-of-sight imaging (Heide et al. 2014a; Kadambi et al. 2016; Tsai et al. 2017) and recover the shape of transparent and translucent objects (Shim and Lee 2015; Tanaka et al. 2016). Past work using time-resolved methods (O'Toole et al. 2014a; Wu et al. 2014; Gupta et al. 2015; Gkioulekas et al. 2015) also separates direct and indirect light transports.

Far infrared heating characteristics are known to have the temporal transience. Thompson et al. (2013) model the temporal effects of heating during infrared neural stimulation. Ito et al. (2000) describe the heating effects in skin during continuous wave near infrared spectroscopy. We define far infrared light transport components as light and heat, from the law of thermodynamics and simulation.

7.2.2. *Computational thermal imaging*

Thermal imaging approaches have traditionally not been widely used to solve computer vision problems. Saponaro et al. (2015) estimate the material from the water permeation and heating/cooling process of the object. Miyazaki et al. (2002) resolve the ambiguity regarding polarization-based shape reconstruction using a thermal image. Eren et al. (2009) recover the transparent shape by triangulation using laser beam spot heating and thermal imaging. In this study, we firstly propose a light transport model of thermal imaging.

7.2.3. *Photometric stereo*

Photometric stereo has been a broad interest in the computer vision field. The Lambertian photometric stereo (Woodham 1980) is a standard way to recover the surface normal by assuming Lambert reflection, no optical effect such as shadow and scattering, orthogonal projection, and parallel lights. To apply the Lambertian photometric stereo for a non-Lambert surface, other optical components need to be separated by pattern projection (Nayar et al. 2006), polarization (Nayar et al. 1997) and fluorescence (Treibitz et al. 2012). Similar to these approaches, we apply the Lambertian photometric stereo after extracting the diffuse component.

Inoshita et al. (2014) improve the photometric stereo for translucent objects using surface normal deconvolution, Ngo et al. (2015) use a polarization cue to recover a smooth surface, and Murez et al. (2017) develop photometric stereo in a scattering media that consider the blur depending on the distance. While these methods jointly compensate for the global light transport in their solutions, we aim to separate the far infrared light transport.

7.3. Far infrared light transport

We begin by establishing a few basic principles pertaining to thermal and far infrared light transports. In particular, we will establish a theory of light transport for *thermal cameras*.

A thermal camera is a unique camera that is designed to measure the intensity of far infrared light ($8 - 14$ μm). A typical sensor as shown in Figure 7.4 measures the intensity of far infrared light, which corresponds to the temperature. When the object is a black body, the temperature and the intensity of far infrared light are governed by the Stefan–Boltzmann law (Howell et al. 2015), which represents a one-to-one correspondence between temperature and intensity:

$$E = \sigma T^4, \tag{7.1}$$

where E is the intensity of the radiated far infrared light, σ is the Stefan–Boltzmann constant[1] and T is the thermodynamic temperature. Consider the scene is illuminated by a stable parallel light source of far infrared light and the object is captured by a thermal camera as shown in Figure 7.2. When the object is not a black body, a part of the far infrared light reflects on the surface, while the rest of the light is absorbed and converted to the heat energy, the temperature increases, and far infrared light is emitted corresponding to its temperature. The observation is the sum of these effects, and we term this total energy transport as *far infrared light transport*.

---

1. Stefan–Boltzmann constant: $\sigma = 5.67 \times 10^{-8}$ W.m^{-2}.K^{-4}.

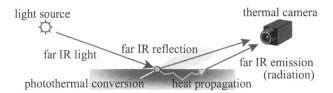

Figure 7.2. *Far infrared light transport. While far infrared light can partially be reflected on the surface, the rest of the light is converted to heat energy, propagates inside the object and is then converted to far infrared light corresponding to the temperature. The composition of all the components is captured by a camera. The observation system is closed in the far infrared light domain. For a color version of this figure, see www.iste.co.uk/funatomi/computational.zip*

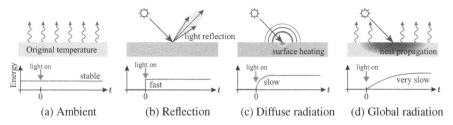

Figure 7.3. *Far infrared light and heat transport components. Similar to the visible light transport, far infrared light transport consists of (a) ambient, which is the original temperature, (b) reflection as light, (c) diffuse radiation and (d) global radiation caused by heat propagation. Because the speed of heat is slower than that of light, every component has distinctive transient properties and hence they are separable. For a color version of this figure, see www.iste.co.uk/funatomi/computational.zip*

Although analogous light transport effects can be observed between ordinary and thermal cameras, the underlying causes can be fundamentally distinct. An image captured by an ordinary camera is the composition of multiple light transport effects, e.g., specular and diffuse reflections, inter-reflection and subsurface scattering. Analogously, the thermal image is a sum of the multiple far infrared light transports, as shown in Figure 7.3. The key difference lies in the optical factors behind these light transport effects.

Here, we consider one such optical factor: the transience of light transport for a planar object. Transient light transport is observable at video frame rates for thermal cameras. This is due to the slow propagation of heat. In contrast, the transient state of ordinary light transport is dependent on the speed of light rather than of heat propagation. Figure 7.5 illustrates a concept of the temperature transition of the far infrared light transport components. Before the light source is turned on, the observation consists of only the ambient component. The reflection appears immediately after the light source is turned on, and diffuse and global radiations slowly appear as the temperature increases. Then, the diffuse radiation reaches the steady state faster than the global radiation.

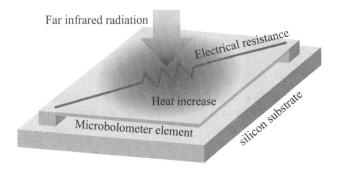

Figure 7.4. *The architecture of a typical thermal sensor, a microbolometer. A microbolometer element converts far infrared radiation to heat, which changes electrical resistance. The intensity of far infrared is captured by measuring the electrical resistance of the element. To prevent the surrounding temperature change, the sensor is covered by vacuum package. For a color version of this figure, see www.iste.co.uk/funatomi/computational.zip*

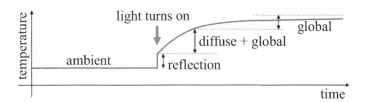

Figure 7.5. *Transient properties of far infrared light transport. Because the temporal responses of the components are significantly different, they can be separated from the thermal video frames. For a color version of this figure, see www.iste.co.uk/ funatomi/computational.zip*

The observed thermal image $I(t)$ at a video frame t can be modeled as

$$I(t) = A(t) + S(t) + D(t) + G(t), \qquad [7.2]$$

where A, S, D, G are the ambient, reflection, diffuse radiation and global radiation components, respectively. The time parameter t is referenced such that $t = 0$ refers to the time that the light is turned on. In what follows, we define the four components and provide supporting physical justifications.

DEFINITION 7.1.– *The original temperature T_{obj} of the object is defined as the temperature when the light source is turned off, i.e., $t < 0$. The ambient component*

A(t) is the radiation emitted by the object due to the original temperature T_{obj} of the object, as shown in Figure 7.3a.

CLAIM II.1.– The ambient component, $A(t)$, is a time-invariant function.

PROOF 7.1.– The Stefan–Boltzmann law (equation [7.1]) can be rewritten in terms of the ambient component, such that:

$$A(t) = \epsilon \sigma T_{obj}^4, \tag{7.3}$$

where ϵ is the emissivity of the non-black-body object. Because no heat source or sink is present when the light source is turned off, T_{obj} is constant. Then, since T_{obj} is constant, it follows from equation [7.3] that $A(t)$ is time-invariant. □

DEFINITION 7.2.– *The reflection component $S(t)$ is the far infrared light reflected off of the object when irradiated by an external source, as shown in Figure 7.3b.*

CLAIM II.2.– The reflection component $S(t)$ is a time-invariant function of the thermal reflectance.

PROOF 7.2.– $S(t)$ refers to the reflective component of long-wave infrared (LWIR) light radiation. Similar to visible light, the LWIR bidirectional reflectance distribution function (BRDF) is assumed to be time-invariant. As shown in Figure 7.3, the incident light is directional along $S(t)$ to be written as a single integral over the reflectance angles:

$$S(t) = L_0 \int_{2\pi} R_i(\Omega) \, d\Omega, \tag{7.4}$$

where $d\Omega$ is the differential solid angle over which reflectance varies, $R_i(\cdot)$ is the reflectance function and L_0 is the incident light. Since neither L_0 nor $R_i(\cdot)$ varies with time, $S(t)$ is time-invariant. □

COROLLARY 7.1.– *The previous proof assumes that the reflectance function is time-invariant. This is a standard assumption, though there are edge cases where the reflectance may vary with time (Sunkavalli et al. 2007). These edge cases are not unique to thermal light transport, and are therefore not a special consideration for this study.*

With the preceding two claims, we have established that neither the ambient nor reflective components exhibit transience. In what follows, we define two additional components of thermal light transport, which do exhibit transience.

DEFINITION 7.3.– *The diffuse radiation component $D(t)$ is the radiation emitted as a result of local surface heating occurring at the point of irradiance, as shown in Figure 7.3c.*

DEFINITION 7.4.– *The global radiation component $G(t)$ is the radiation emitted as a result of global heat transfer, as shown in Figure 7.3d.*

Similar to visible light (Nayar et al. 2006), we assume that the radiation due to heating is the sum of diffuse and global components. In the case of visible light, it has been reported that the diffuse reflection and subsurface scattering are the same physical phenomenon (Hanrahan and Krueger 1993; Jensen et al. 2001; Tanaka et al. 2017); the light scatters on or beneath the surface and eventually bounces off of the material in random directions. Diffuse reflection represents the total intensity of light close to the incident point on the surface, and the subsurface scattering represents the light at a distance away from the incident point on the surface. Although there is an intermediate state between diffuse reflection and subsurface scattering (Mukaigawa et al. 2010), modeling using two components, i.e., diffuse reflection and subsurface scattering, is consistent with prior work in light transport (Nayar et al. 2006). Inspired by previous two-component simplifications, we define two analogous components for thermal radiation. According to definitions 7.3 and 7.4, the difference between the diffuse component $D(t)$ and the global component $G(t)$ is whether heating is local or global; diffuse radiation is the heat energy whose heating point is local and global radiation is the heat that is propagated to the other points.

CLAIM II.3.– The diffuse radiation component exhibits a transience and increases with time.

JUSTIFICATION.– From definition 7.3, the diffuse radiation component is due to the local surface heating. The phenomenon of local surface heating occurs as follows. When an object is irradiated with far infrared light, a part of energy is absorbed (Howell et al. 2015), which raises the temperature of the object. In accordance with the Stefan–Boltzmann law (equation [7.1]), this increase in temperatures results in the emission of radiation. However, temperature increase is a function of time (Howell et al. 2015), and therefore the emitted radiation varies with time.   □

CLAIM II.4.– The global radiation component exhibits a transience and increases with time.

JUSTIFICATION.– From definition 7.4, the global radiation component is due to the global heat transfer. The phenomenon of global heat transfer occurs as follows. When an object is irradiated for a prolonged duration of time, the absorbed energy is transferred across the spatial profile of the object, resulting in a temperature increase away from the point of irradiance (see Unsworth and Duarte 1979). The temperature gradually increases until it reaches an equilibrium state (Howell et al. 2015). In accordance with the Stefan–Boltzmann law (equation [7.1]), the emitted radiation has an increasing transience according to the globally increasing temperature.   □

CLAIM II.5.– The rate of increase for global radiation is slower than that of diffuse radiation.

JUSTIFICATION.– From definitions 7.3 and 7.4, the difference between global and diffuse radiations is only in the transfer of heat energy. Hence, the rate of increase in global radiation is slower than direct radiation due to the additional time taken for the heat transfer to occur across a larger spatial region.   □

COROLLARY 7.2.– *The exact functions of these transiences are not obvious. Prior work in heat transfer literature (Unsworth and Duarte 1979; Carslaw and Jaeger 1959) suggests that the equation takes the form of an exponential. Keeping the same functional form (i.e. an exponential), we simplify the model to include constants, such that:*

$$\begin{cases} D(t) & = R_\infty (1 - e^{-\sigma_d t}) d_\infty \\ G(t) & = R_\infty (1 - e^{-\sigma_g t}) g_\infty, \end{cases} \tag{7.5}$$

where σ_d and σ_g ($\sigma_g \ll \sigma_d$) represent the coefficients of the transient speed of diffuse and global radiations, respectively, d_∞ and g_∞ represent the ratios of diffuse and the global radiation components at the steady state to the total radiation, respectively, and R_∞ is the steady state of the radiation components.

Validation. To validate our model, we render the radiation transience by physical simulation and confirm that our model fits the ground truth.

Heat transfer is described using differential heat equations (Carslaw and Jaeger 1959) for a hypothetical 1D object. Let $T(x, t)$ be the temperature function where x is the variable in space and t is the variable in time. The heat equation is then

$$\frac{\partial T}{\partial t} - k \left(\frac{\partial^2 T}{\partial^2 x} \right) = 0, \tag{7.6}$$

where k is the thermal diffusivity. From the literature (Unsworth and Duarte 1979), this equation is theoretically generalizable for a 3D object as well. This formula has the exact solution for certain initial and boundary conditions, where the input heat is spatio-temporally impulse and the surface is infinite plane. In our case, the object is illuminated over time and area. Unfortunately, there is no closed-form solution for the general case, especially for curved surfaces. Instead, we render the radiation transience and confirm how equation [7.5] approximates the ground-truth heat transfer.

Transience of the radiation can be simulated by solving the heat equation numerically. We perform this by applying the forward time central space (FTCS) method (Crank and Nicolson 1947; Becker and Kaus 2016) to solve equation [7.6]. We can estimate the change in temperature as a function of space and time for any thermal diffusivity objects. Using this simulation, we can obtain the ground-truth temperature curve, which can be used for validating our model.

To confirm that our model approximates the radiation transience, we fit the proposed double exponential model, as shown in equation [7.5], which is the sum of direct and global components, to the simulated temperature curve by FTCS, as shown in Figure 7.6. Our model fits the FTCS curves of different thermal diffusivities of 99.89% on average, which shows the validity of our proposed model.

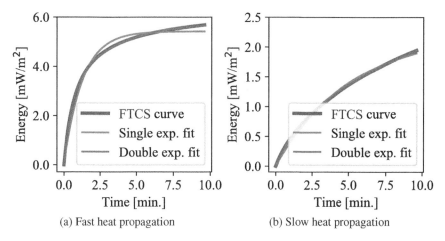

(a) Fast heat propagation   *(b) Slow heat propagation*

Figure 7.6. *Double exponential fitting result to the FTCS curves of different parameters. Double exponential is the sum of two exponentials, as shown in equation [7.5], representing both the direct and global components. Our model is a good approximation of the radiation transience. For a color version of this figure, see www.iste.co.uk/funatomi/computational.zip*

7.4. Decomposition and application

Here, we develop an inverse problem to separate the different components of transient thermal light transport. These results also motivate a first attempt at a form of *thermal photometric stereo* (TPS).

7.4.1. *Far infrared light transport decomposition*

In separating different light transport components, we make the following assumptions:

– the ambient component is observed before the light source is turned on;

– transient state of increasing temperature is observed until the temperature becomes steady.

In what follows, we expand on the methodology of separating the different components.

7.4.2. *Separating the ambient component*

The ambient component is the observation before the light source is turned on, and is determined as

$$A = I(0). \tag{7.7}$$

The transient observation $T_r(t)$ is the rest of the observation, given as

$$T_r(t) = I(t) - A. \tag{7.8}$$

7.4.3. *Separating reflection and radiation*

The reflection component is the reflection of light and has no transient state; hence, it can be obtained as the increase immediately after the light source is turned on. The reflection component S is obtained as

$$S = T_r(\epsilon), \tag{7.9}$$

where ϵ is an infinitesimal time duration.

The rest is the radiation, which has a temporal transient state. The radiation $R(t)$ can be obtained as

$$R(t) = T_r(t) - S. \tag{7.10}$$

7.4.4. *Separating diffuse and global radiations*

We fit the radiation components $R(t)$ to the model defined in equation [7.5] as

$$\hat{\sigma}_d, \hat{d_\infty}, \hat{\sigma}_g, \hat{g_\infty} = \underset{\sigma_d, d_\infty, \sigma_g, g_\infty}{\operatorname{argmin}} \ \|R(t) - D(t) - G(t)\|_2^2$$

$$\text{s.t.} \qquad \min_t \frac{-\log\left(R_\infty - R(t)\right)}{t} \leq \sigma_g < \sigma_d$$

$$0 \leq d_\infty \leq 1$$

$$0 \leq g_\infty \leq 1$$

$$d_\infty + g_\infty = 1. \tag{7.11}$$

The first constraint represents that the time duration to the steady state of each component is smaller than the time for the observation to reach the steady state. Because the diffuse radiation is faster than the global radiation, σ_g is much less than σ_d

($\sigma_g \ll \sigma_d$). The second and third constraints represent that the intensity of the diffuse and global radiations is smaller than that of the total radiation. The last constraint represents that the total radiation is a sum of diffuse and global radiations, which reduces one degree of freedom. Fitting these parameters is not a convex problem, so we use a grid search to find the global optimum. This does not involve a large computational cost because there are only three variables and the boundaries of the parameters can be predicted by the radiation profile $R(t)$.

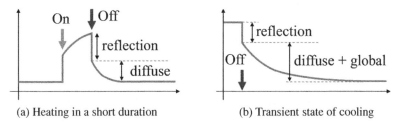

(a) Heating in a short duration   (b) Transient state of cooling

Figure 7.7. *Other viable approaches. (a) By turning on and off the light source in a sufficiently short time, the reflection and diffuse radiation can be directly obtained. (b) Transient state after turning off the light source contains similar information. For a color version of this figure, see www.iste.co.uk/funatomi/computational.zip*

7.4.5. *Other options*

Another viable approach is to use the decrease in temperature after the light source is turned off. By switching on and off the light source over a short duration, the reflection and diffuse radiation can be directly obtained, as shown in Figure 7.7(a), because the effect of heat propagation is negligible over a very short time. However, the diffuse radiation does not reach the steady state; hence, it may suffer from extremely low SNR. Extending the heating time could improve the SNR; however, the global radiation cannot be ignored.

The cooling process is also useful to analyze far infrared light transport, as shown in Figure 7.7(b). Because heating and cooling are the reverse phenomena, light transport decomposition can be achieved in a very similar way. Because this takes twice as long time, we chose to analyze only the heating process.

7.4.6. *Thermal photometric stereo*

TPS is different from ordinary photometric stereo because it relies on the slow transience of heat propagation. The assumptions of TPS are as follows:

– parallel far infrared light source or the light placed sufficiently far from the object is used;

– far infrared light transport decomposition is performed for each light source direction with the same ambient temperature.

When the object is heated by a narrow beam, the point absorbs the energy and radiates far infrared light according to the increased temperature. The absorbed energy follows the cosine law (Howell et al. 2015) as is observed for the light irradiance. Since the increased temperature due to heating can be observed from any camera position, the diffuse component at the steady state can be represented as

$$D(\infty) = R_\infty d_\infty = R_\infty \rho i^\top n, \qquad [7.12]$$

where ρ is the albedo of far infrared corresponding to the absorptivity, and $i \in \mathbb{R}^3$ and $n \in \mathbb{R}^3$ represent the light direction and surface normal, respectively.

Because the diffuse radiation and diffuse reflection follow the same cosine law, the ordinary photometric stereo can be applied for diffuse radiation. The ordinary photometric stereo is not applicable for black body, transparent objects and translucent objects that do not have diffuse reflection or are governed by other light transports. However, the diffuse radiation is a phenomenon of energy absorption and emission, so the surface normal of much more objects can be uniformly obtained using diffuse radiation.

As shown in equation [7.12], the decomposed diffuse radiation follows the cosine law; hence, it can be directly used for the Lambertian photometric stereo. The estimated diffuse radiation component $\hat{d}_\infty$ can be simply represented as

$$\hat{d}_\infty = \rho i^\top n, \qquad [7.13]$$

where ρ is the albedo of far infrared. When multiple light sources are placed at different positions, multiple observations can be obtained that can be superposed in a matrix form as

$$d = \rho I n, \qquad [7.14]$$

where d is the superposed diffuse component vector and $I = [i_1^\top \ i_2^\top \ \cdots]^\top$ is the superposed light source direction matrix. When the light direction matrix is a full-rank matrix, the surface normal and albedo can be obtained as

$$\begin{cases} n & = \frac{I^\dagger d}{\rho}, \\ \rho & = \left\| I^\dagger d \right\|_2, \end{cases} \qquad [7.15]$$

where $I^\dagger$ is a pseudo-inverse matrix of I.

7.5. Experiments

The experimental setup is shown in Figure 7.8. The target object is illuminated by far infrared spot lights (Exo Terra Heat-Glo 100W) and measured by a thermal camera

(InfRec R500). The ambient component is observed before the light source is turned on. Then, the light source is turned on and the change of temperature is captured as a video.

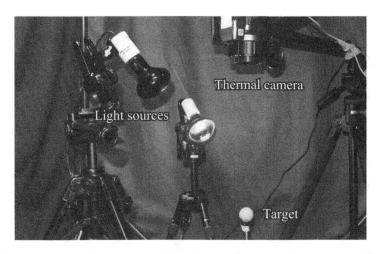

Figure 7.8. *Experimental setup. The object is illuminated by far infrared light and captured by a thermal camera. For a color version of this figure, see www.iste.co.uk/funatomi/computational.zip*

The real light bulb is not stable immediately after turning on and requires a warm-up period in practice. In our experiments, the bulb is warmed up outside the experiment room and brought in under a cover. Removal of the cover is the actual meaning of the light being turned on. The wall of the room is heated over the experiment time, and it could become a heat source. To avoid this effect, we place the object far from the wall and the room is actively cooled using an air conditioner.

7.5.1. *Decomposition result*

A black-painted wooden sphere as shown in Figure 7.9a is measured. A frame of the measured thermal video is shown in Figure 7.9b. The ambient component is the measured temperature before turning on the light source, and reflection component is the increased intensity immediately after the light source is turned on. The radiation components are the rest and are shown in Figure 7.9f. The radiation components are not fitted well by a single exponential curve because this is a sum of the diffuse and global radiations. Figure 7.9g shows the decomposed diffuse and global radiations. The sum of these fits well to the observation.

This procedure is applied for all the pixels, and the decomposed images are shown in Figure 7.9c – 7.9e. The reflection component represents the reflection of the light source on the surface, the diffuse radiation represents the reasonable shading and the global radiation represents the warming of the entire object.

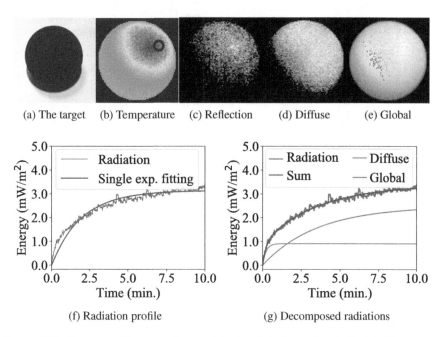

(a) The target   (b) Temperature   (c) Reflection   (d) Diffuse   (e) Global

(f) Radiation profile   (g) Decomposed radiations

Figure 7.9. *Decomposition result for a black-painted wooden ball. (a) The scene. (b) One of the thermal video frames. Transient profiles of a point, indicated by a black circle, are shown. (c–e) Decomposed images of reflection, diffuse and global radiations, respectively. (f) Radiation profile. Ambient and reflection components are subtracted. (g) Decomposed diffuse and global radiations. For a color version of this figure, see www.iste.co.uk/funatomi/computational.zip*

To show that our dual exponential approximation is valid, experiments were performed on objects of four different surface properties. As shown in Figure 7.10, four spherical objects made of (a) wood, (b) glass, (c) plastic and (d) marble were chosen for this experiment, which are shown in Figure 7.13. The objects were chosen so as to cover a wide range of variability in surface properties. Infrared light was irradiated and energy measurements were collected for each of the four objects. In each case, it can be observed that a dual exponential model is a better fit for the experimental measurement of energy.

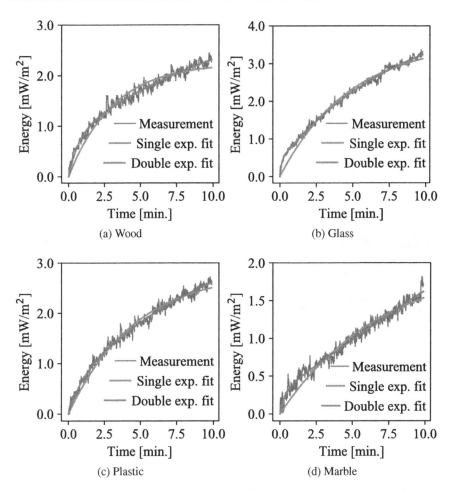

Figure 7.10. *Exponential fitting results. Double exponential curves fit the observation better than single exponential curves. For a color version of this figure, see www.iste.co.uk/funatomi/computational.zip*

7.5.2. *Surface normal estimation*

By using multiple light source positions and separating each diffuse radiation, we can apply the Lambertian photometric stereo. Figure 7.11 shows the result of the TPS for the same object as shown in Figure 7.9. A normal of the sphere is obtained as shown in Figure 7.11d. The result is compared with the result without light transport decomposition (composition of reflection, diffuse, and global) and radiation (composition of diffuse and global) as shown in Figure 7.12. As the temperature is not increased around $t = 0$, the compared results are noisy. The error increases owing to the global radiation at a longer time. As the best result, the angular errors of the result without decomposition and that of radiation are 7.71 and 6.50 degrees, respectively, while our method achieves a better result and the angular error is 5.85 degrees. This result shows the effectiveness of the separation of diffuse radiation.

We apply our method to other materials, including crystal glass, translucent plastic and translucent marble. The decomposed diffuse component and estimated surface normal are shown in Figure 7.13. Because our method is based on the diffuse radiation, materials that are difficult to measure with the ordinary vision techniques, e.g., transparent and translucent objects, can be measured in the same way. A plastic ornament is also measured, and the result shows the feasibility of our method to complex shaped objects.

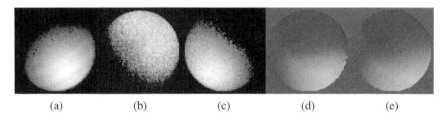

(a)   (b)   (c)   (d)   (e)

Figure 7.11. *Results of the thermal photometric stereo. (a–c) Decomposed diffuse radiation at different light positions. (d) Estimated surface normal. (e) The ground-truth normal. For a color version of this figure, see www.iste.co.uk/funatomi/computational. zip*

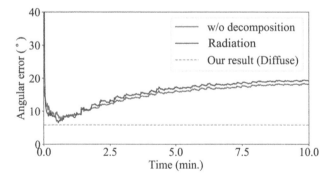

Figure 7.12. *The effectiveness of decomposition. Photometric stereo result without decomposition, result using radiation components and comparison with our method. Our method is time invariant, and the accuracy is shown as a dotted line. The angular error of our method is 5.85 degrees, which shows that our decomposition is effective for the separation of diffuse radiation. For a color version of this figure, see www.iste.co.uk/funatomi/computational.zip*

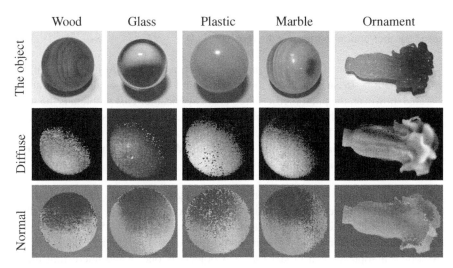

Figure 7.13. *Results on various materials. Spheres made of wood, glass, plastic and marble are measured, which are challenging objects for ordinary computer vision techniques. Our method uniformly recovers the surface normal for many materials. A complex shape is also measured, and our method recovers the normal appropriately. For a color version of this figure, see www.iste.co.uk/funatomi/computational.zip*

7.6. Conclusion

This study presents a novel technique for the time-resolved decomposition of far infrared light transport. We describe the far infrared light transport model and its transient properties and that the ordinary vision techniques can be applied to decomposed thermal images. Because this is a first work of far infrared light transport, we believe that this work will spur further researches using far infrared light transport, including thermal reflectance analysis, thermal structured light (Erdozain et al. 2020), thermal non-line-of-sight imaging (Kaga et al. 2019; Maeda et al. 2019), material classification, and so on.

We also propose a time-resolved far infrared light transport decomposition and TPS as its application. Separating diffuse components of far infrared radiations, the surface normal of any objects that absorb the incident light, including transparent, translucent and black objects as well as matte objects, are recovered.

While the effectiveness of our method is shown by some real-world experiments, some limitations are also encountered. First, the result is noisy owing to noisy observations and pixel-wise calculation. One reason is that far infrared sensors are not developed for measuring small temperature changes, hence low SNR. Naturally, the quality of the sensor will be improved in the future and it will directly improve our results. A global optimization that considers smoothness or simply using a smoothing filter is another option to improve the results.

Another limitation is that some materials, such as metals, do not exhibit much diffuse radiation. In such a case, the ambient and reflection components can be separated; however, the photometric stereo is not applicable. This problem is the same as that encountered with visible light observation, e.g., photometric stereo suffers from mirror surface objects. In contrast, the absorption of many objects, including glass, is high, hence the potential applicability of our method is relatively higher than visible light observation techniques.

We only model a simple far infrared light transport. Inhomogeneous, multi-material, thin and/or locally spiny shape has complicated heat transport; thus, it might not exhibit the exponential transience as in equation [7.5]. Therefore, our method works only for locally planar, homogeneous and sufficiently thick objects. Developing a sophisticated model for such objects is one of the interesting future directions of this research.

7.7. References

Becker, T.W. and Kaus, B.J. (2016). Numerical modeling of Earth Systems: An introduction to computational methods with focus on solid Earth applications of continuum mechanics. Lecture Notes, University of Southern California.

Bhandari, A., Feigin, M., Izadi, S., Rhemann, C., Schmidt, M., Raskar, R. (2014). Resolving multipath interference in Kinect: An inverse problem approach. *IEEE SENSORS*, pp. 614–617.

Carslaw, H.S. and Jaeger, J.C. (1959). *Conduction of Heat in Solids*, 2nd edition. Clarendon Press, Oxford.

Crank, J. and Nicolson, P. (1947). A practical method for numerical evaluation of solutions of partial differential equations of the heat-conduction type. *Mathematical Proceedings of the Cambridge Philosophical Society*, 43(1), 50–67.

Dorrington, A.A., Godbaz, J.P., Cree, M.J., Payne, A.D., Streeter, L.V. (2011). Separating true range measurements from multi-path and scattering interference in commercial range cameras. *SPIE 7864, Three-Dimensional Imaging, Interaction, and Measurement*, p. 786404.

Erdozain, J., Ichimaru, K., Maeda, T., Kawasaki, H., Raskar, R., Kadambi, A. (2020). 3D imaging for thermal cameras using structured light. *2020 IEEE International Conference on Image Processing (ICIP)*, pp. 2795–2799.

Eren, G., Aubreton, O., Meriaudeau, F., Sanchez Secades, L.A., Fofi, D., Naskali, A.T., Truchetet, F., Ercil, A. (2009). Scanning from heating: 3D shape estimation of transparent objects from local surface heating. *Optics Express*, 17(14), 11457–11468.

Freedman, D., Krupka, E., Smolin, Y., Leichter, I., Schmidt, M. (2014). SRA: Fast removal of general multipath for ToF sensors. *Proc. European Conference on Computer Vision (ECCV)*, pp. 234–249.

Fuchs, S. (2010). Multipath interference compensation in time-of-flight camera images. *Proc. International Conference on Pattern Recognition*. IEEE, pp. 3583–3586.

Gkioulekas, I., Levin, A., Durand, F., Zickler, T. (2015). Micron-scale light transport decomposition using interferometry. *ACM Transactions on Graphics (ToG)*, 34(4), 37:1–14.

Godbaz, J.P., Cree, M.J., Dorrington, A.A. (2012). Closed-form inverses for the mixed pixel/multipath interference problem in AMCW Lider. *SPIE 8296, Computational Imaging X*, p. 909778.

Gupta, M., Nayar, S.K., Hullin, M.B., Martin, J. (2015). Phasor imaging: A generalization of correlation-based time-of-flight imaging. *ACM Transactions on Graphics (ToG)*, 34(5), 156:1–18.

Hanrahan, P. and Krueger, W. (1993). Reflection from layered surfaces due to subsurface scattering. *Proceedings of SIGGRAPH*. ACM Press, pp. 165–174.

Heide, F., Hullin, M.B., Gregson, J., Heidrich, W. (2013). Low-budget transient imaging using photonic mixer devices. *ACM Transactions on Graphics (ToG)*, 32(4), 45:1–10.

Heide, F., Xiao, L., Heidrich, W., Hullin, M.B. (2014a). Diffuse mirrors: 3D reconstruction from diffuse indirect illumination using inexpensive time-of-flight sensors. *Proceedings Computer Vision and Pattern Recognition (CVPR)*, pp. 3222–3229.

Heide, F., Xiao, L., Kolb, A., Hullin, M.B., Heidrich, W. (2014b). Imaging in scattering media using correlation image sensors and sparse convolutional coding. *Optics Express*, 22(21), 26338–26350.

Howell, J.R., Mengüç, M.P., Siegel, R. (2015). *Thermal Radiation Heat Transfer*, 6th edition. CRC Press.

Inoshita, C., Mukaigawa, Y., Matsushita, Y., Yagi, Y. (2014). Surface normal decomposition: Photometric stereo for optically thick translucent objects. *Proceedings of European Conference on Computer Vision (ECCV)*, pp. 346–359.

Ito, Y., Kennan, R.P., Watanabe, E., Koizumi, H. (2000). Assessment of heating effects in skin during continuous wave near-infrared spectroscopy. *Journal of Biomedical Optics*, 5(4), 383–391.

Jensen, H.W., Marschner, S.R., Levoy, M., Hanrahan, P. (2001). A practical model for subsurface light transport. *Proceedings of SIGGRAPH*, ACM Press, pp. 511–518.

Jimenez, D., Pizarro, D., Mazo, M., Palazuelos, S. (2012). Modelling and correction of multipath interference in Time of Flight cameras. *Proceedings of Computer Vision and Pattern Recognition (CVPR)*. IEEE, pp. 893–900.

Kadambi, A., Whyte, R., Bhandari, A., Streeter, L., Barsi, C., Dorrington, A., Raskar, R. (2013). Coded Time of Flight cameras: Sparse deconvolution to address multipath interference and recover time profiles. *ACM Transactions on Graphics (ToG)*, 32(6), 167:1–10.

Kadambi, A., Schiel, J., Raskar, R. (2016a). Macroscopic interferometry: Rethinking depth estimation with frequency-domain time-of-flight. *Proceedings of Computer Vision and Pattern Recognition (CVPR)*, pp. 893–902.

Kadambi, A., Zhao, H., Shi, B., Raskar, R. (2016b). Occluded imaging with time-of-flight sensors. *ACM Transactions on Graphics (ToG)*, 35(2), 15:1–12.

Kaga, M., Kushida, T., Takatani, T., Tanaka, K., Funatomi, T., Mukaigawa, Y. (2019). Thermal non-line-of-sight imaging from specular and diffuse reflections. *IPSJ Transactions on Computer Vision and Applications (TCVA)*, 11(1), 8.

Kirmani, A., Benedetti, A., Chou, P.A. (2013). SPUMIC: Simultaneous phase unwrapping and multipath interference cancellation in time-of-flight cameras using spectral methods. *IEEE International Conference on Multimedia and Expo (ICME)*, pp. 1–6.

Kitano, K., Okamoto, T., Tanaka, K., Aoto, T., Kubo, H., Funatomi, T., Mukaigawa, Y. (2017). Recovering temporal PSF using ToF camera with delayed light emission. *IPSJ Transactions on Computer Vision and Applications*, 9(15), 1–6.

Lamond, B., Peers, P., Debevec, P. (2007). Fast image-based separation of diffuse and specular reflections. *Proceedings of SIGGRAPH Sketches*, p. 74.

Lee, S. and Shim, H. (2015). Skewed stereo time-of-flight camera for translucent object imaging. *Image and Vision Computing*, 43(C), 27–38.

Maeda, T., Wang, Y., Raskar, R., Kadambi, A. (2019). Thermal non-line-of-sight imaging. *Proc. International Conference on Computational Photography (ICCP)*. IEEE, pp. 1–11.

Miyazaki, D., Saito, M., Sato, Y., Ikeuchi, K. (2002). Determining surface orientations of transparent objects based on polarization degrees in visible and infrared wavelengths. *Journal of the Optical Society of America A*, 19(4), 687–694.

Mukaigawa, Y., Raskar, R., Yagi, Y. (2010). Analysis of light transport in scattering media. *Proceedings of Computer Vision and Pattern Recognition (CVPR)*, pp. 153–160.

Murez, Z., Treibitz, T., Ramamoorthi, R., Kriegman, D.J. (2017). Photometric stereo in a scattering medium. *IEEE Transactions on Pattern Analysis and Machine Intelligence*, 39(9), 1880–1891.

Naik, N., Zhao, S., Velten, A., Raskar, R., Bala, K. (2011). Single view reflectance capture using multiplexed scattering and time-of-flight imaging. *ACM Transactions on Graphics (ToG)*, 30(6), 171:1–10.

Naik, N., Kadambi, A., Rhemann, C., Izadi, S., Raskar, R., Bing Kang, S. (2015). A light transport model for mitigating multipath interference in time-of-flight sensors. *Proceedings of Computer Vision and Pattern Recognition (CVPR)*, pp. 73–81.

Nayar, S.K., Fang, X.-S., Boult, T. (1997). Separation of reflection components using color and polarization. *International Journal of Computer Vision (IJCV)*, 21(3), 163–186.

Nayar, S.K., Krishnan, G., Grossberg, M.D., Raskar, R. (2006). Fast separation of direct and global components of a scene using high frequency illumination. *ACM Transactions on Graphics (ToG)*, 25(3), 935–944.

Ngo, T.T., Nagahara, H., Taniguchi, R. (2015). Shape and light directions from shading and polarization. *Proceedings of Computer Vision and Pattern Recognition (CVPR)*.

Nguyen, T., Vo, Q.N., Yang, H.-J., Kim, S.-H., Lee, G.-S. (2014). Separation of specular and diffuse components using tensor voting in color images. *Applied Optics*, 53(33), 7924–7936.

O'Toole, M., Raskar, R., Kutulakos, K.N. (2012). Primal-dual coding to probe light transport. *ACM Transactions on Graphics (ToG)*, 31(4), 39:1–11.

O'Toole, M., Heide, F., Xiao, L., Hullin, M.B., Heidrich, W., Kutulakos, K.N. (2014a). Temporal frequency probing for 5D transient analysis of global light transport. *ACM Transactions on Graphics (ToG)*, 33(4), 87:1–11.

O'Toole, M., Mather, J., Kutulakos, K.N. (2014b). 3D shape and indirect appearance by structured light transport. *Proceedings of Computer Vision and Pattern Recognition (CVPR)*. IEEE, pp. 3246–3253.

O'Toole, M., Achar, S., Narasimhan, S.G., Kutulakos, K.N. (2015). Homogeneous codes for energy-efficient illumination and imaging. *ACM Transactions on Graphics (ToG)*, 34(4), 35:1–13.

O'Toole, M., Heide, F., Lindell, D., Zang, K., Diamond, S., Wetzstein, G. (2017). Reconstructing transient images from single-photon sensors. *Proceedings of Computer Vision and Pattern Recognition (CVPR)*.

Qiao, H., Lin, J., Liu, Y., Hullin, M.B., Dai, Q. (2015). Resolving transient time profile in ToF imaging via log-sum sparse regularization. *Optics Letters*, 40(6), 918–21.

Ren, W., Tian, J., Tang, Y. (2017). Specular reflection separation with color-lines constraint. *IEEE Transactions on Image Processing*, 26(5), 2327–2337.

Saponaro, P., Sorensen, S., Kolagunda, A., Kambhamettu, C. (2015). Material classification with thermal imagery. *Proceedings of Computer Vision and Pattern Recognition (CVPR)*, pp. 4649–4656.

Sato, Y. and Ikeuchi, K. (1994). Temporal-color space analysis of reflection. *Journal of the Optical Society of America A*, 11(7), 2990–3002.

Shafer, S.A. (1985). Using color to separate reflection components. *Color Research & Application*, 10(4), 210–218.

Shim, H. and Lee, S. (2015). Recovering translucent object using a single time-of-flight depth camera. *IEEE Transactions on Circuits and Systems for Video Technology*, 26(5), 841–854.

Sunkavalli, K., Belhumeur, P.N., Ramamoorthi, R., Sun, B., Nayar, S.K. (2007). Time-varying BRDFs. *IEEE Transactions on Visualization & Computer Graphics*, 13(03), 595–609.

Tanaka, K., Mukaigawa, Y., Matsushita, Y., Yagi, Y. (2013). Descattering of transmissive observation using parallel high-frequency illumination. *Proceedings of International Conference on Computational Photography (ICCP)*. IEEE, pp. 96–103.

Tanaka, K., Mukaigawa, Y., Kubo, H., Matsushita, Y., Yagi, Y. (2016). Recovering transparent shape from time-of-flight distortion. *Proceedings of Computer Vision and Pattern Recognition (CVPR)*, pp. 4387–4395.

Tanaka, K., Mukaigawa, Y., Kubo, H., Matsushita, Y., Yagi, Y. (2017). Recovering inner slices of layered translucent objects by multi-frequency illumination. *IEEE Transactions on Pattern Analysis and Machine Intelligence (TPAMI)*, 39(4), 746–757.

Tanaka, K., Ikeya, N., Takatani, T., Kubo, H., Funatomi, T., Mukaigawa, Y. (2018). Time-resolved light transport decomposition for thermal photometric stereo. *Proceedings of Computer Vision and Pattern Recognition (CVPR)*.

Tanaka, K., Ikeya, N., Takatani, T., Kubo, H., Funatomi, T., Ravi, V., Kadambi, A., Mukaigawa, Y. (2021). Time-resolved far infrared light transport decomposition for thermal photometric stereo. *IEEE Transactions on Pattern Analysis and Machine Intelligence (TPAMI)*, 43(6), 2075–2085.

Thompson, A.C., Wade, S.A., Cadusch, P.J., Brown, W.G., Stoddart, P.R. (2013). Modeling of the temporal effects of heating during infrared neural stimulation. *Journal of Biomedical Optics*, 18(3), 035004.

Treibitz, T. and Schechner, Y.Y. (2009). Active polarization descattering. *IEEE Transactions on Pattern Analysis and Machine Intelligence (TPAMI)*, 31(3), 385–399.

Treibitz, T., Murez, Z., Mitchell, B.G., Kreigman, D. (2012). Shape from fluorescence. *Proceedings of European Conference on Computer Vision (ECCV)*.

Tsai, C.-Y., Kutulakos, K.N., Narasimhan, S.G., Sankaranarayanan, A.C. (2017). The geometry of first-returning photons for non-line-of-sight imaging. *Proceedings of Computer Vision and Pattern Recognition (CVPR)*.

Unsworth, J. and Duarte, F. (1979). Heat diffusion in a solid sphere and Fourier theory: An elementary practical example. *American Journal of Physics*, 47(11), 981–983.

Wolff, L.B. and Boult, T.E. (1991). Constraining object features using a polarization reflectance model. *IEEE Transactions on Pattern Analysis and Machine Intelligence (TPAMI)*, 13(7), 635–657.

Woodham, R.J. (1980). Photometric method for determining surface orientation from multiple images. *Optical Engineering*, 19(1), 139–144.

Wu, D., Velten, A., O'Toole, M., Masia, B., Agrawal, A., Dai, Q., Raskar, R. (2014). Decomposing global light transport using time of flight imaging. *International Journal of Computer Vision (IJCV)*, 107(2), 123–138.

8

Synthetic Wavelength Imaging: Utilizing Spectral Correlations for High-Precision Time-of-Flight Sensing

Florian WILLOMITZER

Wyant College of Optical Sciences, University of Arizona, Tucson, USA

8.1. Introduction

Optical three-dimensional (3D) imaging and ranging techniques have been used in academia and industry for many years with great success. Fields of application include medical imaging, autonomous navigation, industrial inspection, forensics or virtual reality. The great success of these techniques is no coincidence, as high-quality 3D object or scene representations contain a distinctive higher information content than simple 2D images: compared to 2D images, 3D representations are invariant against translation and rotation of the object as well as variations in surface texture or external illumination conditions. This fact has made 3D imaging an established tool in optical metrology and, more recently, in computer vision.

Although the sheer number of available 3D sensing principles is immense, existing approaches can be broadly categorized into three groups: (1) triangulation-based approaches (including active stereo, passive stereo or focus search) (Takeda and Mutoh 1983; Srinivasan et al. 1984; Schaffer et al. 2010; Willomitzer et al. 2010; Ettl

Computational Imaging for Scene Understanding,
coordinated by Takuya FUNATOMI and Takahiro OKABE.
© ISTE Ltd 2024.

et al. 2013; Arold et al. 2014; Schönberger and Frahm 2016; Willomitzer and Häusler 2017), (2) reflectance-based approaches that measure the surface gradient (including photometric stereo and deflectometry) (Woodham 1980; Horn 1990; Knauer et al. 2004; Faber et al. 2012; Huang et al. 2018; Willomitzer et al. 2020; Wang et al. 2021) and (3) approaches that measure the "time of flight" (ToF) or travel distance of light. This last group includes interferometry and so-called "ToF cameras". We will see later that the third group can be again subdivided into ToF imaging on *smooth* and *rough* surfaces, although this distinction will not be discussed in detail (see Häusler and Willomitzer (2022) for more information).

This book chapter focuses on *interferometric multi-wavelength ToF imaging techniques* – a subset of the third category. The chapter should serve as a gentle introduction and is intended for computational imaging scientists and students new to this fascinating topic. Technical details (such as detector or light source specifications) will be largely omitted. Instead, the similarities between different methods will be emphasized to "draw the bigger picture".

In computer vision, ToF cameras have become increasingly popular over the last years. Current techniques either use pulsed waveform modulation/LIDAR (Collis 1970; Weitkamp 2006) (which has already been discussed in previous chapters of this book) or continuous-wave amplitude modulation (CWAM) (Schwarte et al. 1997; Lange and Seitz 2001; Foix et al. 2011). CWAM ToF cameras illuminate the scene with a light source whose intensity is modulated over time, for example, as a sinusoid. The detector pixels in these devices behave as homodyne receivers which accumulate a charge proportional to the phase difference $\phi(\lambda_{mod})$ between the modulated emitted light and the modulated irradiance received at the sensor. For each pixel, the scene distance z can then be estimated via

$$z = \frac{1}{2} \frac{\lambda_{mod} \cdot \phi(\lambda_{mod})}{2\pi} \qquad [8.1]$$

where λ_{mod} is the modulation wavelength of the temporal intensity modulation of the light source. The depth precision δz is directly proportional to the modulation wavelength λ_{mod}: the smaller the wavelength, the better the depth precision (Lange and Seitz 2001; Li et al. 2021):

$$\delta z \sim \lambda_{mod} \qquad [8.2]$$

However, current limitations in silicon manufacturing technology (Schwarte et al. 1997; Lange and Seitz 2001; Li et al. 2018) restrict the minimal achievable modulation wavelengths for CWAM ToF cameras to roughly $\lambda_{mod} \gtrsim 3m$, which results in a depth precision not better than centimeters. This might be sufficient for simple object detection tasks (i.e. "is there an object or not?") or rough depth estimations (i.e. "is the object in the foreground or background?") but leaves much room for improvement for more sophisticated computer vision or metrology applications, where precise measurement of the 3D geometry is important (e.g. in medical imaging or industrial inspection).

Alternatives with higher precision reside on the other end of the spectrum of ToF principles: *optical interferometers* (Born and Wolf 2013; Wyant 2015). As mentioned above, ToF cameras and optical interferometers are closely related. Both systems evaluate the difference in pathlength or travel time between the signal reflected or scattered off the object under test and a reference signal. This difference manifests as a phase shift in the detected signal. Instead of using an intensity-modulated light source or a pulse, optical interferometry exploits the oscillations of the physical wave light field itself, meaning that it works with very small optical wavelengths down to several hundred nanometers. For well-adjusted interferometers, this leads to depth precisions better than nm on specular surfaces, which by far surpasses the precision that can be reached with a ToF camera. However, this very high precision of single-wavelength interferometry systems comes with a trade-off: the inability to measure rough surfaces. If coherent light is scattered off a rough surface, the optical phase of the backscattered light field is randomized. The resulting "speckle field" does seemingly not contain any information about the macroscopic optical pathlength anymore, and range information cannot be extracted from one single (monochromatic) speckle field.

This chapter will discuss how techniques known from multi-wavelength interferometry can be applied to compensate for the problem of phase randomization in a speckle field which is formed when a rough surface is illuminated with coherent light. This becomes possible, as multi-wavelength interferometry exploits information from an additional modality: *spectral diversity*, acquired via additional measurements at different optical wavelengths. It will be shown how to use this additional modality for "synthetic wavelength imaging" concepts, which eventually leads to the development of interferometric time-of-flight (ToF) cameras with μm-range precision for the measurement of macroscopic objects with rough surfaces (Li et al. 2018, 2019; Wu et al. 2020; Li et al. 2021). Subsequently, the concept of "synthetic waves" will be combined with digital holography to develop a novel approach for non-line-of-sight (NLoS) imaging, i.e. to image hidden objects around corners or through scattering media with sub-mm resolution (Willomitzer et al. 2019a, 2019b, 2021a).

8.2. Synthetic wavelength imaging

In the following, the term "synthetic wavelength imaging" collectively groups imaging principles that use well-known techniques from optical metrology (such as interferometry or holography) but perform these techniques at a so-called "synthetic wavelength" instead of optical wavelengths. In many of the introduced methods, the synthetic wave field only "lives" in the computer, and the respective operations (such as propagation, superposition, etc.) are performed purely computationally.

As mentioned, synthetic wavelength imaging exploits concepts known from multi-wavelength interferometry, a technique that has been widely used in optical metrology (see exemplary references (Polhemus 1973; Cheng and Wyant 1985; De Groot and McGarvey 1992; De Groot 1994; Tiziani et al. 1996; Falaggis et al. 2009; Zhou et al. 2022)). For many implementations of multi-wavelength interferometry, the additional spectral information is used to increase the unambiguous measurement range for the interferometric measurement of optically smooth surfaces with large height variations. The resulting increase in dynamic range is available for the high-precision measurements of optical components or technical parts. However, it turns out that the utilization of spectral diversity in interferometry has another significant benefit: it enables interferometric measurements of *optically rough surfaces*. This is very surprising at first glance: it was just discussed that illuminating an optically rough surface with coherent light at wavelength λ_1 produces a speckle field, which is randomized in phase and intensity (Häusler 2004; Goodman 2007) and optical pathlength information cannot be inferred. However, by probing the scene with a second wavelength λ_2, slightly different from the initial wavelength λ_1, additional information can be exploited. The basic process is explained in Figure 8.1: the speckle field $E(\lambda_1)$, emerging from a tilted planar rough surface which is illuminated with a spatially and temporally coherent beam at wavelength λ_1, is randomized in intensity and phase (see Figure 8.1(a) and (b)). Illuminating the surface with a second beam at another wavelength λ_2 produces a second speckle field $E(\lambda_2)$ (see Figure 8.1(c) and (d)). If the two beams at λ_1 and λ_2 originate from the same source (e.g. from the same fiber tip of the illumination unit), the respective fields *undergo the exact same pathlength variations* before reaching the detector. For closely spaced wavelengths λ_1 and λ_2, this results in two speckle fields with a largely similar intensity distribution (see Figure 8.1(a) and (c)) – the speckle fields are "spectrally correlated" (Goodman 2007).

The complex valued speckle fields $E(\lambda_1)$ and $E(\lambda_2)$ incident on the detector can be expressed as

$$E(\lambda_1) = A_1 \cdot e^{i\phi(\lambda_1)}, E(\lambda_2) = A_2 \cdot e^{i\phi(\lambda_2)}, \qquad [8.3]$$

with A_1, A_2 being the respective amplitudes and $\phi(\lambda_1), \phi(\lambda_2)$ the respective (speckled) phase maps at the given wavelengths λ_1, λ_2. After measuring and storing $E(\lambda_1)$ and $E(\lambda_2)$, a so-called "synthetic field" $E(\Lambda)$ can be calculated. One possibility to calculate $E(\Lambda)$ is *computational mixing* of $E(\lambda_1)$ and $E(\lambda_2)$:

$$E(\Lambda) = E(\lambda_1) \cdot E^*(\lambda_2) = A_1 A_2 \cdot e^{i(\phi(\lambda_1) - \phi(\lambda_2))} = A_1 A_2 \cdot e^{i\phi(\Lambda)}, \qquad [8.4]$$

where $E^*(\lambda_2)$ denotes the complex conjugate of $E(\lambda_2)$. Other options to calculate the synthetic field can be found in Li et al. (2018, 2021). It can be seen in equation [8.4] that the phase of $E(\Lambda)$ is composed of the *difference of the two optical phase maps*.

This means that the synthetic field $E(\Lambda)$ contains information that is on the order of a "synthetic wavelength" Λ, which is the beat wavelength of λ_1 and λ_2:

$$\Lambda = \frac{\lambda_1 \cdot \lambda_2}{|\lambda_1 - \lambda_2|} \tag{8.5}$$

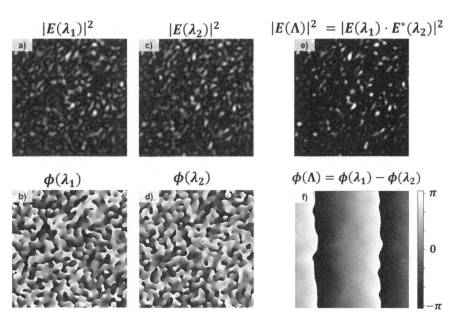

Figure 8.1. *Formation of the synthetic wave field, explained at the example of the measurement of a tilted planar rough surface (experiment): the surface is subsequently illuminated with a coherent beam at wavelength $\lambda_1 = 854.39$ nm and $\lambda_2 = 854.47$ nm, and the complex speckle fields $E(\lambda_1)$ and $E(\lambda_2)$ are captured with an interferometer. The intensity of both fields (a and c) shows the typical distribution of a speckle pattern, and the phase of both fields (b and d) is randomized. The synthetic field $E(\Lambda)$ is formed (in this example) via mixing (equation [8.4]) and the synthetic wavelength calculates to $\Lambda \approx 9$ mm. The microscopic pathlength variations cancel each other out, and the phase of the synthetic field (f) is not subject to phase randomization anymore. Macroscopic information about the surface (shape, tilt, etc.) can be extracted from (f)*

With a *closely spaced* selection of λ_1 and λ_2, the synthetic wavelength can be chosen orders of magnitudes larger than the two optical wavelengths or the microscopic roughness[1] of the surface. For example, the selection of the two optical wavelengths $\lambda_1 = 550$ nm and $\lambda_2 = 550.1$ nm leads to a synthetic wavelength of

---

1. For the sake of brevity and a gentle introduction, an exact definition of the term "surface roughness" is omitted at this early point of the chapter. The topic will be described in detail in section 8.5, when the theoretical limits of the introduced methods are discussed.

$\Lambda \approx 3\ mm$. For surfaces with roughness $\ll \Lambda$, the fields $E(\lambda_1)$ and $E(\lambda_2)$ are *highly correlated* (Goodman 2007). Remaining phase fluctuations in the synthetic phase map $\phi(\Lambda) = \phi(\lambda_1) - \phi(\lambda_2)$ are negligible, and $\phi(\Lambda)$ becomes largely equivalent to the phase map *that would be measured for a source that emits electromagnetic radiation at the much larger wavelength* Λ, meaning that it displays no speckle artifacts (see Figure 8.1f). This is possible, since the phase perturbations, imparted for each field by the microscopic pathlength variations, cancel each other out and information about the macroscopic pathlength variation is now visible in the resulting phase map $\phi(\Lambda)$. It should be emphasized that this is only the case for the *phase map* and not the intensity term $|E(\Lambda)|^2 = |E(\lambda_1) \cdot E^*(\lambda_2)|^2$, which still displays the spatial structure of the two speckle fields at the optical wavelengths λ_1 and λ_2 (see Figure 8.1e).

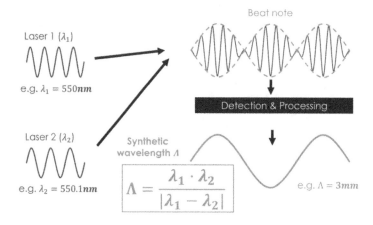

Figure 8.2. *Synthetic wavelength imaging exploits spectral diversity, i.e. encoding at different optical illumination wavelengths. The superposition of two or more optical fields at slightly different wavelengths (in this example, $\lambda_1 = 550$ nm and $\lambda_1 = 550.1$ nm) produces a high-frequency wave with a low-frequency beat note. The beat wave can be isolated by special detection and processing methods and forms the "synthetic wave", where the synthetic wavelength (in this example, $\Lambda = 3\ mm$) is calculated by equation [8.5]. For the majority of results shown in this chapter, the synthetic field is obtained via mixing of the two optical fields (see equation [8.4]). For a color version of this figure, see www.iste.co.uk/funatomi/computational.zip*

Obtaining the largely speckle-free synthetic phase map $\phi(\Lambda)$ is the first step of further computational processing. To evaluate the final result, synthetic wavelength imaging procedures often use well-known methods of computational optics (such as computational wave propagation) and perform these methods "at the synthetic wavelength". This is possible, as the computationally constructed synthetic wave behaves much like an optical wave (Willomitzer et al. 2021a, 2021b), and hence can be processed using the same algorithms.

In the following, two examples of this striking similarity are introduced: *synthetic wavelength interferometry* for the precise "ToF" 3D measurement of macroscopic objects with rough surfaces (Li et al. 2018, 2019; Wu et al. 2020; Li et al. 2021), and *synthetic wavelength holography* which can be used to image hidden objects around corners or through scattering media (Willomitzer et al. 2019a, 2019b, 2021a). In both examples, the calculation of the synthetic field $E(\Lambda)$ helps to circumvent the deleterious effects of speckle arising at the optical wavelengths. However, it should be emphasized that the fields at the optical wavelengths are still the carrier of the information, and inexpensive sensors for visible light (no mm-wave detectors) can be used for their detection. For the sake of brevity, the methods are mainly introduced with work performed by the author's research group.

8.3. Synthetic wavelength interferometry

Interferometric imaging methods obtain 3D shape information by measuring the phase of a light field reflected off the surface under test with respect to a reference field (Hecht 2012; Wyant 2015). As common for principles measuring the standoff distance of an object via optical pathlength differences (such as interferometry or ToF imaging), the unique measurement range ΔZ is defined by 1/2 of the wavelength λ_{mod} at which the signal is modulated ($\Delta Z = \frac{\lambda_{mod}}{2}$). The distance value z is calculated by equation [8.1] after measuring the respective phase map $\phi(\lambda_{mod})$. As discussed, CWAM ToF cameras exploit intensity modulation of an (in most cases) incoherent light source (Schwarte et al. 1997; Lange and Seitz 2001), while optical interferometry uses the modulation of the physical wave light field itself, meaning that $\lambda_{mod} = \lambda$ is within or close to the visible wave band. This reveals an interesting trade-off: The extremely high precision of optical (single-wavelength) interferometry comes at the price of an extremely small unique measurement range $\Delta Z = \frac{\lambda}{2}$ (several hundred nm for visible light), and phase wrapping occurs for surface height variations exceeding ΔZ. Phase unwrapping methods can be used to extend the unique measurement range ΔZ and to disambiguate the captured phase map $\phi(\lambda)$. The probably most prominent example is multi-frequency phase unwrapping (Huntley and Saldner 1993), which requires the measurement of a second phase map at a different wavelength.

Although the exact details will be discussed later, it should be briefly mentioned here that synthetic wavelength imaging can also be seen as a special form of multi-frequency phase unwrapping. This is because the technique is able to disambiguate the random phase fluctuations in speckled interferograms. The microscopic pathlength variations in the two optical interferograms at λ_1 and λ_2 cancel each other out, and the final synthetic phase map $\phi(\Lambda)$ contains only information about the macroscopic pathlength variations on the order of the synthetic wavelength Λ. As discussed, $\phi(\Lambda)$ is virtually speckle-free if the distance between the two optical wavelengths λ_1 and λ_2 is chosen sufficiently small (i.e. the synthetic wavelength Λ is sufficiently large). In this case, the surface height can be simply determined by

$$z = \frac{1}{2} \frac{\Lambda \cdot \phi(\Lambda)}{2\pi}. \qquad [8.6]$$

Note that this expression is equivalent to equation [8.1], with the synthetic wavelength being now the utilized modulation wavelength. It should also be emphasized that surface measurements at the synthetic wavelength do not exclude the possibility of phase wrapping. Depending on the ratio between surface height variations and unique measurement range $\Delta Z = \frac{\Lambda}{2}$, the synthetic phase map $\phi(\Lambda)$ could be wrapped. In analogy to classical interferometry, unwrapping can be performed. Besides the described multi-frequency phase unwrapping, which now would use a second *synthetic* phase map acquired at a different synthetic wavelength (Li et al. 2021), other unwrapping methods could be applied as well. Examples are spatial phase unwrapping (Willomitzer 2019), or data-driven phase unwrapping procedures (Wang et al. 2019; Yin et al. 2019).

Synthetic wavelength interferometry is a long-known and well-established principle in optical surface metrology. Over the years, many different flavors of the synthetic wavelength idea have been published. The exemplary references (Fercher et al. 1985; Vry and Fercher 1986; Dändliker et al. 1988; De Groot and McGarvey 1992; Li et al. 2021; Kotwal et al. 2022; Zhou et al. 2022) should give the reader an idea about the diversity of approaches and their applications. While the method of processing the synthetic interferograms is fairly similar to most approaches, variations exist in how the complex-valued speckle fields at the two or more optical wavelengths are acquired. Common procedures include simultaneous and sequential capturing of $E(\lambda_1)$ and $E(\lambda_2)$, and rely on, for example, spatial heterodyning, frequency heterodyning or phase shifting of the reference beam with respect to the object beam. The latter can be performed, for example, by moving a mirror in the reference arm. An overview of different methods is given in Kreis et al. (1997), Kim (2010) and Zhou et al. (2022).

The remainder of this section focuses on a very specific application of synthetic wavelength interferometry, which recently has been investigated by the author's research group: the precise 3D acquisition of macroscopic objects with rough surfaces for the specific application to problems in computer vision, medical imaging, automotive or virtual reality. Or, in other words: *Using synthetic wavelength interferometry to build a "high-precision ToF camera"*. As before, explanations will be kept to a high level, and technical details will be largely omitted. Further details can be found in Li et al. (2018, 2019, 2021) and Wu et al. (2020).

Figure 8.3a displays the schematic setup of the "high precision ToF camera", which consists of a dual-wavelength Michelson interferometer, equipped with a lens that images the object onto a high-resolution pixel array of a camera (herein referred to as "focal plane array" (FPA) sensor). The wavelength of at least one laser in the setup is tunable so that different wavelength spacings $|\lambda_1 - \lambda_2|$, and hence different synthetic wavelengths Λ can be realized. Amongst the various detection methods mentioned above, the author's research group has predominantly used heterodyne (Fercher et al. 1985) and superheterodyne (Dändliker et al. 1988) detection, by using the frequency modulation of additional acousto-optic modulators (AOMs) integrated with a fiber-based setup. To the best of its knowledge, the group was the first team who paired heterodyne/superheterodyne interferometry with high-resolution FPA detectors for full-field 3D measurements (Li et al. 2021). Raster-scanning-based superheterodyne interferometers equipped with a single-pixel detector have been studied as well (Li et al. 2018).

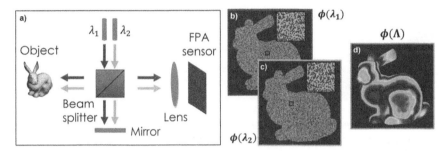

Figure 8.3. *High-precision ToF imaging with synthetic wavelength interferometry (simulation): a) Schematic setup of a dual-wavelength Michelson interferometer, which serves as "high-precision ToF camera". b) and c) Phase maps $\phi(\lambda_1)$ and $\phi(\lambda_2)$ acquired at the two optical wavelengths. Both phase maps are subject to heavy phase randomization. d) Resulting speckle-free synthetic phase map $\phi(\Lambda)$, which allows for the extraction of depth information. For a color version of this figure, see www.iste.co.uk/funatomi/computational.zip*

The simulated results in Figure 8.3b,c display exemplary phase maps acquired from a macroscopic object with optically rough surface (bunny). As discussed before, it can be seen that $\phi(\lambda_1)$ and $\phi(\lambda_2)$ are subject to heavy phase randomization due to speckle, while $\phi(\Lambda)$ (Figure 8.3d) is virtually speckle-free. In the realized example of an FPA-based heterodyne interferometer, the fields at both optical wavelength $E(\lambda_1), E(\lambda_2)$ are acquired in a sequential fashion to form the synthetic field $E(\Lambda)$ via equation [8.4].

The demonstrated realization of the FPA-based superheterodyne interferometer (Li et al. 2021) uses simultaneous illumination with both lasers and is able to measure the synthetic phase map $\phi(\Lambda)$ directly but at the cost of ≥ 3 phase shifts of the AOM modulated signal. Figure 8.4(b)–(d) shows full-field synthetic phase measurements of a plaster bust, captured with the tunable dual-wavelength superheterodyne interferometer (Li et al. 2021). Synthetic phase maps acquired for different synthetic wavelengths are shown, and the field of view is approximately 10 cm × 10 cm.

It can be seen that the phase maps at smaller synthetic wavelengths are subject to serious phase wrapping. After unwrapping (in the particular case of Figure 8.4 via multi-frequency phase unwrapping), the respective depth maps can be calculated (see Figure 8.4(e)–(g)). As expected, the noise level decreases for smaller synthetic wavelengths.

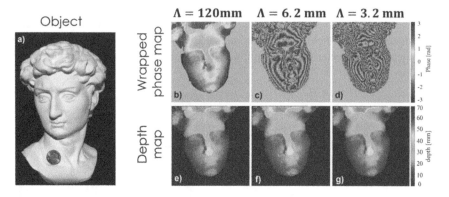

Figure 8.4. *Measurement of a plaster bust with a focal plane array (FPA)-based superheterodyne interferometer (experiment) (Li et al. 2021). a) Image of the bust with US penny for size comparison. b–d) Measured synthetic phase maps for different synthetic wavelengths (120 mm, 6.2 mm and 3.2 mm). e–g) Respective depth maps after phase unwrapping. For a color version of this figure, see www.iste.co.uk/funatomi/computational.zip*

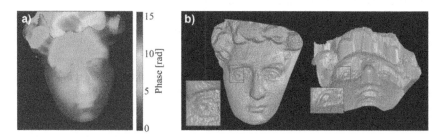

Figure 8.5. *Measurement of a plaster bust with a flutter-shutter camera-based tunable dual-wavelength heterodyne interferometer (experiment) (Li et al. 2021). The measurement was captured at a synthetic wavelength of $\Lambda \approx 43$ mm, and the depth precision was evaluated to $\delta z < 380\mu m$. a) Measured phase map. b) Calculated 3D model shown from two different perspectives. For a color version of this figure, see www.iste.co.uk/funatomi/computational.zip*

Figure 8.5 shows the 3D model of the same plaster bust, now acquired with a flutter-shutter camera-based tunable dual-wavelength heterodyne interferometer (Li et al. 2021). The 3D model was captured at a synthetic wavelength of $\Lambda \approx 43\ mm$, and the precision of the measurement was evaluated to $\delta z < 380\mu m$. Measurements at smaller synthetic wavelengths reach higher precision. A comprehensive precision evaluation of the method is given in Li et al. (2021), and the related fundamental limits are discussed in section 8.5.

8.4. Synthetic wavelength holography

It was discussed above that interferometry at the synthetic wavelength has a long tradition in optical metrology. This is also true for synthetic wavelength holography, which is used, for example, for surface profiling of technical parts or industrial inspection (see Javidi et al. (2005), Mann et al. (2008), Fu et al. (2009), Yamagiwa et al. (2018), Fratz et al. (2021) and Hase et al. (2021) for exemplary references). Similar to synthetic wavelength interferometry, the required synthetic phase maps can be captured in multiple ways, including methods that use spatial heterodyning, frequency heterodyning or phase shifting (Kreis et al. 1997; Kim 2010; Zhou et al. 2022).

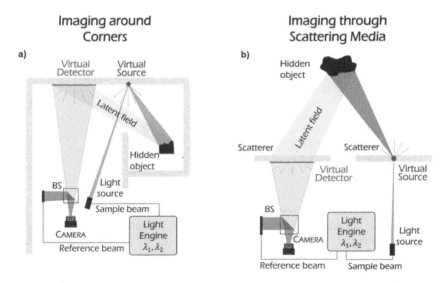

Figure 8.6. *Schematic setups for NLoS imaging around corners (a) and NLoS imaging through scatterers (b) with synthetic wavelength holography: A spot on the wall/scatterer (the "virtual source" (VS)) is illuminated by the sample beam. The VS scatters light towards the hidden object. A small fraction of the light incident on the object is scattered back to the wall/scatterer, where it hits the "virtual detector" (VD). By imaging the VD with a camera, a synthetic hologram is captured at the VD surface. Details about the light engine are specified in Willomitzer et al. (2021b). For a color version of this figure, see www.iste.co.uk/funatomi/computational.zip*

This section will focus on a specific novel application of synthetic wavelength holography which has been recently published in Willomitzer et al. (2021a): Imaging hidden objects around corners or through scattering media (which can be collectively referred to as "non-line-of-sight" (NLoS) imaging) at high resolution. As before, *spectral diversity*, necessary to form the synthetic field $E(\Lambda)$, is used to mitigate the phase randomization in a speckle field. However, one of the new and important insights that have been demonstrated in Willomitzer et al. (2019a, 2019b, 2021a) is that the synthesis of a synthetic field $E(\Lambda)$ from two speckled optical fields is still possible if *more than one* scattering process (e.g. in form of more than one optically rough surface) is involved. In this case, *all* introduced microscopic pathlength variations contribute to the spectral decorrelation, and the synthetic wavelength has to be adjusted accordingly (see section 8.5).

SWH Image Formation

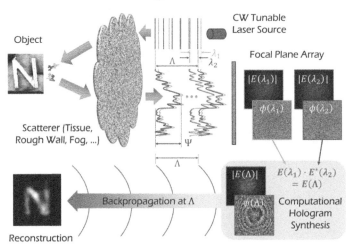

Figure 8.7. *Synthetic wavelength holography (SWH) image formation: The scene/object is illuminated with a continuous-wave tunable laser at two slightly different wavelengths λ_1 and λ_2. Each field $E(\lambda_1)$, $E(\lambda_2)$ undergoes multiple scattering processes in or at the scatterer (which could be a wall, tissue, fog, etc.) and the rough object surface. The introduced pathlength variation Ψ leads to a randomization of the captured fields $E(\lambda_1)$ and $E(\lambda_2)$ and their respective phase maps $\phi(\lambda_1)$ and $\phi(\lambda_2)$. However, computational mixing of the speckled fields via equation [8.4] yields a complex-valued hologram $E(\Lambda)$ of the object at a synthetic wavelength Λ (equation [8.5]). The object is reconstructed by backpropagating $E(\Lambda)$ with the synthetic wavelength Λ. For a color version of this figure, see www.iste.co.uk/funatomi/computational.zip*

Over the years, many techniques have been proposed to image "hidden" objects, obscured from direct view (Freund 1990; Yoo and Alfano 1990; Wang et al. 1991;

Dunsby and French 2003; Yaqoob et al. 2008; Ntziachristos 2010; Vellekoop et al. 2010; Xu et al. 2011; Bertolotti et al. 2012; Mosk et al. 2012; Singh et al. 2014; Hoshi and Yamada 2016; Doktofsky et al. 2020; Yoon et al. 2020). Now, the problem is enjoying renewed attention. A solution can lead to potential applications in autonomous navigation, industrial inspection, planetary exploration or early-warning systems for first-responders (Katz et al. 2012, 2014; Velten et al. 2012; Heide et al. 2014; Gariepy et al. 2016; O'Toole et al. 2018; Faccio 2019; Lindell et al. 2019; Liu et al. 2019; Sava and Asphaug 2019; Faccio et al. 2020; Lindell and Wetzstein 2020; Metzler et al. 2021; Batarseh et al. n.d.). Potential application scenarios are imaging through deep turbulence or fog, imaging through optically opaque barriers like a skull, face identification around corners and many more.

8.4.1. *Imaging around corners with synthetic wavelength holography*

The experimental setup used to image hidden objects around corners is schematically depicted in Figure 8.6a. A reflective scatterer such as a wall is used to scatter light towards the hidden object and to intercept the back-scattered light. The respective portions of the wall are identified as "virtual source" (VS) and "virtual detector" (VD). The nomenclature "VS" and "VD" allude to the fact that the method indeed synthesizes a virtual computational holographic camera (with source and detector) on the wall. The position of this virtual camera on the wall is chosen in a way that the hidden object resides *in direct line of sight* of this virtual camera. The FPA camera is focused on the VD portion of the wall, where the backscattered fields $E(\lambda_1)$ and $E(\lambda_2)$ are recorded. Eventually, the synthetic field $E(\Lambda)$ can be assembled, for example, by mixing (see equation [8.4]). As $E(\Lambda)$ is captured at the VD position on the wall and the light waves are subject to an additional propagation between the hidden object and the wall, $E(\Lambda)$ represents now a *hologram of the hidden object at the synthetic wavelength* Λ. Again, if Λ is chosen sufficiently large (see discussion in next section), the synthetic hologram $E(\Lambda)$ is not affected by speckle and shows a clear structure. A three-dimensional representation of the hidden object can be reconstructed by numerically backpropagating the assembled synthetic wavelength hologram $E(\Lambda)$ with the synthetic wavelength Λ. Figure 8.7 depicts the generalized image formation process.

Figure 8.8(a)–(d) displays the phase $\phi(\Lambda)$ of the computationally assembled synthetic wavelength hologram, for a specific set of synthetic wavelengths Λ. For each measurement, it is possible to recover phase information, despite the pronounced multiple scattering at the object surface and wall. A 2D image of the final reconstruction is shown in Figures 8.8(e)–(h). The images show the squared magnitude of the backpropagated synthetic holograms at the standoff distance of the object (character "N"). As expected, this is the backpropagation distance that produces the "sharpest" image of the character for the respective synthetic wavelength. The experiments demonstrate the ability to recover an image of a small character "N" (dimensions $15\ mm \times 20\ mm$, see Figure 8.9(a)) despite being obscured from direct view. Furthermore, the *phase information* encapsulated in the synthetic hologram allows for an evaluation of the depth location of the hidden object within the obscured volume.

As it can be seen in Figure 8.8(e)–(h), the lateral resolution of the reconstruction improves with decreasing synthetic wavelength Λ. This behavior is in complete agreement with results from classical holography. It confirms again that the synthetic wave, although a computational construct, has distinct characteristics that it shares with a physical wave at the respective wavelength Λ. The fundamental resolution limits of synthetic wavelength holography will be discussed in the next section.

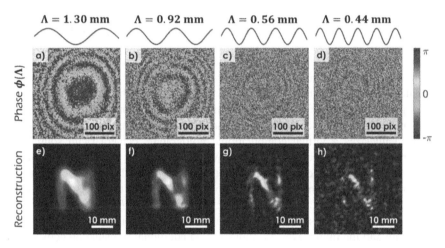

Figure 8.8. *Imaging around corners with synthetic wavelength holography (experiment). The character "N" (dimensions $\sim 15\ mm \times 20\ mm$, photo shown in Figure 8.9a) is imaged around the corner at four different synthetic wavelengths Λ. The setup schematic is shown in Figure 8.6a. a–d) Phase maps of synthetic holograms captured at the VD surface. e–h) Respective reconstructions, displaying the squared magnitude of the backpropagated synthetic fields at the standoff distance of the character. The lateral resolution of the reconstructions increases with decreasing synthetic wavelength. However, the speckle artifacts increase as well due to the decorrelation of the two optical fields at λ_1 and λ_2. For a color version of this figure, see www.iste.co.uk/funatomi/computational.zip*

8.4.2. *Imaging through scattering media with synthetic wavelength holography*

In addition to imaging hidden objects around corners, synthetic wavelength holography can also be used to image hidden objects through a scattering medium, like skin, or fog. Although such a scenario is often considered the transmissive equivalent to the reflective "imaging around the corner" scattering problem, the scattering can be much more severe in the transmissive case. Compared to the two to three distinct surface scattering processes for imaging around corners, light is commonly subject to much larger pathlength variations when penetrating through a volume scatterer. This makes imaging through strongly scattering media a much harder computational imaging problem.

The set of experiments shown in Figure 8.10 demonstrates the versatility of the synthetic wavelength holography principle by recovering holograms of objects hidden behind a scattering medium. The schematic setup is illustrated in Figure 8.6b. The setup is equivalent to the setup depicted in Figure 8.6a, but now adjusted to transmissive scattering.

In a first experiment, a small character "U" (dimensions $15\ mm \times 20\ mm$) is imaged through a 220 grit diffuser (Figure 8.9(b) left). The diffuser scatters light only at one of its surfaces. This means that this first experiment is indeed the transmissive equivalent to the "imaging around corners" scenario discussed above, as the total scattering for this scenario can be again described as two to three distinct surface scattering processes. Similar to the "imaging around corners" configuration, illumination and FPA camera are focused on the surface of the scatterer and form a VS and VD. Again, the synthetic wavelength hologram $E(\Lambda)$ is captured at the VD surface, and the object is reconstructed via numerical backpropagation at the synthetic wavelength Λ. The holographic reconstructions of the character "U" are shown in Figure 8.10(a)–(d). Again, the lateral resolution of the reconstruction increases with decreasing synthetic wavelength Λ.

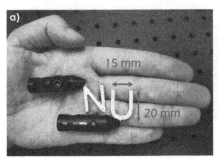

Figure 8.9. *a) Photo of the two objects (character "N" and "U") used for the shown synthetic wavelength holography experiments. Each character has the dimensions $\sim 15\ mm \times 20\ mm$ (see hand for size comparison). b) Scatterers obscuring the "U" character for the imaging through scattering media experiment shown in Figure 8.10: A 220-grit ground glass diffuser and a milky white plastic plate of $\sim 4\ mm$ thickness, both placed $\sim 1\ cm$ over a printed checker pattern to demonstrate the degradation in visibility. For a color version of this figure, see www.iste.co.uk/funatomi/computational.zip*

In a second experiment, the ground glass diffuser in the imaging path is swapped with a $4\ mm$ thick milky plastic plate (Figure 8.9(b) right). The plate can be considered a strong volume scatterer, as the light undergoes several scattering processes during its transmission. This becomes apparent in Figure 8.9(b) by comparing the degraded visibility of a checkerboard pattern, which is viewed through the plastic plate and the 220 grit diffuser for comparison. However, despite pronounced scattering in the plastic plate, the character "U" can be reconstructed for synthetic wavelengths exceeding $360\mu m$, as shown in Figures 8.10(e)–(h). This confirms the ability to recover image information at visibility levels far below the perceptual threshold. A comparison of the reconstructions for the plastic plate and the diffuser reveals only a marginal change in the smallest achievable synthetic wavelength. This observation can be explained by the fact that the visibility of ballistic light paths decays exponentially with the propagation distance through a scattering volume (in accordance with Beer's law 1852), whereas the lateral resolution for synthetic wavelength holography is linearly related to the choice of Λ, as discussed in the next section.

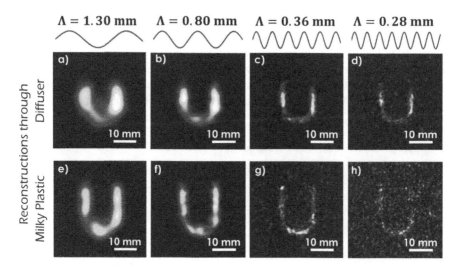

Figure 8.10. *Imaging through scatterers with synthetic wavelength holography (experiment). The imaged object is a character "U" (dimensions $\sim 15\ mm \times 20\ mm$, photo shown in Figure 8.9a). a–d) Reconstructions of measurements taken through a 220-grit ground glass diffuser (shown in Figure 8.9b) for different synthetic wavelengths Λ. e–h) Reconstructions of measurements taken through a milky white plastic plate with $\sim 4\ mm$ thickness (shown in Figure 8.9b) for different synthetic wavelength Λ. In analogy to Figure 8.8, smaller synthetic wavelengths Λ deliver higher lateral resolution but increased speckle artifacts. The larger pathlength difference in the plastic plate leads to greater decorrelation. For a color version of this figure, see www.iste.co.uk/funatomi/computational.zip*

8.4.3. *Discussion and comparison with the state of the art*

As mentioned before, the problem of "non-line-of-sight" imaging, which is here collectively referred to as the task of imaging around corners and imaging through scattering media, has recently enjoyed renewed attention. Besides the introduced synthetic wavelength holography technique, existing active methods are either based on ToF imaging ("transient techniques"; see also earlier chapters of this book) or exploit spatial correlations in the scattered optical fields, i.e. the so-called spatial (or angular) "memory effect" (Freund et al. 1988; Goodman 2007).

Recent work in the area of ToF-based techniques using transients has demonstrated results with centimeter-scale lateral resolution over a $\sim 1m \times 1m \times 1m$ working volume, and in select cases providing near-real-time reconstructions. However, many approaches rely on raster scanning large areas on VS and/or VD whose dimensions are comparable to the obscured volume (Velten et al. 2012; Heide et al. 2014; O'Toole et al. 2018; Faccio 2019; Lindell et al. 2019; Liu et al. 2019; Nam et al. 2021).

Spatial correlation-based techniques allow for the highest lateral resolution of object reconstructions ($< 100\mu m$ at $1m$ standoff). Moreover, the probing area on the intermediary VS/VD surface can be less than a few centimeters. These benefits, however, come at the expense of a highly restricted angular field of view ($< 2°$), as determined by the angular decorrelation of scattered light. This angular memory effect limits not only the field of view but also the maximal possible size of the measured object, which is not allowed to exceed the respective working volume (Katz et al. 2014; Singh et al. 2014; Edrei and Scarcelli 2016; Balaji et al. 2017; Viswanath et al. 2018; Rangarajan et al. 2019; Metzler et al. 2020).

The wide disparity in achievable field of view and resolution of other NLoS imaging techniques can limit their usability. In contrast, the introduced synthetic wavelength holography technique allows for a combination of capabilities that is, to the author's knowledge, currently unmatched by the state of the art (Willomitzer et al. 2021a). The respective attributes (also shown in Figure 8.11) are:

Small probing area: many transient-based NLoS schemes require probing areas (VD or VS sizes) with dimensions around $1m \times 1m$, which limits their ability to detect hidden objects in confined spaces. Synthetic wavelength holography provides the ability to image obscured objects by simultaneously illuminating and observing a small area ($58\ mm \times 58\ mm$ for the shown experiments);

Wide angular field of view: angular memory effect-based approaches are limited to highly restricted fields of view ($< 2°$ for drywall). As a holographic method, synthetic wavelength holography provides the ability to recover obscured objects over a nearly hemispherical field of view that far exceeds the limited angular extent of the memory effect;

High spatial resolution: transient or ToF camera-based approaches generally produce rather low spatial resolutions ($\sim$ cm), due to the long modulation wavelengths. Synthetic wavelength holography provides the ability to resolve small features on obscured objects (up to $< 1mm$ in the shown experiments);

High temporal resolution: many transient-based approaches rely on point-wise raster-scanning. Synthetic wavelength holography is able to recover full field holograms of the obscured object using off-the-shelf FPA technology. The synthetic wavelength holography principle even allows for single-shot acquisition.

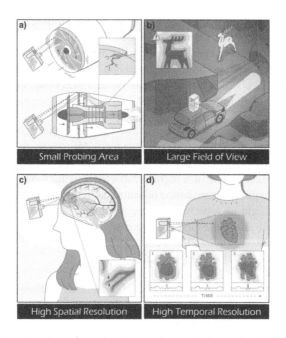

Figure 8.11. *Key attributes of synthetic wavelength holography (SWH) and potential future applications. The SWH approach combines four key attributes, each shown in a potential future NLoS application example: in each scenario, a scattering surface or medium is used to indirectly illuminate and intercept light scattered by the hidden object. a) A small probing area allows for the inspection of defects in tightly confined spaces, for example, in running aircraft engines. b) A wide angular field of view allows the measurement/detection of hidden objects without previous knowledge of their position as, for example, important when navigating in degraded visual environments. c) High spatial resolution allows for the measurement of small structures, such as non-invasive imaging of brain vessels through the skull. d) High temporal resolution allows the imaging of objects in motion, for example, to potentially discern cardiac arrhythmia through the chest. Synthetic wavelength holography combines all of these four attributes in one single approach to NLoS imaging. For a color version of this figure, see www.iste.co.uk/funatomi/computational.zip*

This concludes the description of synthetic wavelength holography. The following section will discuss an important topic that has been largely omitted so far: *fundamental performance limits of synthetic wavelength imaging* in general. It already became clear from the preceding explanations that the size of the synthetic wavelength Λ, as well as the severity of scattering imparted by the rough surfaces and the scattering media, plays an important role in this discussion. It will be explained how these parameters and other system parameters like the VD size influence the resolution of the respective methods, and the related trade-off space will be explored.

8.5. Fundamental performance limits of synthetic wavelength imaging

It has been mentioned at the beginning of this chapter that the depth precision of ToF-based methods like interferometry or CWAM ToF imaging is directly proportional to the modulation wavelength (equation [8.2]). The discussion has been further backed up by the notion that a generalized wave concept can be indeed used to discuss general trends of the performance (such as the trend of increasing precision with decreasing modulation wavelength) (Gupta et al. 2015; Reza et al. 2019; Willomitzer et al. 2021a). For this general discussion, the *origin* of the wave (complex electromagnetic field, optical beat, amplitude modulation, etc.) became of secondary importance. The aim for higher depth precision and/or lateral resolution was one of the main motivation points to switch from a meter-sized amplitude modulated wave (as used in CWAM ToF cameras) to a synthetic wave with a much smaller wavelength Λ. However, it also became clear that Λ cannot become indefinitely small: As seen in single-wavelength interferometry, measurements at a very small (i.e. optical) wavelength are subject to phase randomization in a speckle field. The inevitable question:

How small can the synthetic wavelength Λ become, and what are the related limits?

A well-known model to describe the severity of scatter and the related formation of a speckle field for *optical waves* is the so-called "Rayleigh quarter wavelength criterion" (Rayleigh 1879). According to this criterion, an optical field that is reflected off a rough surface or surpasses a scattering medium forms speckle if the maximal pathlength variation Ψ introduced by the scatterer and the geometry exceeds $1/4$ of the wavelength λ within one object-sided diffraction disk[2]:

$$\Psi < \frac{\lambda}{4} \quad \text{for speckle-free imaging} \qquad [8.7]$$

This criterion is intuitively understandable: A light field that exhibits random pathlength variations of maximal $\Psi = \lambda/4$ while propagating from the source to

---

2. The size of the object-sided diffraction is the size of the image-sided diffraction disk divided by the magnification of the optical imaging system. It can be seen as the projection of an image-sided diffraction disk onto the object surface.

the object, and another $\Psi = \lambda/4$ for propagating from the object back to the detector is subject to spatially varying random phase shifts of maximal π. A phase shift of π is just large enough for fully destructive interference, i.e. an interference pattern at full contrast (Häusler 2004; Häusler and Ettl 2011).

It should be noted for the sake of completeness that the Rayleigh quarter wavelength criterion is also frequently used to define whether a surface is optically rough or not. In this case, a surface can be defined as "rough" if the pathlength variations within one object-sided diffraction disk exceed $\lambda/4$. This leads to the interesting fact that the roughness definition depends not only on the surface itself but also on the geometry (tilt) and the resolving power of the optical imaging apparatus. The direct follow-up question is *which statistical definition of "roughness" should be used*. Strictly spoken, the first destructive interferences (dark speckles) should appear if the peak-to-valley surface roughness R_p exceeds $\lambda/4$. However, as the microscopic height values of most surfaces are normal distributed, this will happen very infrequently, which is the reason why most definitions use the RMS surface roughness σ_h instead. The RMS surface roughness σ_h will also be used here for further definitions.

So far, the discussion has been focused on imaging at the optical wavelength λ. Although some analogies between the synthetic wavelength and the optical wavelength have been drawn already, related limit considerations for the synthetic wavelength Λ seem less intuitive. This is particularly true for such realizations of synthetic wavelength imagers that acquire both optical fields $E(\lambda_1)$ and $E(\lambda_2)$ in a sequential fashion (e.g. dual-wavelength heterodyne interferometers). In this case, the two optical fields never physically beat together, and the synthetic wave becomes a purely computational construct that is generated in the computer via post-processing (equation [8.4]). Nevertheless, it has been shown in Willomitzer et al. (2021b) that the Rayleigh quarter wavelength criterion *can still be applied* to measurements at the synthetic wavelength. For artifact-free imaging, this means that the smallest possible synthetic wavelength Λ can be estimated via

$$\Psi < \frac{\Lambda}{4} \quad \Rightarrow \quad \Lambda > 4 \cdot \Psi. \qquad [8.8]$$

Reconstructions obtained from measurements at Λ close to or smaller than 4Ψ start to exhibit speckle-like artifacts (see, for example, Figure 8.8(h) or Figure 8.10(g) and (h)) which become more severe with decreasing Λ. Indeed, these artifacts can be interpreted as "synthetic speckle!"

This is a ***remarkable analogy*** which requires further explanation: It has been discussed that an artifact-free synthetic phase map $\phi(\Lambda)$ can be obtained if the two optical (speckle) fields $E(\lambda_1)$ and $E(\lambda_2)$ are sufficiently correlated. For the synthetic wavelength imaging methods introduced here, the two object beams originate from the same point (same fiber tip), and the respective speckle patterns change their appearance with different wavelengths. Hence, this correlation can be understood as a *spectral correlation*. The related effect of spectral decorrelation shows a strong analogy to the "memory effect" for angular decorrelation (Freund et al. 1988; Goodman 2007). In fact, it can be understood as "spectral memory effect" (Goodman 2007; Willomitzer et al. 2021a). The transition from correlated to uncorrelated fields is, of course, a fluent and not binary process, meaning that different criteria exist regarding whether two fields can be treated as "correlated" or not (Goodman 2007). Applying similar decorrelation criteria used to define the isoplanatic angle for the angular memory effect leads to a maximal wavelength separation $|\lambda_1 - \lambda_2|$, which is dependent on the maximal pathlength variation Ψ and the "starting" wavelength λ_1 (Goodman 2007). Combined with equation [8.5], this results in equation [8.8]. The concrete calculations and further details can be found in Willomitzer et al. (2021b).

What does the limitation of equation [8.8] mean for synthetic wavelength interferometry and synthetic wavelength holography? In both cases, the maximal pathlength variation Ψ introduced by the scatterer restricts the smallest possible synthetic wavelength Λ.

For synthetic wavelength interferometry on pure surface scatterers, the pathlength variations are introduced by the surface roughness and the geometry. A beam that hits the rough surface from an oblique angle is subject to different pathlength variations than a beam at normal incidence. In the limit case of scanning the surface at a single point with beam diameter $\lambda_1 \approx \lambda_2$ and normal incidence, the pathlength variation converges into the RMS surface roughness ($\Psi_{Int} \rightarrow \sigma_h$) (Willomitzer et al. 2021b). This results in the criterion

$$\Lambda > 4 \cdot \Psi_{Int} > 4\sigma_h. \tag{8.9}$$

For imaging around the corner with synthetic wavelength holography, the coherent field exhibits *at least* two scattering processes on a rough surface (the wall). Assuming again no subsurface scattering, normal incidence and a VS/VD diameter of $\lambda_1 \approx \lambda_2$, the pathlength variation converges to $\Psi_{Hol1} \rightarrow 2 \cdot \sigma_h$ (Willomitzer et al. 2021b), which means that

$$\Lambda > 4 \cdot \Psi_{Hol1} > 8 \cdot \sigma_h. \tag{8.10}$$

For imaging through scattering media with synthetic wavelength holography, it is assumed that the light traverses two times (round trip) through a volume scattering medium with thickness L and transport mean free path l^*. In the described limit case, the pathlength variation converges to $\Psi_{Hol2} \rightarrow 2 \cdot L^2/l^*$ (Willomitzer et al. 2021b), leading to

$$\Lambda > 4 \cdot \Psi_{Hol2} > 8 \cdot \frac{L^2}{l^*} \tag{8.11}$$

Having established these bounds, it can finally be discussed how the size of the smallest possible synthetic wavelength is tied to the resolution of the respective method. It is known from metrology literature that well-calibrated single-wavelength interferometers can reach impressive depth precisions of $\delta z = \lambda/100$ (or even better) on specular surfaces. A depth error of, for example, $\lambda/100$ means that the optical phase can be determined with a precision of $\delta\phi = 2\pi/100$. However, caution is advised in extrapolating this result to synthetic wavelength interferometry on rough surfaces and assuming that precisions of $\delta z = \Lambda/100$ can easily be reached as well! For every point in the camera image, waves are accumulated from a small surface area on the object (the object-sided diffraction disk). Per previous definition, the surface topography of rough surfaces varies significantly within this object-sided diffraction disk. The consequence is a large uncertainty in the measured optical phase, which translates to the synthetic phase. This is the reason why the precision of multi-wavelength interferometry on rough surfaces is ultimately bound by the surface roughness (Dresel et al. 1992; Häusler and Neumann 1993; Häusler 1994; Ettl et al. 1998) and can typically not reach the very high phase precision $\delta\phi$ of single-wavelength interferometry on smooth surfaces. Refer to Häusler and Willomitzer (2022) for further information about this topic. Li et al. (2021) contains a precision evaluation of the introduced synthetic wavelength interferometer for macroscopic objects with rough surfaces.

For the introduced technique of synthetic wavelength holography to image around corners or through scattering media, the lateral resolution and longitudinal localization uncertainty are influenced by the geometry of the setup and the size of the synthetic wavelength Λ . To reconstruct an image of the hidden object, the field captured at the VD is back-propagated in the hidden volume. In an analogy to light focusing through a lens, we can approximate a numerical aperture by the radius of the virtual detector $D_{VD}/2$ divided by the standoff distance z. Analog to classical optics, the lateral resolution δx can then be approximated as (Willomitzer et al. 2021a)

$$\delta x \approx \frac{\Lambda z}{D_{VD}}. \qquad\qquad [8.12]$$

Figure 8.12 displays experimental results to evaluate the lateral resolution δx around the corner. A surface patch on the rough wall (the VD) was illuminated by a point-like light source (a fiber tip), which is located in the hidden volume. The patch has a diameter of $D_{VD} = 58\ mm$ and the fiber tip is located at a standoff distance of $z = 95\ mm$. After illuminating the VD with two wavelengths λ_1 and λ_2 and forming the synthetic hologram, the point-like light source was reconstructed by numerical back-propagation at the synthetic wavelength Λ. It can be seen that the diameter of the "synthetic diffraction disk" decreases with decreasing synthetic wavelength and that the general trend closely follows the theoretical expectation from equation [8.12]. For the measurement at $\Lambda = 0.28\ mm$, the experimentally evaluated lateral resolution around the corner is *below* 1 mm (Figure 8.12c,f).

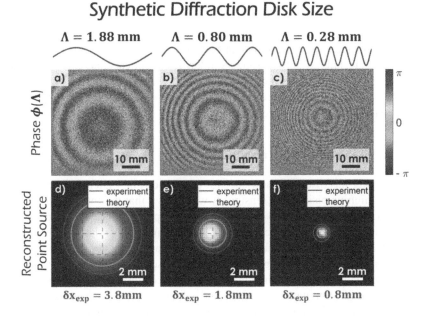

Figure 8.12. *Measurement of the "synthetic diffraction disk" via reconstruction of an obscured point source for three different synthetic wavelengths (experiment). a–c) Phase of the synthetic holograms captured at the VD surface. The diameter of the VD is $D_{VD} = 58\ mm$ and the standoff distance of the point source (fiber tip) is $z = 95\ mm$. d–e) Reconstruction of the "synthetic diffraction disk". In agreement with classical optics (equation [8.12]), the disk size varies linearly with the wavelength (in this case, the synthetic wavelength). The experimental value is close to the theoretical expectation. For $\Lambda = 0.28\ mm$ (f), the point source is reconstructed with sub-mm lateral resolution. For a color version of this figure, see www.iste.co.uk/funatomi/computational.zip*

Besides the discussed lateral resolution δx, the longitudinal localization uncertainty δz behaves also as expected from classical optics and goes proportional to $\delta z \sim \Lambda z^2 / D_{VD}^2$. The exact value is also strongly dependent on the noise properties of the system. This is, of course, the case for all of the introduced precision and resolution measures, although it was not discussed in detail in this chapter. The reader is referred to Willomitzer et al. (2021a) and Li et al. (2021) and future publications of the author's research group.

Similar to what has been discussed for synthetic wavelength interferometry, the longitudinal localization uncertainty of synthetic wavelength holography can be improved by the use of *multiple* synthetic wavelengths. A related experiment ("synthetic pulse holography") is described in Willomitzer et al. (2021a): The hidden scene is interrogated at multiple synthetic wavelengths, and the respective back-propagated (complex) fields are superpositioned coherently in the computer. As this procedure mimics the computational synthesis of a "synthetic pulse train", the longitudinal localization uncertainty can be significantly improved. It should also be noted that this technique shows striking similarities to optical coherence tomography (OCT) or white light interferometry (WLI), which will be further investigated in the future. Again, the reader is referred to Willomitzer et al. (2021a, 2019b) for additional information and further references.

8.6. Conclusion and future directions

This chapter has discussed how spectral correlations in scattered light fields can be used for high-precision ToF sensing. It was shown that the introduced synthetic wavelength imaging techniques are able to extract phase information from optical speckle fields, which are subject to heavy scattering. The related techniques of synthetic wavelength interferometry and synthetic wavelength holography have long been known in optical metrology and have been used, for example, for industrial inspection and surface testing. This chapter has outlined how to apply these methods to novel problems in computational imaging and computer vision, such as high-precision ToF imaging for AR/VR and medical applications, or the problem of imaging objects around corners and through scattering media. It has been demonstrated that the introduced techniques can achieve very high depth precision and lateral resolution (in many cases much higher than established methods), and show a unique combination of other valuable attributes, such as a wide field of view, or high temporal resolution.

However, the discussions in the last section revealed that the introduced synthetic wavelength imaging approaches are not without limitations. The good news: Physical limitations often come in the shape of uncertainty products! This makes it possible to optimize a technique towards a specific quantity (e.g. speed or resolution) by trading in information less critical for the targeted application. These limitations and their trade-off spaces will be further explored and exploited by the author's research group in the future.

8.7. Acknowledgment

Research is always a team effort! Hence, it should be emphasized that the presented results and insights are the product of a continuous process that involves many students and colleagues and have not been produced and derived by the author alone. Amongst others, these students and colleagues are Muralidhar M. Balaji, Fengqiang Li, Marc P. Christensen, Manuel Ballester and Heming Wang. From the involved colleagues, the author would particularly thank Oliver Cossairt, Prasanna Rangarajan and Gerd Häusler for proofreading this chapter, for their valuable comments and for the always exciting discussions.

8.8. References

Arold, O., Ettl, S., Willomitzer, F., Häusler, G. (2014). Hand-guided 3D surface acquisition by combining simple light sectioning with real-time algorithms. *arXiv:1401.1946*.

Balaji, M.M., Rangarajan, P., MacFarlane, D., Corliano, A., Christensen, M.P. (2017). Single-shot holography using scattering surfaces. *Imaging and Applied Optics 2017 (3D, AIO, COSI, IS, MATH, pcAOP)*. Optical Society of America, p. CTu2B.1 [Online]. Available at: http://www.osapublishing.org/abstract.cfm? URI=COSI-2017-CTu2B.1.

Batarseh, M., Sukov, S., Shen, Z., Gemar, H., Rezvani, R., Dogariu, A. (2018). Passive sensing around the corner using spatial coherence. *Nature Communications*, 9, 3629.

Beer, A. (1852). Bestimmung der absorption des rothen lichts in farbigen fluessigkeiten. *Annalen der Physik*, 162(5), 78–88 [Online]. Available at: https:// onlinelibrary.wiley.com/doi/abs/10.1002/andp.18521620505.

Bertolotti, J., Putten, E., Blum, C., Lagendijk, A., Vos, W., Mosk, A. (2012), Non-invasive imaging through opaque scattering layers. *Nature*, 491, 232–234.

Born, M. and Wolf, E. (2013). *Principles of Optics: Electromagnetic Theory of Propagation, Interference and Diffraction of Light*. Elsevier.

Cheng, Y.-Y. and Wyant, J.C. (1985). Multiple-wavelength phase-shifting interferometry. *Applied Optics*, 24(6), 804–807.

Collis, R. (1970). Lidar. *Applied Optics*, 9(8), 1782–1788.

Dändliker, R., Thalmann, R., Prongué, D. (1988). Two-wavelength laser interferometry using superheterodyne detection. *Optics Letters*, 13(5), 339–341 [Online]. Available at: http://ol.osa.org/abstract.cfm?URI=ol-13-5-339.

De Groot, P.J. (1994). Extending the unambiguous range of two-color interferometers. *Applied Optics*, 33(25), 5948–5953.

De Groot, P.J. and McGarvey, J. (1992). Chirped synthetic-wavelength interferometry. *Optics Letters*, 17(22), 1626–1628.

Doktofsky, D., Rosenfeld, M., Katz, O. (2020). Acousto optic imaging beyond the acoustic diffraction limit using speckle decorrelation. *Communications Physics*, 3, 5.

Dresel, T., Häusler, G., Venzke, H. (1992). Three-dimensional sensing of rough surfaces by coherence radar. *Applied Optics*, 31(7), 919–925 [Online]. Available at: http://ao.osa.org/abstract.cfm?URI=ao-31-7-919.

Dunsby, C. and French, P. (2003). Techniques for depth-resolved imaging through turbid media including coherence-gated imaging. *Journal of Physics D: Applied Physics*, 36, 207–227.

Edrei, E. and Scarcelli, G. (2016). Optical imaging through dynamic turbid media using the Fourier-domain shower-curtain effect. *Optica*, 3(1), 71–74 [Online]. Available at: http://www.osapublishing.org/optica/abstract.cfm?URI= optica-3-1-71.

Ettl, P., Schmidt, B.E., Schenk, M., Laszlo, I., Häusler, G. (1998). Roughness parameters and surface deformation measured by coherence radar. *International Conference on Applied Optical Metrology*, vol. 3407. SPIE, pp. 133–140.

Ettl, S., Rampp, S., Fouladi-Movahed, S., Dalal, S.S., Willomitzer, F., Arold, O., Stefan, H., Häusler, G. (2013). Improved EEG source localization employing 3D sensing by "flying triangulation". *Videometrics, Range Imaging, and Applications XII; and Automated Visual Inspection*, vol. 8791. SPIE, pp. 194–200.

Faber, C., Olesch, E., Krobot, R., Häusler, G. (2012). Deflectometry challenges interferometry: The competition gets tougher! *Proceedings of SPIE*, 8493 [Online]. Available at: https://doi.org/10.1117/12.957465.

Faccio, D. (2019). Non-line-of-sight imaging. *Optics & Photonics News*, 30(1), 36–43 [Online]. Available at: http://www.osa-opn.org/abstract.cfm?URI=opn-30-1-36.

Faccio, D., Velten, A., Wetzstein, G. (2020). Non-line-of-sight imaging. *Nature Reviews Physics*, 2, 318–327.

Falaggis, K., Towers, D.P., Towers, C.E. (2009). Multiwavelength interferometry: Extended range metrology. *Optics Letters*, 34(7), 950–952.

Fercher, A.F., Hu, H.Z., Vry, U. (1985). Rough surface interferometry with a two-wavelength heterodyne speckle interferometer. *Applied Optics*, 24(14), 2181–2188 [Online]. Available at: http://ao.osa.org/abstract.cfm?URI=ao-24-14-2181.

Foix, S., Alenya, G., Torras, C. (2011). Lock-in time-of-flight (ToF) cameras: A survey. *IEEE Sensors Journal*, 11(9), 1917–1926.

Fratz, M., Seyler, T., Bertz, A., Carl, D. (2021). Digital holography in production: An overview. *Light: Advanced Manufacturing*, 2(3), 283–295.

Freund, I. (1990). Looking through walls and around corners. *Physica A: Statistical Mechanics and its Applications*, 168(1), 49–65 [Online]. Available at: http://www.sciencedirect.com/science/article/pii/037843719090357X.

Freund, I., Rosenbluh, M., Feng, S. (1988). Memory effects in propagation of optical waves through disordered media. *Physical Review Letters*, 61, 2328–2331.

Fu, Y., Pedrini, G., Hennelly, B.M., Groves, R.M., Osten, W. (2009). Dual-wavelength image-plane digital holography for dynamic measurement. *Optics and Lasers in Engineering*, 47(5), 552–557.

Gariepy, G., Tonolini, F., Henderson, R., Leach, J., Faccio, D. (2016). Detection and tracking of moving objects hidden from view. *Nature Photonics*, 10, 23–26.

Goodman, J. (2007). *Speckle Phenomena in Optics: Theory and Applications*. Roberts and Company Publishers.

Gupta, M., Nayar, S.K., Hullin, M.B., Martin, J. (2015). Phasor imaging: A generalization of correlation-based time-of-flight imaging. *ACM Transactions on Graphics (ToG)*, 34(5), 1–18.

Hase, E., Tokizane, Y., Yamagiwa, M., Minamikawa, T., Yamamoto, H., Morohashi, I., Yasui, T. (2021). Multicascade-linked synthetic-wavelength digital holography using a line-by-line spectral-shaped optical frequency comb. *Optics Express*, 29(10), 15772–15785.

Häusler, G. (1994). Range sensing of the first, the second, and the third kind. *Proceedings of EOS TOPICAL MEETING Optical Metrology and Nanotechnology*, pp. 27–30.

Häusler, G. (2004). Speckle and coherence. *Encyclopedia of Modern Optics*. Academic Press, Oxford.

Häusler, G. and Ettl, S. (2011). Limitations of optical 3D sensors. *Optical Measurement of Surface Topography*. Springer.

Häusler, G. and Neumann, J. (1993). Coherence radar: An accurate 3D sensor for rough surfaces. *Optics, Illumination, and Image Sensing for Machine Vision VII*, vol. 1822. SPIE, pp. 200–205.

Häusler, G. and Willomitzer, F. (2022). Reflections about the holographic and non-holographic acquisition of surface topography: Where are the limits? *Light: Advanced Manufacturing*, 3(2), 1–10.

Hecht, E. (2012). *Optics*. Pearson Education India.

Heide, F., Xiao, L., Heidrich, W., Hullin, M.B. (2014). Diffuse mirrors: 3D reconstruction from diffuse indirect illumination using inexpensive time-of-flight sensors. *Proceedings of the IEEE Conference on Computer Vision and Pattern Recognition*, pp. 3222–3229.

Horn, B.K. (1990). Height and gradient from shading. *International Journal of Computer Vision*, 5(1), 37–75.

Hoshi, Y. and Yamada, Y. (2016). Overview of diffuse optical tomography and its clinical applications. *Journal of Biomedical Optics*, 21(9), 1–11 [Online]. Available at: https://doi.org/10.1117/1.JBO.21.9.091312.

Huang, L., Idir, M., Zuo, C., Asundi, A. (2018). Review of phase measuring deflectometry. *Optics and Lasers in Engineering*, 107, 247–257 [Online]. Available at: http://www.sciencedirect.com/science/article/pii/S0143816618300599.

Huntley, J.M. and Saldner, H. (1993). Temporal phase-unwrapping algorithm for automated interferogram analysis. *Applied Optics*, 32(17), 3047–3052.

Javidi, B., Ferraro, P., Hong, S.-H., De Nicola, S., Finizio, A., Alfieri, D., Pierattini, G. (2005). Three-dimensional image fusion by use of multiwavelength digital holography. *Optics Letters*, 30(2), 144–146.

Katz, O., Small, E., Silberberg, Y. (2012). Looking around corners and through thin turbid layers in real time with scattered incoherent light. *Nature Photonics*, 6, 549553.

Katz, O., Heidmann, P., Fink, M., Gigan, S. (2014). Non-invasive single-shot imaging through scattering layers and around corners via speckle correlations. *Nature Photonics*, 8(10), 784.

Kim, M.K. (2010). Principles and techniques of digital holographic microscopy. *SPIE Reviews*, 1(1), 018005.

Knauer, M.C., Kaminski, J., Häusler, G. (2004). Phase measuring deflectometry: A new approach to measure specular free-form surfaces. *Proceedings of SPIE*, 5457 [Online]. Available at: https://doi.org/10.1117/12.545704.

Kotwal, A., Levin, A., Gkioulekas, I. (2022). Swept-angle synthetic wavelength interferometry. *arXiv:2205.10655*.

Kreis, T.M., Adams, M., Jüptner, W.P. (1997). Methods of digital holography: A comparison. *Optical Inspection and Micromeasurements II*, vol. 3098. International Society for Optics and Photonics, pp. 224–233.

Lange, R. and Seitz, P. (2001). Solid-state time-of-flight range camera. *IEEE Journal of Quantum Electronics*, 37(3), 390–397.

Li, F., Willomitzer, F., Rangarajan, P., Gupta, M., Velten, A., Cossairt, O. (2018). SH-ToF: Micro resolution time-of-flight imaging with superheterodyne interferometry. *2018 IEEE International Conference on Computational Photography (ICCP)*, pp. 1–10.

Li, F., Willomitzer, F., Rangarajan, P., Cossairt, O. (2019). Mega-pixel time-of-flight imager with GHz modulation frequencies. *Imaging and Applied Optics 2019 (COSI, IS, MATH, pcAOP)*, Optical Society of America, p. CTh2A.2 [Online]. Available at: http://www.osapublishing.org/abstract.cfm?URI=COSI-2019-CTh2A.2.

Li, F., Willomitzer, F., Balaji, M.M., Rangarajan, P., Cossairt, O. (2021). Exploiting wavelength diversity for high resolution time-of-flight 3D imaging. *IEEE Transactions on Pattern Analysis and Machine Intelligence*, 43(7), 2193–2205.

Lindell, D.B. and Wetzstein, G. (2020). Three-dimensional imaging through scattering media based on confocal diffuse tomography. *Nature Communications*, 11(1), 1–8.

Lindell, D.B., Wetzstein, G., O'Toole, M. (2019). Wave-based non-line-of-sight imaging using fast f-k migration. *ACM Transactions on Graphics (SIGGRAPH)*, 38(4), 116.

Liu, X., Guillen, I., Manna, M.L., Nam, J.H., Reza, S.A., Le, T.H., Jarabo, A., Gutierrez, D., Velten, A. (2019). Non-line-of-sight imaging using phasor-field virtual wave optics. *Nature*, 572(7771), 620–623.

Mann, C.J., Bingham, P.R., Paquit, V.C., Tobin, K.W. (2008). Quantitative phase imaging by three-wavelength digital holography. *Optics Express*, 16(13), 9753–9764.

Metzler, C.A., Heide, F., Rangarajan, P., Balaji, M.M., Viswanath, A., Veeraraghavan, A., Baraniuk, R.G. (2020). Deep-inverse correlography: Towards real-time high-resolution non-line-of-sight imaging. *Optica*, 7(1), 63–71.

Metzler, C.A., Lindell, D.B., Wetzstein, G. (2021). Keyhole imaging: Non-line-of-sight imaging and tracking of moving objects along a single optical path. *IEEE Transactions on Computational Imaging*, 7, 1–12.

Mosk, A., Lagendijk, A., Lerosey, G., Fink, M. (2012). Controlling waves in space and time for imaging and focusing in complex media. *Nature Photonics*, 6, 283–292.

Nam, J.H., Brandt, E., Bauer, S., Liu, X., Renna, M., Tosi, A., Sifakis, E., Velten, A. (2021). Low-latency time-of-flight non-line-of-sight imaging at 5 frames per second. *Nature Communications*, 12(1), 1–10.

Ntziachristos, V. (2010). Going deeper than microscopy: The optical imaging frontier in biology. *Nature Methods*, 7, 603–614.

O'Toole, M., Lindell, D., Wetzstein, G. (2018). Confocal non-line-of-sight imaging based on the light-cone transform. *Nature*, 555, 338–341.

Polhemus, C. (1973). Two-wavelength interferometry. *Applied Optics*, 12(9), 2071–2074.

Rangarajan, P., Willomitzer, F., Cossairt, O., Christensen, M.P. (2019). Spatially resolved indirect imaging of objects beyond the line of sight. In *Unconventional and Indirect Imaging, Image Reconstruction, and Wavefront Sensing*, Dolne, J.J., Spencer, M.F., Testorf, M.E. (eds). International Society for Optics and Photonics, SPIE [Online]. Available at: https://doi.org/10.1117/12.2529001.

Rayleigh, L. (1879). Reprinted in his *Scientific Papers* (Cambridge University Press, 1899), vol. 1, pp. 432–435.). *Philosophical Magazine*, 8, 403.

Reza, S.A., La Manna, M., Bauer, S., Velten, A. (2019). Phasor field waves: A Huygens-like light transport model for non-line-of-sight imaging applications. *Optics Express*, 27(20), 29380–29400.

Sava, P. and Asphaug, E. (2019). Seismology on small planetary bodies by orbital laser Doppler vibrometry. *Advances in Space Research*, 64(2), 527–544 [Online]. Available at: http://www.sciencedirect.com/science/article/pii/S0273117719302777.

Schaffer, M., Grosse, M., Kowarschik, R. (2010). High-speed pattern projection for three-dimensional shape measurement using laser speckles. *Applied Optics*, 49(18), 3622–3629.

Schönberger, J.L. and Frahm, J.-M. (2016). Structure-from-motion revisited. *Conference on Computer Vision and Pattern Recognition (CVPR)*.

Schwarte, R., Xu, Z., Heinol, H.-G., Olk, J., Klein, R., Buxbaum, B., Fischer, H., Schulte, J. (1997). New electro-optical mixing and correlating sensor: Facilities and applications of the photonic mixer device (PMD). *Proceedings of SPIE*, 3100, 245–254.

Singh, A.K., Naik, D.N., Pedrini, G., Takeda, M., Osten, W. (2014). Looking through a diffuser and around an opaque surface: A holographic approach. *Optics Express*, 22(7), 7694–7701 [Online]. Available at: http://www.opticsexpress.org/abstract.cfm?URI=oe-22-7-7694.

Srinivasan, V., Liu, H.-C., Halioua, M. (1984). Automated phase-measuring profilometry of 3-D diffuse objects. *Applied Optics*, 23(18), 3105–3108.

Takeda, M. and Mutoh, K. (1983). Fourier transform profilometry for the automatic measurement of 3-D object shapes. *Applied Optics*, 22(24), 3977–3982 [Online]. Available at: http://ao.osa.org/abstract.cfm?URI=ao-22-24-3977.

Tiziani, H., Rothe, A., Maier, N. (1996). Dual-wavelength heterodyne differential interferometer for high-precision measurements of reflective aspherical surfaces and step heights. *Applied Optics*, 35(19), 3525–3533.

Vellekoop, I., Lagendijk, A., Mosk, A. (2010). Exploiting disorder for perfect focusing. *Nature Photonics*, 4, 320–322.

Velten, A., Willwacher, T., Gupta, O., Veeraraghavan, A., Bawendi, M., Raskar, R. (2012). Recovering three-dimensional shape around a corner using ultrafast time-of-flight imaging. *Nature Communications*, 1–8.

Viswanath, A., Rangarajan, P., MacFarlane, D., Christensen, M.P. (2018). Indirect imaging using correlography. *Imaging and Applied Optics 2018 (3D, AO, AIO, COSI, DH, IS, LACSEA, LS&C, MATH, pcAOP)*. Optical Society of America, p. CM2E.3 [Online]. Available at: http://www.osapublishing.org/abstract.cfm?URI=COSI-2018-CM2E.3.

Vry, U. and Fercher, A.F. (1986). Higher-order statistical properties of speckle fields and their application to rough-surface interferometry. *Journal of the Optical Society of America A*, 3(7), 988–1000 [Online]. Available at: http://josaa. osa.org/abstract.cfm?URI=josaa-3-7-988.

Wang, L., Ho, P.P., Liu, C., Zhang, G., Alfano, R.R. (1991). Ballistic 2-D imaging through scattering walls using an ultrafast optical Kerr gate. *Science*, 253(5021), 769–771 [Online]. Available at: https://science.sciencemag.org/content/253/5021/769.

Wang, K., Li, Y., Kemao, Q., Di, J., Zhao, J. (2019). One-step robust deep learning phase unwrapping. *Optics Express*, 27(10), 15100–15115.

Wang, J., Xu, B., Wang, T., Lee, W.J., Walton, M., Matsuda, N., Cossairt, O., Willomitzer, F. (2021). *VR Eye-Tracking using Deflectometry. Computational Optical Sensing and Imaging*, Optical Society of America, pp. CF2E–3.

Weitkamp, C. (2006). *Lidar: Range-Resolved Optical Remote Sensing of the Atmosphere*, vol. 102. Springer Science & Business.

Willomitzer, F. (2019). Single-shot 3D sensing close to physical limits and information limits. Dissertation, Springer Theses.

Willomitzer, F. and Häusler, G. (2017). Single-shot 3D motion picture camera with a dense point cloud. *Optics Express*, 25(19), 23451–23464 [Online]. Available at: http://www.opticsexpress.org/abstract.cfm?URI=oe-25-19-23451.

Willomitzer, F., Yang, Z., Arold, O., Ettl, S., Häusler, G. (2010). 3D face scanning with "flying triangulation". *DGaO Proc.*, 111, 18.

Willomitzer, F., Li, F., Balaji, M.M., Rangarajan, P., Cossairt, O. (2019a). High resolution non-line-of-sight imaging with superheterodyne remote digital holography. *Computational Optical Sensing and Imaging*, Optical Society of America, pp. CM2A–2.

Willomitzer, F., Rangarajan, P.V., Li, F., Balaji, M.M., Christensen, M.P., Cossairt, O. (2019b). Synthetic wavelength holography: An extension of Gabor's holographic principle to imaging with scattered wavefronts. *arXiv:1912.11438*.

Willomitzer, F., Yeh, C.-K., Gupta, V., Spies, W., Schiffers, F., Katsaggelos, A., Walton, M., Cossairt, O. (2020). Hand-guided qualitative deflectometry with a mobile device. *Optics Express*, 28(7), 9027–9038 [Online]. Available at: http://www.opticsexpress.org/abstract.cfm?URI=oe-28-7-9027.

Willomitzer, F., Rangarajan, P.V., Li, F., Balaji, M.M., Christensen, M.P., Cossairt, O. (2021a). Fast non-line-of-sight imaging with high-resolution and wide field of view using synthetic wavelength holography. *Nature Communications*, 12(1), 1–11.

Willomitzer, F., Rangarajan, P.V., Li, F., Balaji, M.M., Christensen, M.P., Cossairt, O. (2021b). Supplementary material to: Fast non-line-of-sight imaging with high-resolution and wide field of view using synthetic wavelength holography. *Nature Communications*, 12(1), 1–11.

Woodham, R.J. (1980). Photometric method for determining surface orientation from multiple images. *Optical Engineering*, 19(1), 139–144.

Wu, Y., Li, F., Willomitzer, F., Veeraraghavan, A., Cossairt, O. (2020). Wished: Wavefront imaging sensor with high resolution and depth ranging. *2020 IEEE International Conference on Computational Photography (ICCP)*.

Wyant, J.C. (2015). Interferometric optical metrology. *OSA Century of Optics* [Online]. Available at: https://opg.optica.org/books/bookshelf/osa-century-optics.cfm.

Xu, X., Liu, H., Wang, L. (2011). Time-reversed ultrasonically encoded optical focusing into scattering media. *Nature Photonics*, 5, 154.

Yamagiwa, M., Minamikawa, T., Trovato, C., Ogawa, T., Ibrahim, D.G.A., Kawahito, Y., Oe, R., Shibuya, K., Mizuno, T., Abraham, E. et al. (2018). Multicascade-linked synthetic wavelength digital holography using an optical-comb-referenced frequency synthesizer. *Optics Express*, 26(20), 26292–26306.

Yaqoob, Z., Psaltis, D., Feld, M., Yang, C. (2008). Optical phase conjugation for turbidity suppression in biological samples. *Nature Photonics*, 2, 110–115.

Yin, W., Chen, Q., Feng, S., Tao, T., Huang, L., Trusiak, M., Asundi, A., Zuo, C. (2019). Temporal phase unwrapping using deep learning. *Scientific Reports*, 9(1), 1–12.

Yoo, K.M. and Alfano, R.R. (1990). Time-resolved coherent and incoherent components of forward light scattering in random media. *Optics Letters*, 15(6), 320–322.

Yoon, S., Kim, M., Jang, M., Choi, Y., Choi, W., Kang, S., Choi, W. (2020). Deep optical imaging within complex scattering media. *Nature Reviews Physics*, 2(3), 141–158.

Zhou, H., Hussain, M.M., Banerjee, P.P. (2022). A review of the dual-wavelength technique for phase imaging and 3D topography. *Light: Advanced Manufacturing*, 3(1), 1–21.

PART 3

Polarimetric Imaging and Processing

9

Polarization-Based Shape Estimation

Daisuke Miyazaki

Department of Intelligent Systems, Hiroshima City University, Japan

9.1. Fundamental theory of polarization

Light is an electromagnetic wave, and therefore it oscillates. Polarization is a phenomenon in which the direction of light oscillation is biased. Although there is linear polarization and circular polarization, this section only explains linear polarization. Light oscillating in a single direction is called perfectly linear polarized; while light oscillating isotropically is denoted as unpolarized (Figure 9.1). The intermediate state of perfectly polarized and unpolarized can be expressed as partially polarized. DOP (degree of polarization) is a measure that represents the polarization state of light, which ranges from 0 to 1. The DOP of perfectly polarized light is 1 and that of unpolarized light is 0. Light penetrating through the linear polarizer becomes perfectly linear polarized. Light passes through the linear polarizer when its orientation is the same as the orientation of the oscillation of perfectly polarized light. On the contrary, light is blocked when two directions of polarizers are orthogonal.

The brightness of the transmitted light varies depending on the rotation angle of the polarizer if we observe the partially polarized light with a linear polarizer. The polarizer should be set so that the light strikes orthogonally to the polarizer plane. As for the brightness observed when the polarizer is rotated, the maximum brightness is denoted as I_{max}, while the minimum brightness is denoted as I_{min}. Let us define the

Computational Imaging for Scene Understanding,
coordinated by Takuya Funatomi and Takahiro Okabe.
© ISTE Ltd 2024.

two-dimensional coordinates (x-axis and y-axis) on the plane including the polarizer. The polarizer angle v can be defined as the angle between the polarizer axis and the $+x$-axis, where its angle is defined as the angle that starts from the $+x$-axis and aims for the $+y$-axis. Note that the polarizer axis is the orientation that passes the light oscillating in the same orientation. Since the cycle of the polarizer is $180°$, the polarizer angle also ranges from $0°$ to $180°$. Let us define the phase angle ψ as the polarizer angle v at which the maximum brightness $I_{\max}$ is observed. As a result, the observed brightness I while the polarizer rotated can be expressed as follows:

$$I = \frac{I_{\max} + I_{\min}}{2} + \frac{I_{\max} - I_{\min}}{2} \cos(2v - 2\psi). \qquad [9.1]$$

Figure 9.2(a) is a typical illustration of a linear polarizer. Here, the unpolarized light enters the polarizer, while the transmitted light oscillates only vertically.

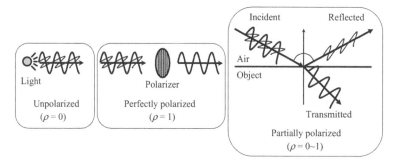

Figure 9.1. *Polarization*

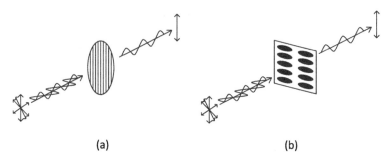

(a)   (b)

Figure 9.2. *(a) Linear polarizer illustrated as a circle with a line grid inside it, and (b) linear polarizer absorbing the light whose orientation is the same as the ingredients*

Typical artificial polarizers are either dichroic or wire-grid polarizers. Figure 9.2(b) shows that light transmitted through a dichroic polarizing film is vertically polarized when the long axis of the absorbing molecules in the film is horizontal. In this case, the horizontal component of the light is absorbed by these aligned molecules. Similarly, for wire-grid polarizers, if the wire grid is aligned horizontally, the transmitted light becomes vertically polarized.

If the polarization state of the light is measured using the camera on which the linear polarizer is mounted, the camera gamma should be 1. We should be careful not to make a gap between the polarizing filter and the lens of the camera.

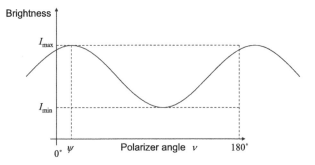

Figure 9.3. *Sinusoid of the brightness while rotating the polarizer*

If we rotate the linear polarizer around the optical axis of the camera direction, the brightness of the image will change sinusoidally according to equation [9.1] (Figure 9.3). The period of the sinusoid is 180°. Three or more images with different rotating angle of the polarizer enable us to calculate the polarization parameter of linear polarization. Equation [9.1] is also represented as follows:

$$I = a \sin \beta + b \cos \beta + c, \qquad [9.2]$$

where

$$a \sin \beta + b \cos \beta = \sqrt{a^2 + b^2} \sin(\beta + \alpha), \qquad [9.3]$$

$$\sin \alpha = \frac{b}{\sqrt{a^2 + b^2}}, \qquad [9.4]$$

$$\cos \alpha = \frac{a}{\sqrt{a^2 + b^2}}, \qquad [9.5]$$

$$c = \frac{I_{\max} + I_{\min}}{2}, \qquad [9.6]$$

$$\sqrt{a^2 + b^2} = \frac{I_{\max} - I_{\min}}{2}, \qquad [9.7]$$

$$\beta = 2\nu, \qquad [9.8]$$

$$\alpha = \pi/2 - 2\psi. \qquad [9.9]$$

If we substitute $0°$, $45°$, $90°$ and $135°$ into equation [9.1], we obtain the following:

$$I_0 = \frac{I_{max} + I_{min}}{2} + \frac{I_{max} - I_{min}}{2} \cos(2\psi),$$ [9.10]

$$I_{45} = \frac{I_{max} + I_{min}}{2} + \frac{I_{max} - I_{min}}{2} \sin(2\psi),$$ [9.11]

$$I_{90} = \frac{I_{max} + I_{min}}{2} - \frac{I_{max} - I_{min}}{2} \cos(2\psi),$$ [9.12]

$$I_{135} = \frac{I_{max} + I_{min}}{2} - \frac{I_{max} - I_{min}}{2} \sin(2\psi).$$ [9.13]

The DOP ρ for linear polarization is defined as follows:

$$\rho = \frac{I_{max} - I_{min}}{I_{max} + I_{min}}.$$ [9.14]

If we capture four images with $0°$, $45°$, $90°$ and $135°$ polarizing angles, the DOP and the phase angle can be calculated as follows:

$$\rho = \frac{\sqrt{(I_0 - I_{90})^2 + (I_{45} - I_{135})^2}}{(I_0 + I_{45} + I_{90} + I_{135})/2},$$ [9.15]

$$\psi = \frac{1}{2}\text{atan2}(I_{45} - I_{135}, I_0 - I_{90}).$$ [9.16]

Here, atan2 represents the arctangent function.

$$\text{atan2}(y, x) = \tan^{-1}\frac{y}{x}.$$ [9.17]

If we capture three images with $0°$, $45°$ and $90°$ polarizing angles, the DOP and the phase angle can be calculated as follows:

$$\rho = \frac{\sqrt{(I_0 - I_{90})^2 + (2I_{45} - I_0 - I_{90})^2}}{I_0 + I_{90}},$$ [9.18]

$$\psi = \frac{1}{2}\text{atan2}(2I_{45} - I_0 - I_{90}, I_0 - I_{90}).$$ [9.19]

If we capture three or more images with ν_1, ν_2, ..., ν_N polarizing angles, the polarization parameters a, b and c shown in equation [9.2] can be calculated from the pixel intensities I_1, I_2, ..., I_N as follows:

$$\begin{pmatrix} I_1 \\ I_2 \\ \vdots \\ I_N \end{pmatrix} = \begin{pmatrix} \sin\beta_1 & \cos\beta_1 & 1 \\ \sin\beta_2 & \cos\beta_2 & 1 \\ \vdots & \vdots & \vdots \\ \sin\beta_N & \cos\beta_N & 1 \end{pmatrix} \begin{pmatrix} a \\ b \\ c \end{pmatrix}.$$ [9.20]

The closed-form solution can be obtained from the known polarizer angles ν_1, ν_2, ..., ν_N, where $\beta_1 = 2\nu_1$, $\beta_2 = 2\nu_2$, ..., $\beta_N = 2\nu_N$ hold.

The polarization imaging camera with assorted pixels (Figure 9.4) is now available from various manufacturers. The Phoenix polarization camera (LUCID Vision Labs, Inc., Canada) uses the SONY Polarsens CMOS imaging sensor (Yamazaki et al. 2016). This CMOS sensor detects four types of polarization states of light with $0°$, $45°$, $90°$ and $135°$ polarizer angles.

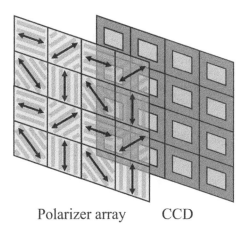

Polarizer array   CCD

Figure 9.4. *Assorted pixels of polarization imaging camera*

9.2. Reflection component separation

The reflection on the object surface can be generally represented by a dichromatic reflection model, i.e. the reflection can be expressed as a combination of diffuse reflection and specular reflection. Figure 9.5 explains the separation of the diffuse reflection component and the specular reflection component by setting the linear polarizer in front of both the camera and the light source. Light passing through the linear polarizer set in front of the light source becomes perfectly polarized and hits the object surface. Since specular reflection instantly reflects at the interface between the air and the object, the specularly reflected light remains perfectly polarized light. Diffuse reflection is unpolarized since the light randomly reflects in every direction inside the object due to the non-uniformity of the IOR (index of refraction) of the material. In this situation, we can separate the diffuse reflection component and the specular reflection component by rotating the polarizer set in front of the camera.

The maximum brightness observed I_{max} and the minimum brightness observed I_{min} are represented as follows, where specular reflection brightness and diffuse reflection brightness are represented as I_s and I_d, respectively:

$$I_{\text{min}} = \frac{1}{2} I_d \qquad\qquad [9.21]$$

$$I_{\text{max}} = \frac{1}{2} I_d + I_s \; . \qquad\qquad [9.22]$$

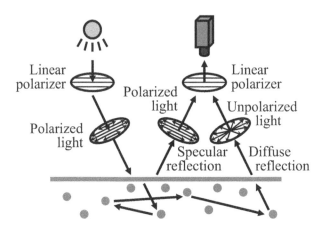

Figure 9.5. *Separation of the specular reflection component and the diffuse reflection component using two pieces of linear polarizers*

9.3. Phase angle of polarization

Let us assume that the object surface is optically smooth, and it is an isotropic structure that is a transparent dielectric. Assume that the unpolarized light is illuminated from medium 1 to medium 2, as shown in Figure 9.6. The angle between the surface normal and the incident light is denoted as θ_1, and the angle between the surface normal and the reflected light is denoted as θ_1'. Here, $\theta_1 = \theta_1'$ holds since this section assumes the optically smooth surface. The angle between the surface normal and the transmitted light is denoted as θ_2. Snell's law holds for the incident angle and the transmitted angle:

$$n_1 \sin \theta_1 = n_2 \sin \theta_2. \qquad\qquad [9.23]$$

Here, n_1 is the IOR of medium 1 and n_2 is the IOR of medium 2.

We define the plane including the incident light vector and the surface normal vector as the POI (plane of incidence). Since the surface is smooth, the POI also includes the reflected light vector and the transmitted light vector. Assume that the reflected light vector is the z-axis. Let us define the POI angle φ as the angle between the $+x$-axis and the surface normal vector projected onto the (x, y)-plane. This angle is defined as starting from the $+x$-axis and going in the direction of the $+y$-axis.

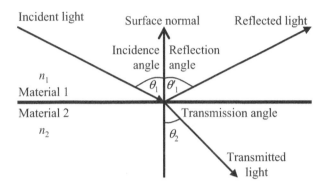

Figure 9.6. *Reflection, transmission and refraction*

We use the subscript $\parallel$ for the component parallel to the POI, and the subscript $\perp$ for the component perpendicular to the POI. The intensity reflectivity R and the intensity transmissivity T are represented as follows:

$$R_{\parallel} = \frac{\tan^2(\theta_1 - \theta_2)}{\tan^2(\theta_1 + \theta_2)} \tag{9.24}$$

$$R_{\perp} = \frac{\sin^2(\theta_1 - \theta_2)}{\sin^2(\theta_1 + \theta_2)} \tag{9.25}$$

$$T_{\parallel} = \frac{\sin 2\theta_1 \sin 2\theta_2}{\sin^2(\theta_1 + \theta_2)\cos^2(\theta_1 - \theta_2)} \tag{9.26}$$

$$T_{\perp} = \frac{\sin 2\theta_1 \sin 2\theta_2}{\sin^2(\theta_1 + \theta_2)}. \tag{9.27}$$

These values are shown in Figure 9.7, where the relative IOR $n = n_2/n_1$ is 1.5. The horizontal axis represents the incident angle $\theta_1 (0° \leq \theta_1 < 90°)$, and the vertical angle represents the intensity reflectivity/transmissivity. The following property holds for the intensity reflectivity/transmissivity:

$$R_{\parallel} \leq R_{\perp} \tag{9.28}$$

$$T_{\parallel} \geq T_{\perp}. \tag{9.29}$$

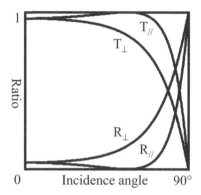

Figure 9.7. *Intensity reflectivity and intensity transmissivity*

9.4. Surface normal estimation from the phase angle

The unpolarized light specularly reflected at the object surface becomes partially polarized. The angle at which $I_{\min}$ is observed represents the angle of the plane that includes the surface normal. This property can be used to estimate the surface normal. Minimum brightness is observed when the polarizer angle coincides with the POI, i.e. the phase angle ψ is orthogonal to the POI angle ($\varphi = \psi + 90°(\mathrm{mod}\,360°)$ or $\varphi = \psi - 90°(\mathrm{mod}\,360°)$).

As for diffuse reflection, the angle at which $I_{\max}$ is observed includes the surface normal. The angle of diffuse reflection is $90°$ rotated from the angle of specular reflection. If we know that the object surface is either specular-dominant or diffuse-dominant, i.e. if we know that the object surface produces either a specular reflection stronger than its diffuse reflection or a diffuse reflection stronger than its specular reflection, we can obtain a constraint on the surface normal from the polarization of light.

If we illuminate unpolarized light onto the object surface, the specularly reflected light will be partially polarized. We can obtain the azimuth angle of the object surface from the polarization angle when $I_{\min}$ is observed. The surface normal cannot be uniquely determined from a single view (Figure 9.8 (a)). Wolff and Boult (1991) presented that the surface normal can be uniquely determined if we use another additional view (Figure 9.8 (b)). Using this property, the surface shape can be obtained when observed from two views. However, polarization data of two views should be analyzed at the same point on the object surface. Miyazaki et al. (2016) predetermined the abstract shape using a space carving technique to analyze the polarization state of the reflected light at the corresponding points when observed from multiple viewpoints.

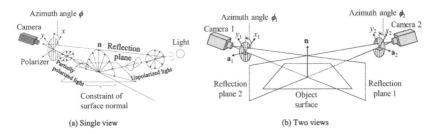

Figure 9.8. *Relationship between the surface normal and the POI: (a) single viewpoint and (b) two viewpoints*

Here, ϕ_k represents the azimuth angle of the surface point observed by the camera $k = (1, 2, \cdots, K)$, and $\mathbf{a}_k$ represents the vector orthogonal to the POI under the coordinate system of the camera k. Because $\mathbf{a}_k$ is orthogonal to the POI, we obtain equation [9.30] using the phase angle ψ_k or the azimuth angle ϕ_k:

$$\mathbf{a}_k = \begin{pmatrix} \cos(\phi_k + 90°) \\ \sin(\phi_k + 90°) \\ 0 \end{pmatrix} = \begin{pmatrix} \cos\psi_k \\ \sin\psi_k \\ 0 \end{pmatrix}. \qquad [9.30]$$

The rotation matrix $\mathbf{R}_k$ represents the transformation from the world coordinate system to the local coordinate system of the camera denoted by k. The transformation from the local coordinate system of the camera k to the world coordinate system is the transpose of $\mathbf{R}_k$. Because the transformed vector becomes orthogonal to the surface normal $\mathbf{n} = (n_x, n_y, n_z)$, equation [9.31] holds:

$$(\mathbf{R}_k^\top \mathbf{a}_k) \cdot \mathbf{n} = 0, \quad (k = 1, 2, \cdots, K). \qquad [9.31]$$

If we concatenate equation [9.31] for K cameras, we obtain equation [9.32]:

$$\begin{pmatrix} \mathbf{a}_1^\top \mathbf{R}_1 \\ \mathbf{a}_2^\top \mathbf{R}_2 \\ \vdots \\ \mathbf{a}_K^\top \mathbf{R}_K \end{pmatrix} \begin{pmatrix} n_x \\ n_y \\ n_z \end{pmatrix} = \begin{pmatrix} 0 \\ 0 \\ \vdots \\ 0 \end{pmatrix},$$

$$\mathbf{A}\mathbf{n} = \mathbf{0}. \qquad [9.32]$$

The surface normal $\mathbf{n}$, which satisfies equation [9.32] in the least-squares sense, can be estimated using SVD. The $K \times 3$ matrix $\mathbf{A}$ can be decomposed by SVD as follows:

$$\begin{pmatrix} \mathbf{a}_1^\top \mathbf{R}_1 \\ \mathbf{a}_2^\top \mathbf{R}_2 \\ \vdots \\ \mathbf{a}_K^\top \mathbf{R}_K \end{pmatrix} = \mathbf{U}\mathbf{W}\mathbf{V}^\top = \mathbf{U} \begin{pmatrix} w_1 & & \\ & w_2 & \\ & & 0 \end{pmatrix} \begin{pmatrix} \mathbf{v}_1 \\ \mathbf{v}_2 \\ \mathbf{v}_3 \end{pmatrix}. \qquad [9.33]$$

Here, $\mathbf{U}$ is a $K{\times}3$ orthogonal matrix, $\mathbf{W}$ is a $3{\times}3$ diagonal matrix with non-negative values and $\mathbf{V}^{\top}$ is a $3{\times}3$ orthogonal matrix. The diagonal item w_i of the matrix $\mathbf{W}$ is the singular value of the matrix $\mathbf{A}$ and the singular vector corresponding to w_i is $\mathbf{v}_i$. Owing to the relationship between the surface normal and the POIs, the rank of matrix $\mathbf{A}$ is at most 2; thus, one of the three singular values becomes 0. The surface normal $\mathbf{n}$ can be represented as equation [9.34], which can be calculated from the singular vector that has the smallest singular value, i.e. the third row of $\mathbf{V}^{\top}$ in equation [9.33]:

$$\mathbf{n} = s\mathbf{v}_3^{\top}. \tag{9.34}$$

In the general case, s is an arbitrary scalar coefficient. However, since the surface normal and the singular vectors are normalized vectors, s would be either $+1$ or -1. Whether s must be positive or negative can be easily determined to ensure that the surface normal is facing towards the camera.

Unlike a conventional photometric stereo or multiview stereo, which cannot estimate the shape of a black specular object, this method (Miyazaki et al. 2016) estimates the surface normal and 3D coordinates of black specular objects via polarization analysis and space carving (Figure 9.9).

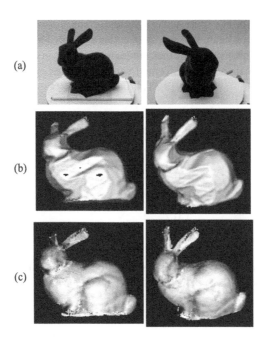

Figure 9.9. *Shape from polarization and space carving: (a) target object, (b) shape calculated from space carving and (c) shape calculated from polarization and space carving*

This method (Miyazaki et al. 2016) is not suitable for concave objects since it depends on space carving. Miyazaki et al. (2020) modified this method to estimate the surface normal of concave objects. The target object has a specular surface without diffuse reflection. Factories in industrial fields have a high demand for estimating the shape of cracks since it is quite important for the quality control of products. Therefore, there is a great demand for estimating the shape of concave objects of highly specular surfaces since it is a challenging task. Their method estimates the surface normal of a black specular object with a concave shape by analyzing the polarization state of the reflected light in which the target object is observed from multiple views. The camera parameters can be estimated a priori using known corresponding points. Since the target object is assumed to be almost planar, the corresponding points between multiple views can be obtained. Since each point on the surface corresponds, the surface normal of the object is uniquely determined from the polarization data of multiple views. Figure 9.10 shows the algorithm flow. The result of Figure 9.11 is shown in Figures 9.12 and 9.13. The result of Figure 9.14 is shown in Figure 9.15.

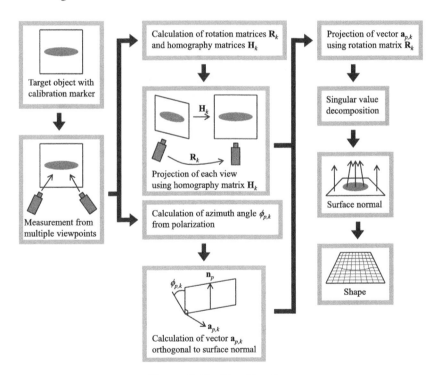

Figure 9.10. *Algorithm flow*

Figure 9.11. *Target object (ellipsoid)*

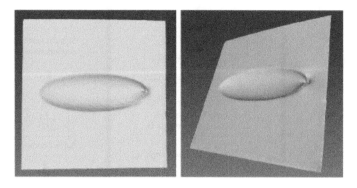

Figure 9.12. *Estimated shape (ellipsoid)*

Ground truth

Proposed method

Figure 9.13. *Intersection shape (ellipsoid)*

Figure 9.14. *Target object (stripe)*

Figure 9.15. *Estimated shape (stripe)*

9.5. Degree of polarization

Let us assume that medium 1 is the air and medium 2 is the object. Assume that the reflected light vector is the z-axis in the camera coordinate system. We set the linear polarizer in front of the camera and observe the reflected light. We consider the case that the incident light is unpolarized. We assume the IOR of the air n_1 as 1 and represent the IOR of the object as $n = n_2/n_1 = n_2$. The surface normal can be represented in polar coordinates, where the azimuth angle is ϕ and the zenith angle is θ. The zenith angle is defined as the angle from the $+z$-axis to the surface normal, and the azimuth angle is defined as the angle from the $+x$-axis to the $+y$-axis. The azimuth angle ϕ coincides with the POI angle φ ($\phi = \varphi$). The DOP is defined as equation [9.14]. If we observe the specular reflection, the following holds from equation [9.28]:

$$I_{\max} = \frac{R_\perp}{R_\parallel + R_\perp} I_s \tag{9.35}$$

$$I_{\min} = \frac{R_\parallel}{R_\parallel + R_\perp} I_s \tag{9.36}$$

Here, I_s represents the brightness of the reflected light. Substituting equation [9.35] and equation [9.36] into equation [9.14] results in the following:

$$\rho = \frac{R_\perp - R_\parallel}{R_\perp + R_\parallel} \qquad [9.37]$$

From equation [9.24], equation [9.25], equation [9.23] and equation [9.37], the DOP is expressed as follows using the reflected angle $\vartheta \equiv \theta_1 = \theta_1'$:

$$
\begin{aligned}
\rho &= \frac{\cos^2(\theta_1 - \theta_2) - \cos^2(\theta_1 + \theta_2)}{\cos^2(\theta_1 - \theta_2) + \cos^2(\theta_1 + \theta_2)} \\
&= \frac{\sqrt{\sin^4 \vartheta \cos^2 \vartheta (n^2 - \sin^2 \vartheta)}}{\left(\sin^4 \vartheta + \cos^2 \vartheta (n^2 - \sin^2 \vartheta)\right)/2}.
\end{aligned}
\qquad [9.38]
$$

Here, ϑ is the zenith angle θ of the surface normal ($\theta = \vartheta$). The DOP when the IOR is 1.5 is shown in Figure 9.16.

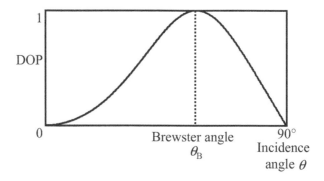

Figure 9.16. *Degree of polarization of the reflected light*

An object radiates infrared light when it is heated. This phenomenon is called thermal radiation (black-body radiation). Figure 9.17 shows the light inside the object emitting into the air.

Thermal radiation is unpolarized when it is inside the object since it is randomly generated and randomly reflected inside the object. The light becomes partially polarized when it is transmitted through the interface of the object and the air (Figure 9.17). The observed brightness is expressed as follows if we set the infrared linear polarizer in front of the infrared camera:

$$I_{\max} = \frac{T_\parallel}{T_\parallel + T_\perp} I_t \qquad [9.39]$$

$$I_{\min} = \frac{T_\perp}{T_\parallel + T_\perp} I_t \qquad [9.40]$$

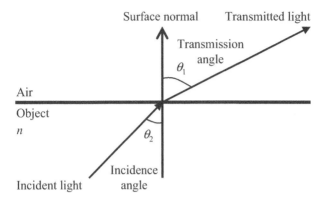

Figure 9.17. *Emission*

Here, I_t represents the brightness of thermal radiation. Substituting equation [9.39] and equation [9.40] into equation [9.14] results in the following:

$$\rho = \frac{T_\| - T_\perp}{T_\| + T_\perp} \tag{9.41}$$

The emitted angle θ_1 is the angle between the surface normal vector and the observation vector, and the incident angle θ_2 is the angle between the incident light vector and the surface normal vector. Here, the IOR of medium 1 (the air) is $n_1 = 1$, and the IOR of medium 2 (the object) is $n_2 = n$. The DOP is expressed as follows using equation [9.26], equation [9.27], equation [9.23], equation [9.41] and the emitted angle $\vartheta \equiv \theta_1$.

$$
\begin{aligned}
\rho &= \frac{1 - \cos^2(\theta_1 - \theta_2)}{1 + \cos^2(\theta_1 - \theta_2)} \\
&= \frac{\sin^2 \vartheta (\cos \vartheta - \sqrt{n^2 - \sin^2 \vartheta})^2}{2n^2 - \sin^2 \vartheta (\cos \vartheta - \sqrt{n^2 - \sin^2 \vartheta})^2} \\
&= \frac{(n - 1/n)^2 \sin^2 \vartheta}{2 + 2n^2 - (n + 1/n)^2 \sin^2 \vartheta + 4 \cos \vartheta \sqrt{n^2 - \sin^2 \vartheta}},
\end{aligned}
\tag{9.42}
$$

The emitted angle ϑ is the zenith angle θ of the surface normal ($\theta = \vartheta$). The DOP when the IOR is 1.5 is shown in Figure 9.18. There is a one-to-one correspondence between the DOP and the zenith angle.

Diffuse reflection can also be regarded as the emission from the inner part of the object. Therefore, the DOP of diffuse reflection can be represented as equation [9.42] when the object surface is optically smooth. When the surface is rough, the DOP becomes lower than equation [9.42].

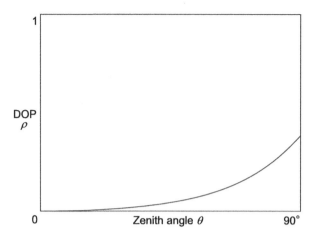

Figure 9.18. *Degree of polarization of thermal radiation*

9.6. Surface normal estimation from the degree of polarization

Candidates of the zenith angle of the surface normal can be calculated from the DOP. Two candidates of the zenith angle are obtained from specular reflection (Figure 9.16), while one candidate of the zenith angle is obtained from diffuse reflection (Figure 9.18). Regarding thermal radiation of infrared light, Miyazaki et al. (2002) estimated the surface shape using polarization. They observed the DOP of specular reflection in the visible light domain (Figure 9.16) and the DOP of thermal radiation in the infrared light domain (Figure 9.18). The reliable zenith angle is obtained from equation [9.38] (Figure 9.16), and the unique zenith angle is obtained from equation [9.42] (Figure 9.18).

We have two candidates if we want to estimate the zenith angle of the surface normal from the DOP of specular reflection (equation [9.38] and Figure 9.16). One of the candidate zenith angles is true and the other is wrong. In order to solve the ambiguity of the zenith angle, we need additional data. Miyazaki et al. (2004) used a geometrical invariant to match the corresponding points from two views to estimate the surface normal of a transparent object.

9.7. Stokes vector

As shown in the previous sections, light can be expressed as the brightness, DOP and phase angle (Shurcliff 1962; Hecht 1998; Born and Wolf 2013). Another representation for the polarization state of light is the coherence matrix (Born and Wolf 2013), the Mueller calculus (Shurcliff 1962; Hecht 1998) and the Jones calculus (Shurcliff 1962; Hecht 1998). These four types of representation can be converted to each other. The Mueller calculus is relatively simple to understand, and thus it is often used in the field of computer vision. The Mueller calculus represents the polarization state of light using the 4D vector $s = (s_0, s_1, s_2, s_3)$, which is called the Stokes vector. Here, s_0 represents the brightness, s_1 represents the strength of the linear polarization along the x-axis, s_2 represents the strength of the linear polarization along the $+45°$ diagonal orientation between the x-axis and the y-axis, and s_3 represents the strength of the right circular polarization (Figure 9.19).

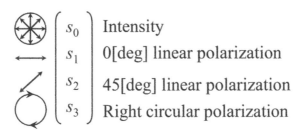

$$\begin{pmatrix} s_0 \\ s_1 \\ s_2 \\ s_3 \end{pmatrix} \quad \begin{array}{l} \text{Intensity} \\ \text{0[deg] linear polarization} \\ \text{45[deg] linear polarization} \\ \text{Right circular polarization} \end{array}$$

Figure 9.19. *Stokes vector*

9.8. Surface normal estimation from the Stokes vector

Miyazaki and Ikeuchi (2007) solved the inverse problem of polarization ray tracing to estimate the surface normal of a transparent object.

The algorithm of the polarization ray-tracing method can be divided into two parts. For the first part, the calculation of the propagation of the ray, they use the same algorithm used in the conventional ray-tracing method. For the second part, the calculation of the polarization state of light, we can use the Mueller calculus. Let us denote the input polarization data as $s_{\mathcal{I}}(x, y)$, where (x, y) represents the pixel position. Polarization ray tracing can render the polarization data from the shape of the transparent object. Let us denote this rendered polarization image as $s_{\mathcal{R}}(x, y)$.

The shape of transparent objects is represented as the height $H(x, y)$ set for each pixel. The rendered polarization image $s_{\mathcal{R}}(x, y)$ depends on the height and the surface normal. The problem is to find the best values to reconstruct a surface $H(x, y)$ that satisfy the following equation:

$$s_{\mathcal{I}}(x, y) = s_{\mathcal{R}}(x, y),$$
[9.43]

for all pixels (x, y). The optimization problem can be described as follows:

$$H(x, y) = \arg \min_{H(x,y)} \iint \|\mathbf{s}_\mathcal{I}(x, y) - \mathbf{s}_\mathcal{R}(x, y)\|^2 \, dx dy. \qquad [9.44]$$

The target object is set inside the center of a plastic sphere whose diameter is 35 cm (Figure 9.20). This plastic sphere is illuminated by 36 incandescent lamps that are almost uniformly distributed spatially around the plastic sphere by a geodesic dome. The plastic sphere diffuses the light that comes from light sources and behaves as a spherical light source, which illuminates the target object from every direction. The target object is observed by a monochrome camera from the top of the plastic sphere, which has a 6 cm diameter hole on the top. A linear polarizer is set in front of the camera. The camera, object and light sources are fixed. The least-squares method calculates $I_{\max}$, $I_{\min}$ and ψ from four images taken by rotating the polarizer at $0°$, $45°$, $90°$ and $135°$.

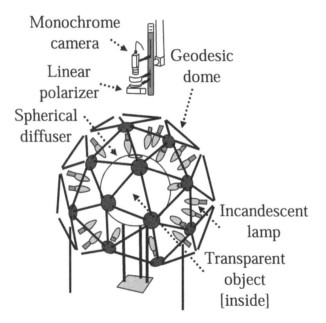

Figure 9.20. *Acquisition system for measuring transparent surfaces based on polarization ray tracing*

A dog-shaped object shown in Figure 9.21(1a) (2a) is made of glass and has a refractive index of 1.5. The shape obtained manually by human operation is used as the initial value for the experiment, which is shown in Figure 9.21(1b) (2b). The estimation results after 10 iterations are shown in Figure 9.21(1c) (2c).

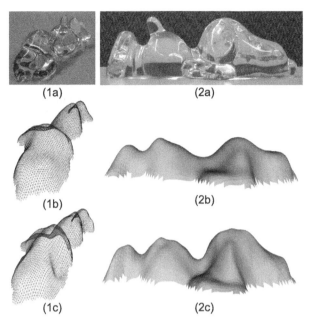

Figure 9.21. *Dog-shaped glass object: (1a) (2a) real image, (1b) (2b) initial state of the shape given manually and (1c) (2c) estimated shape after 10 iterations*

9.9. References

Born, M. and Wolf, E. (2013). *Principles of Optics: Electromagnetic Theory of Propagation, Interference and Diffraction of Light.* Elsevier.

Hecht, E. (1998). *Optics.* Addison Wesley Longman Inc., 1.

Miyazaki, D. and Ikeuchi, K. (2007). Shape estimation of transparent objects by using inverse polarization ray tracing. *IEEE Transactions on Pattern Analysis and Machine Intelligence*, 29(11), 2018–2030.

Miyazaki, D., Saito, M., Sato, Y., Ikeuchi, K. (2002). Determining surface orientations of transparent objects based on polarization degrees in visible and infrared wavelengths. *JOSA A*, 19(4), 687–694.

Miyazaki, D., Kagesawa, M., Ikeuchi, K. (2004). Transparent surface modeling from a pair of polarization images. *IEEE Transactions on Pattern Analysis & Machine Intelligence*, 26(1), 73–82.

Miyazaki, D., Shigetomi, T., Baba, M., Furukawa, R., Hiura, S., Asada, N. (2016). Surface normal estimation of black specular objects from multiview polarization images. *Optical Engineering*, 56(4), 041303.

Miyazaki, D., Furuhashi, R., Hiura, S. (2020). Shape estimation of concave specular object from multiview polarization. *Journal of Electronic Imaging*, 29(4), 041006.

Shurcliff, W.A. (1962). *Polarized Light; Production and Use*. Harvard University Press.

Wolff, L.B. and Boult, T.E. (1991). Constraining object features using a polarization reflectance model. *IEEE Transactions on Pattern Analysis & Machine Intelligence*, 13(7), 635–657.

Yamazaki, T., Maruyama, Y., Uesaka, Y., Nakamura, M., Matoba, Y., Terada, T., Komori, K., Ohba, Y., Arakawa, S., Hirasawa, Y. et al. (2016). Four-directional pixel-wise polarization CMOS image sensor using air-gap wire grid on 2.5-μm back-illuminated pixels. *2016 IEEE International Electron Devices Meeting (IEDM)*, pp. 8.7.1–8.7.4.

10

Shape from Polarization and Shading

Thanh-Trung NGO[1], Hajime NAGAHARA[2], and Rin-ichiro TANIGUCHI[3]

[1] *Hanoi University of Science and Technology, Vietnam*
[2] *Institute for Datability Science, Osaka University, Japan*
[3] *Kyushu University, Fukuoka, Japan*

10.1. Introduction

Three-dimensional (3D) reconstruction is a key research topic in computer vision, for which various approaches have been proposed. Fundamental research works rely on basic physical parameters of light, such as the speed (Foix and Aleny 2011), direction (Woodham 1980), frequency (Huynh et al. 2013) and polarization state (Rahmann 1999), to reconstruct the 3D shape. For example, the photometric stereo method (Woodham 1980; Basri et al. 2006) observes an object from different light directions while the polarization-based method (Saito et al. 1999; Miyazaki et al. 2002; Atkinson and Hancock 2006) observes the object in different polarization states through a polarizer. Naturally, when more parameters of light are used, there is more information with which to recover the shape. For example, the light frequency and polarization are integrated to estimate the shape and refractive index; the speed of light and polarization (Kadambi et al. 2015) or light direction and polarization (Atkinson and Hancock 2007b; Saman et al. 2011; Atkinson 2017) are used to improve the shape estimation by taking advantage of each other's cue. Our research focuses on the direction and polarization state of light to recover surface normals.

Computational Imaging for Scene Understanding,
coordinated by Takuya FUNATOMI and Takahiro OKABE.
© ISTE Ltd 2024.

The photometric stereo method has been studied intensively and has provided promising results recently (Higo et al. 2010; Lu et al. 2013; Chandraker 2014; Inoshita et al. 2014; Santo et al. 2017; Chen et al. 2019). Conventional photometric stereo methods consider light from different directions interacting with the object surface to sense the 3D structure. In most photometric stereo methods, the light directions are known for the reconstruction. The bidirectional reflection distribution function (BRDF), which models the incoming and outgoing light, plays an important role; however, the BRDF is a complex function that depends on the material. Many efforts have been made to overcome the complexity of the BRDF (Higo et al. 2010; Lu et al. 2013; Chandraker 2014; Ngo et al. 2019), yet the approaches taken have typically ignored the refractive index and polarization of the dielectric materials and hence they do not handle polarization images captured with a polarizer. Moreover, the photometric stereo approach usually faces a limitation at the edge with a large observing angle, where the BRDF is not easily modeled (Wolff et al. 1998; Ngan et al. 2005).

Polarization-based shape reconstruction methods consider polarization states of light reflected on the material surface to sense the 3D structure. The Fresnel equations for refraction and reflection (Hecht 2001) in the interaction of light with a smooth dielectric surface are the main theory underlying this approach. Two types of reflection are considered in this approach: specular and diffuse reflections. In the case of specular reflection, we need to obtain the specular reflection for the whole object and therefore require an omnidirectional light source that is hard to handle theoretically. In the case of diffuse reflection, we do not have to use many light sources. In contrast to photometric stereo methods, polarization-based methods do not require manipulation of the light direction and basically do not use information on the light source. However, polarization-based methods only work well at a large zenith angle (e.g. the edge of an object) where the polarization effect is strong. Polarization-based methods usually require us to know the refractive index (Miyazaki et al. 2002; Atkinson and Hancock 2006) or use a homogeneous material with a uniform refractive index (Zhang and Hancock 2013). Owing to these limitations, it is better that a polarization-based method is combined with another method to improve the accuracy. In fact, polarization-based methods are combined with a multiview stereo (Atkinson and Hancock 2007a; Cui et al. 2017; Chen et al. 2018), camera motion (Yang et al. 2018), depth (Kadambi et al. 2015; Zhu and Smith 2019), light color (frequency) (Huynh et al. 2013) or photometric stereo method (Atkinson and Hancock 2007b; Atkinson 2017) to improve the reconstruction. Although the photometric stereo and polarization-based methods have been combined to recover surface normals, they have simply been used as two independent (Atkinson and Hancock 2007b; Saman et al. 2011) or partially independent (Atkinson 2017) steps.

Under unknown light directions, the photometric stereo method faces a serious ambiguity problem, such as the generalized bas-belief ambiguity (Belhumeur et al.

1999), and needs complicated priors to solve the problem (Hayakawa 1994; Zickler et al. 2002; Okabe et al. 2009; Shi et al. 2010; Sunkavalli et al. 2010). However, because the surface normals and light directions are strongly correlated, the ambiguity can be relaxed by constraining all surface normals. Fortunately, we can use such a prior from the polarization of light reflected from the object.

In this chapter, we present a 3D shape estimation method that uses all the directional information (light direction and polarization state) of the light to benefit from the advantages of two conventional (photometric stereo and polarization-based) approaches while overcoming their disadvantages. Our contributions are summarized as follows. We present two constraints: the *shading-stereoscopic constraint* for a pair of polarization images associated with two light directions and the *polarization-stereoscopic constraint* for a pair of polarizer angles. Using the two constraints, we present two algorithms. The first algorithm estimates the surface normal and refractive index of an individual point on an object knowing the light directions. The second algorithm estimates the surface normals, light directions and refractive indexes. This chapter is a summary of our previous publications (Ngo et al. 2015, 2021b). Our datasets are publicly available (Ngo et al. 2021a).

10.2. Related works

10.2.1. *Shading and polarization fusion*

Photometric stereo and polarization-based fusion methods are used in research. Atkinson and Hancock (2007b) first applied a standard polarization method with a known refractive index to estimate zenith and ambiguous azimuth angles of surface normals. Then, from the relationship between the known light directions and the captured image intensity, they selected the correct azimuth angle for each pixel. This method assumes the refractive index and light directions are known. Similarly, Smith et al. (2016, 2019) assumed a known refractive index of a Lambertian object and estimated the surface normal and height of the object. In their method, shading is not used to improve the accuracy of the estimation but to select the correct surface normal from estimated ambiguous surface normals. In contrast, Saman et al. (2011) first used a photometric stereo method to recover surface normals of a Lambertian object and then estimated the refractive indexes of the material by using the normals estimated earlier. However, an actual material with a polarization effect is widely known to be non-Lambertian. Recently, Atkinson (2017) proposed a method that separates a 3D normal vector into components and estimates them separately relying on polarization and shading cues. In this method, the surface normal is not tightly constrained by both polarization and shading cues. Overall, these fusion methods use shading and polarization independently, and they do not use the true advantages of both cues to enhance shape estimation and still require calibrated light sources.

10.2.2. *Shape estimation under uncalibrated light sources*

Uncalibrated photometric stereo methods generally face strong ambiguity. Many works have aimed to reduce the ambiguity. Under orthographic projection, Lambertian reflectance, and an integrability constraint, it has been shown that the general ambiguity is reduced to the group of generalized bas-relief ambiguity with three parameters (Belhumeur et al. 1999). Additional constraints or assumptions are used to reduce the ambiguity. Using specularities of four corresponding different distant point light sources, the generalized bas-relief ambiguity is reduced to the two-degree-of-freedom group of linear transformations (Drbohlav and Sara 2002). Using the constraint that the relative intensities of six light sources or reflectances at six points on a curved surface are known or constant (Hayakawa 1994), the ambiguity is reduced to the group of scaled orthogonal transformations. Moreover, great efforts have been made to reduce the ambiguity to the convex/concave ambiguity (Lu et al. 2013). Further, assuming the object is concave, the inter-reflection constraint completely resolves the ambiguity (Chandraker et al. 2005). Meanwhile, assuming the object is convex, the shadowing constraint completely resolves the ambiguity (Okabe et al. 2009). Under the perspective projection of a calibrated camera, it has been shown that the ambiguity is completely resolved (Papadhimitri and Favaro 2013, 2014).

Although polarization-based methods are basically independent of the calibration of the light sources, they face the surface normal ambiguity. Knowing the refractive index and assuming a diffuse material reflectance, polarization-based methods reduce the general ambiguity to the convex/concave ambiguity (Atkinson and Hancock 2006; Smith et al. 2019). However, in the case of specular reflectance, polarization-based methods face more complicated ambiguity (Miyazaki et al. 2004).

Under uncalibrated light sources for the shading and polarization fusion approach, the polarization cue is used to reduce the ambiguity. Drbohlav and Sara (2001) proposed a solution for conventional photometric stereo using two polarizers simultaneously, one for the light source and one for the camera. Hence, they have more information with which to reduce the general ambiguity to the convex/concave ambiguity. Compared with the conventional polarization-based and photometric stereo methods, this method is much stricter and not practical because the polarization of light incident on the object and the polarization of light reflected to the camera need to be controlled simultaneously. Moreover, polarization information is not used to improve the accuracy of the estimation of the surface normal or to estimate the refractive index.

10.3. Problem setting and assumptions

Similar to most polarization-based methods, we use a monochromatic camera that observes the target object through a polarizer. A world coordinate system is defined in Figure 10.1, where the origin is at the object center, the z-axis coincides with the optical axis of the camera, and the xy plane is translated from the image plane. For each pixel p on the object's image, we will recover its surface normal $n(p)$ in 3D space and refractive index $\eta(p)$. In this coordinate system, $n(p)$ is defined by a zenith angle $\theta(p)$ and an azimuth angle $\phi(p)$.

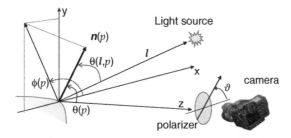

Figure 10.1. *Normal vector $n(p)$ is defined in the world coordinate system by a zenith angle $\theta(p)$ and an azimuth angle $\phi(p)$. For a color version of this figure, see www.iste.co.uk/funatomi/computational.zip*

The following assumptions are made in our proposed method:

1) the object material is heterogeneous with unknown refractive indices or albedo;

2) directional light sources reside on the same side as the camera in relation to the xy plane;

3) the object only exhibits diffuse reflectance and diffuse polarization;

4) the camera projection model is orthogonal.

For the first assumption, we can handle a variety of practical dielectric materials, unlike most existing methods that assume a homogeneous dielectric material (Miyazaki et al. 2002; Atkinson and Hancock 2005, 2006, 2007a; Miyazaki and Ikeuchi 2010; Saman et al. 2011; Mahmoud et al. 2012; Smith et al. 2019). The second assumption states that the intensities of light sources are calibrated and can be considered constant. The third assumption is supported by the second assumption. Since we only consider directional light source, the specular reflectance is very sparse and thus the diffuse reflectance and diffuse polarization are dominant. For the fourth assumption, the camera model is simplified such that the mathematical formulation is similar to that of existing polarization or photometric stereo methods (Woodham 1980; Huynh et al. 2013; Atkinson 2017). In practice, to meet the fourth assumption, we can use a camera with a long-focal-length lens and keep the target object far from the camera.

We propose two constraints, one for a pair of polarizer angles and one for a pair of light directions. To use these constraints, we present two algorithms that estimate the surface normals of the object. In the first algorithm, we assume to know the light directions and then estimate the surface normals and refractive indexes of the object. In the second algorithm, we do not know the light directions and we estimate the surface normals, refractive indexes and light directions simultaneously.

10.4. Shading stereoscopic constraint

In this section, we derive a constraint for two light directions at the same polarizer angle. When a light ray from a direction l is incident on a dielectric surface, it separates into two rays: a reflected ray on the surface and a refracted ray under the surface. Both rays are partially polarized obeying Fresnel theory (Hecht 2001). The reflected ray can only be observed at a specific observing angle and is ignored in our case. The refracted ray is scattered and becomes depolarized and refracts back into the air. The outgoing ray is partially polarized according to Fresnel theory. This is the principle of diffuse reflection on a dielectric material, which is illustrated in Figure 10.2 for a surface point associated with pixel p and normal $n(p)$. The camera observes the diffuse reflection from a wide range of observing directions $o(p)$.

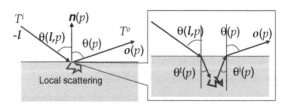

Figure 10.2. *Diffuse reflection at a surface point associated with pixel p and normal vector $n(p)$. All 3D vectors, such as l, $n(p)$ and $o(p)$, are normalized. For a color version of this figure, see www.iste.co.uk/funatomi/computational.zip*

The Fresnel reflection coefficients for perpendicular and parallel polarization components of light reflected on the surface are

$$r_\perp(p) = \frac{\cos\theta(l,p) - \eta(p)\cos\theta^t(p)}{\cos\theta(l,p) + \eta(p)\cos\theta^t(p)}, \tag{10.1}$$

$$r_\parallel(p) = \frac{\eta(p)\cos\theta(l,p) - \cos\theta^t(p)}{\eta(p)\cos\theta(l,p) + \cos\theta^t(p)}, \tag{10.2}$$

where $\theta^t(p)$ is given by Snell's law: $\sin\theta(l,p) = \eta(p)\sin\theta^t(p)$, $\cos\theta(l,p) = l \cdot n(p)$ and $\cos\theta^t(p) = n(p) \cdot o(p)$. Considering the total light energy or intensity, the reflection ratio is computed as

$$F(\theta(p,l),\eta(p)) = r_\perp^2(p) + r_\parallel^2(p), \tag{10.3}$$

and hence, the transmission ratio is $T = 1 - R$. Similarly, we can compute the transmission ratio for the diffuse light coming out of the surface and traveling to the camera at the observing angle $\theta(p)$. This diffuse reflection is modeled using the Wolff reflectance model for a smooth dielectric material (Wolff et al. 1998) without a polarizer in front of the camera:

$$I\big(\theta(l,p), \theta(p), \eta(p)\big) = \rho(p)I(l)T^i(p)T^o(p),\qquad\qquad [10.4]$$

where

$$T^i(p) = \big(1 - F\left(\theta(l,p), \eta(p)\right)\big)\cos\theta(l,p),\qquad\qquad [10.5]$$

$$T^o(p) = 1 - F\left(\arcsin\left(\frac{\sin\theta(p)}{\eta(p)}\right), \eta(p)^{-1}\right),\qquad\qquad [10.6]$$

and $\rho(p)$ is the total diffuse albedo of the surface point associated with p, which is assumed an unknown constant. $I(l)$ is the irradiance from the light source in direction l, which is also an unknown constant according to our assumption. When we capture an image with a polarizer, the rotation is defined by the polarizer angle v. The first transmission term $T^i(p)$ is for the light from the source that transmits through the surface, and it does not depend on any configuration of the polarizer. In contrast, the second transmission term $T^o(p)$ is modulated using v and the image intensity $I\big(\theta(p,l), \theta(p), \eta(p), v\big)$ is captured instead of $I\big(\theta(l,p), \theta(p), \eta(p)\big)$, which we consider in the next section. However, if we fix the polarizer angle at v, the second term is also unchanged.

Assuming that the camera projection is orthogonal, the observation vector $o(p)$ is known, and we re-parameterize the image intensity for p using $I(l, v, p)$. When the polarizer angle is known and the image is captured in different light directions l_1 and l_2, the ratio between image intensities at p can be obtained by removing the common factors:

$$\frac{I(l_1, v, p)}{I(l_2, v, p)} = \frac{\big[1 - F(\theta(l_1, p), \eta(p))\big]\cos\theta(l_1, p)}{\big[1 - F(\theta(l_2, p), \eta(p))\big]\cos\theta(l_2, p)},\qquad\qquad [10.7]$$

where $\theta(l_i, p)$ can be derived from the dot product $l_i \cdot n(p)$, $i = 1, 2$. This direct form of constraint is sensitive to the small denominators. Moreover, if removing the denominators of both sides, the constraint heavily depends on the image intensity.

Considering two two-dimensional vectors constructed from the elements of the left and right sides of equation [10.7], we have

$$s_1 = \big(I(l_1, v, p), I(l_2, v, p)\big)^T\qquad\qquad [10.8]$$

$$s_2 = \big([1 - F(\theta(l_1, p), \eta(p))]\cos\theta(l_1, p),$$

$$[1 - F(\theta(l_2, p), \eta(p))]\cos\theta(l_2, p)\big)^T,\qquad\qquad [10.9]$$

and the constraint equation [10.7] is then transformed into a constraint on the angle between the two vectors, which is much more robust against the drastic change of image intensity. The constraint is then stated as

$$\arccos \frac{s_1.s_2}{\|s_1\| \|s_2\|} = 0. \tag{10.10}$$

This angle is independent of the incident irradiance and the albedo of the object. It is the first constraint on the light directions, surface normal and refractive index of the object.

10.5. Polarization stereoscopic constraint

In this section, we derive a constraint for two different polarizer angles of the same light direction. When we capture an image with a polarizer, the image intensity is manipulated according to the *transmitted radiance sinusoid*:

$$I(l, v, p) = \frac{I_{max}(l, p) + I_{min}(l, p)}{2}$$
$$+ \frac{I_{max}(l, p) - I_{min}(l, p)}{2} \cos(2v - 2\phi(p)), \tag{10.11}$$

where $I_{max}(l, p)$ and $I_{min}(l, p)$ are the maximum and minimum radiances of p for all angles v, respectively. To quantify the polarization effect, the degree of polarization (DOP) is computed as (Hecht 2001)

$$d(p) = \frac{I_{max}(l, p) - I_{min}(l, p)}{I_{max}(l, p) + I_{min}(l, p)}. \tag{10.12}$$

In the case of diffuse reflection, by combining Fresnel transmission coefficients and Snell's law, we have a different formulation of the DOP depending on the refractive index and zenith angle (Atkinson and Hancock 2007a):

$$d(p) = d(\theta(p), \eta(p)) =$$
$$\frac{(\eta(p) - \eta^{-1}(p))^2 \sin^2 \theta(p)}{2 + 2\eta^2(p) - (\eta(p) + \frac{1}{\eta(p)})^2 \sin^2 \theta(p) + 4 \cos \theta(p) \sqrt{\eta^2(p) - \sin^2(\theta(p))}}. \tag{10.13}$$

However, $I_{max}(l, p) + I_{min}(l, p)$ is in fact the image intensity captured without the polarizer (Nayar et al. 1997), $I(l, p)$. Hence, from equation [10.11], equation [10.12] and equation [10.13], we formulate the relationship:

$$I(l, v, p) = \frac{I(l, p)\left[1 + d(\theta(p), \eta(p)) \cos(2v - 2\phi(p))\right]}{2}. \tag{10.14}$$

Therefore, if we capture an image with two different polarizer angles, υ_1 and υ_2, the ratio between image intensities can be computed:

$$\frac{I(l, \upsilon_1, p)}{I(l, \upsilon_2, p)} = \frac{1 + d(\theta(p), \eta(p)) \cos(2\upsilon_1 - 2\phi(p))}{1 + d(\theta(p), \eta(p)) \cos(2\upsilon_2 - 2\phi(p))}.$$ [10.15]

Similar to the above constraint on light directions, considering two two-dimensional vectors, we have

$$\boldsymbol{p_1} = \big(I(l, \upsilon_1, p), I(l, \upsilon_2, p)\big)^T$$ [10.16]

$$\boldsymbol{p_2} = \big(1 + d(\theta(p), \eta(p)) \cos(2\upsilon_1 - 2\phi(p)),$$ [10.17]

$$1 + d(\theta(p), \eta(p)) \cos(2\upsilon_2 - 2\phi(p))\big)^T.$$

Here, the constraint equation [10.15] is transformed into an angle constraint to improve its robustness:

$$\arccos \frac{\boldsymbol{p_1} \cdot \boldsymbol{p_2}}{\|\boldsymbol{p_1}\| \, \|\boldsymbol{p_2}\|} = 0.$$ [10.18]

This is the second constraint that will be used in our reconstruction algorithm. It is independent of the incident irradiance and the albedo of the object.

10.6. Normal estimation with two constraints

In this section, we present two algorithms that reconstruct the object using the proposed constraints. In the first algorithm, we assume that all the light directions are known, as they are in most conventional non-Lambertian photometric stereo methods. Adopting the second algorithm, assuming the light directions are unknown, we recover the surface normals, refractive indexes and light directions.

We define an error function for both constraints. First, for each pair of light directions l_1 and l_2 at polarizer angle υ, the error function based on constraint equation [10.10] for surface point p is defined:

$$E_s(l_1, l_2, \upsilon, p) = \left(\arccos \frac{s_1 \cdot s_2}{\|s_1\| \, \|s_2\|} \right)^2,$$ [10.19]

where s_1 and s_2 are, respectively, defined by equation [10.8] and equation [10.9].

Second, for each pair of polarizer angles υ_1 and υ_2 with light direction l, the error function based on constraint equation [10.18] is defined:

$$E_p(l, \upsilon_1, \upsilon_2, p) = \left(\arccos \frac{\boldsymbol{p_1} \cdot \boldsymbol{p_2}}{\|\boldsymbol{p_1}\| \, \|\boldsymbol{p_2}\|} \right)^2,$$ [10.20]

where $\boldsymbol{p_1}$ and $\boldsymbol{p_2}$ are, respectively, defined by equation [10.16] and equation [10.17].

Considering N surface points, we obtain a total error using both constraints:

$$\mathbb{E}\left(\{\boldsymbol{l}\}, \{p\}, \{v\}\right) = w_s \sum_{\forall \boldsymbol{l}_i \neq \boldsymbol{l}_j, v, p} \alpha(\boldsymbol{l}_i, p)\alpha(\boldsymbol{l}_j, p)E_s(\boldsymbol{l}_i, \boldsymbol{l}_j, v, p) \qquad [10.21]$$

$$+ (1 - w_s) \sum_{\forall \boldsymbol{l}, v_i \neq v_j, p} \alpha(\boldsymbol{l}, p)E_p(\boldsymbol{l}, v_i, v_j, p)$$

$$+ \lambda g(\{\boldsymbol{l}\}, \{p\})$$

where $\alpha(\boldsymbol{l}, p)$ is the visibility coefficient of the light source for direction $\boldsymbol{l}$ and point p. $\alpha(\boldsymbol{l}, p)$ is computed from the actual image intensity of p. If the image intensity is below a threshold (i.e. the image intensity is too dark), then $\alpha(\boldsymbol{l}, p) = 0$; otherwise, $\alpha(\boldsymbol{l}, p) = 1$. w_s and $1 - w_s$ are, respectively, the weights of the shading-stereoscopic and polarization-stereoscopic terms, such that $0 \leq w_s \leq 1$. g is a prior term that represents the prior constraints on the surface normal and light directions, and λ is a weight used to control this term. In the following sections, we describe this prior term specifically for each algorithm.

Given L light directions of the light source and P polarizer angles for each light direction, we have $N_L = L(L-1)/2$ pairs of light directions and $N_P = P(P-1)/2$ pairs of polarizer angles, which means we have N_L equations using equation [10.10] for each polarizer angle and N_P equations using equation [10.18] for each light direction for N points. However, theoretically, there are at most $L-1$ equations for the first constraint and $P-1$ for the second constraint that are effective, and the remaining equations can be derived from those equations. In particular, for polarization, the polarizer angle and image intensity are modulated by equation [10.11], and we need to capture images with only three polarizer angles to recover all the parameters in equation [10.11] (Wolff and Boult 1991). Images for other polarizer angles can be derived from those three images. Moreover, the polarization effect is independent of the light source. We therefore have a condition for the number of unknowns U and number of points N:

$$N\left((L-1)min(P,3) + (min(P,3) - 1)\right) \geq U, \qquad [10.22]$$

In practice, more than three polarization images are usually captured to improve the robustness against image noise (Atkinson and Hancock 2007b; Huynh et al. 2013).

10.6.1. *Algorithm 1: Recovering individual surface points*

Given all light directions, the first algorithm estimates the normal vector for single point associated with pixel p with just three unknowns: $\theta(p), \phi(p), \eta(p)$. In this setting, the surface normal and refractive index of individual pixel p are solved directly by adopting the least-square method:

$$\theta^*(p), \phi^*(p), \eta^*(p) = \arg\min_{\theta(p), \phi(p), \eta(p)} \mathbb{E}\left(\{\boldsymbol{l}\}, p, \{v\}\right). \qquad [10.23]$$

The prior term g is applied to ensure the visibility of surface point. We use a sigmoid-like function (Higo et al. 2010) to make a constraint that favors $\alpha > 0$; an error term is used:

$$\sigma(\alpha, k) = \frac{1 - \alpha}{1 + exp(k\alpha)},$$ [10.24]

where k controls the steepness of this term. Eventually, the prior term is as follows:

$$g = \sum_p \sigma\left(\boldsymbol{n}(p) \cdot \boldsymbol{o}(p), k_n\right)$$ [10.25]

In implementation, we solve the optimization using the Levenberg–Marquardt method. We employ a photometric stereo method (Woodham 1980) to initialize the surface normals, and refractive indexes are initialized to 1. k_n is set to 100 in our experiments.

According to equation [10.22], a minimum configuration of the light direction and polarization angles is $(N, L, P) = (1, 4, 1)$. However, to take advantage of both shading and polarization information, we need at least $(N, L, P) = (1, 3, 2)$.

10.6.2. *Algorithm 2: Recovering shape and light directions*

The second algorithm tackles a much harder situation in which L light directions are unknown. From equation [10.22], we see that the reconstruction cannot work with only one surface point. Instead, we have to observe N points simultaneously. The number of unknowns in this case is $U = 2L + 3N$, where each light direction l can be parameterized by two angles. A minimum configuration for which the algorithm can be applied is $(N, L, P) = (3, 4, 2)$. There are a number of outliers that do not follow the diffuse reflection model owing to specular reflection from the light source or from nearby points (inter-reflection) on the object surface, and large image noise. We simply employ the random sampling method to vote for the light directions with fewer surface points by optimizing equation [10.21]:

$$\{\boldsymbol{l}^*\}, \{\theta^*(p), \phi^*(p), \eta^*(p)\} = \arg\min_{\forall l, p} \mathbb{E}\left(\{\boldsymbol{l}\}, \{p\}, \{v\}\right),$$ [10.26]

and then use *Algorithm 1* to estimate the shape and refractive indexes of the whole object.

To initialize the optimization equation [10.26], we set all surface normals and the light direction towards the camera, which is the direction $(0, 0, 1)$, and the refractive indexes to be 1. Because the optimization function equation [10.21] is nonlinear, the optimization may produce ambiguous solutions for light directions, and we need a constraint for some light directions to guide the optimization function to select the correct ones. We assume that the first and fourth light sources are, respectively, located in the first and second octants of the world coordinate system (i.e. $l_{1,x} > 0, l_{1,y} > 0, l_{1,z} > 0$ and $l_{4,x} < 0, l_{4,y} > 0, l_{4,z} > 0$) and use the above sigmoid-like function, equation [10.24], for this constraint. The prior term is implemented as:

$$g = \sum_{p} \sigma\Big(\boldsymbol{n}(p) \cdot \boldsymbol{o}(p), k_n\Big) + \sum_{l} \sigma\Big(\boldsymbol{l}(p) \cdot \boldsymbol{o}(p), k_l\Big)$$

$$+ \sigma(\boldsymbol{l}_{1,x}, k_l) + \sigma(\boldsymbol{l}_{1,y}, k_l) + \sigma(-\boldsymbol{l}_{4,x}, k_l) + \sigma(\boldsymbol{l}_{4,y}, k_l), \qquad [10.27]$$

where the first and second lines constrain that the surface is visible and the light sources are not in front of the camera, respectively. In our experiments, k_n and k_l are set to 100 and 50, respectively.

10.7. Experiments

We evaluate the proposed algorithms by conducting simulations and real-world experiments. In the case of real-world experiments, we validated the proposed methods using two different types of polarization camera.

We compare the performances of the proposed algorithms with those of state-of-the-art photometric stereo (Woodham 1980), state-of-the-art polarization-based (Atkinson and Hancock 2006; Smith et al. 2019), shading-polarization fusion (Saman et al. 2011) and deep learning-based photometric stereo (Chen et al. 2019) methods. These benchmark methods are, respectively, denoted as *Woodham1980, Atkinson2006, Smith2019_Calib, Smith2019_Uncalib, Saman2011* and *ChenSDPSNet2019*.

In *Saman2011*, light sources are calibrated and surface normals are estimated using *Woodham1980*, and refractive indices are then estimated. Meanwhile, *Atkinson2006* needs a predefined refractive index (e.g. 1.5) and three known light source directions to disambiguate the surface normal. *Smith2019_Calib* and *Smith2019_Uncalib* are proposed by Smith et al. (2019)[1] for shape estimation from at least three polarization images with a single light source and a predefined refractive index (i.e. 1.5). *Smith2019_Calib* assumes known light source direction, while *Smith2019_Uncalib* does not. Because our datasets include eight light sources, these methods produce eight corresponding surface normal maps. For each of these two methods, we use the median normal map as the final map for comparison. In addition, *Smith2019_Uncalib* can only work with objects having uniform albedo in our datasets. *ChenSDPSNet2019*, a deep-learning-based photometric stereo method proposed by Chen et al. (2019)[2],

---

1. Codes: https://github.com/waps101/depth-from-polarisation.

2. Codes: https://github.com/guanyingc/SDPS-Net.

was used. Because this method takes unpolarized 24-bit color object images as the input, we had to crop and convert the images of our datasets to match their input format.

10.7.1. *Simulation experiments with weights for two constraints*

To perform simulation experiments, we created a polarization ray-tracing rendering program that simulates diffuse polarization in the case of an orthogonal camera and distant light source for an opaque dielectric sphere with a refractive index of 1.4553. Examples of original polarization images are shown in Figure 10.3. Gaussian noise with deviation of 0.01 (1% of the highest normalized image intensity) is added randomly to the images. For each light source, we use three polarizer angles, $v \in \{0^o, 45^o, 90^o\}$, to generate polarization images of the object.

Figure 10.3. *Three polarization images taken at a zero polarizer angle for three light directions*

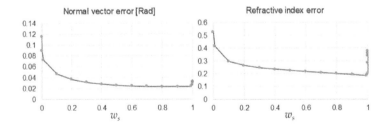

Figure 10.4. *Algorithm 1: Average error for all image pixels and different weights given to the two constraints. For a color version of this figure, see www.iste.co.uk/funatomi/computational.zip*

In this experiment, we analyze the effects of the weights given to the two constraints in equation [10.21]. The results of the experiments are presented in Figure 10.4 for average errors of the surface normal and refractive index and in Figure 10.5 for the error distribution for all image pixels. The polarization-stereoscopic constraint is not used when the weight w_s is 1, and the shading-stereoscopic constraint is not used when w_s is zero. When the shading-stereoscopic constraint is not used, both the surface normal and refractive index are inaccurate. Standard polarization methods face ambiguity of the surface orientation and ambiguity of the relationship between the refractive index and zenith angle. In contrast, when the polarization-stereoscopic constraint is not used, the shading-stereoscopic constraint still works to recover the surface normal but cannot recover the refractive index well. For a weight between those two extremes, $0 < w_s < 1$, the proposed method takes advantage of the two constraints in recovering both the surface normal and refractive index robustly. Moreover, we see that the setting of the weights is relaxed because the proposed method works for a wide range of weights. We choose a balanced weight, $w_s = 0.5$, in the remaining experiments.

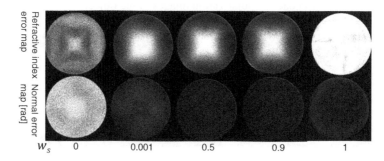

Figure 10.5. *Algorithm 1: Error map of image pixels for different weights given to the two constraints. A brighter pixel indicates larger error*

The distribution of error in Figure 10.5 shows a limitation of the proposed method in that the estimation of the refractive index is inaccurate at the center of the image where the zenith angle is small and the DOP is low. This limitation is not only for the proposed method but also for any method that uses polarization images.

10.7.2. *Real-world experiments*

We carried out experiments in a darkroom with two polarization cameras. For the first experimental setup, we employed a Lumenera INFINITY-3 color camera (8 bits/channel) but only used one image channel. The polarizer was mounted on a Sigma Koki SGSP-120YAW rotation stage. Both camera and rotation stage were controlled automatically with a personal computer, which allowed us to capture polarization images easily. A halogen light source was mounted on a tripod at a distance of about 2.8 m from the target object. All light directions were manually measured, which are visualized in Figure 10.11. The setup of the camera system is shown in Figure 10.7a. Examples of captured images with the eight light directions are shown in Figure 10.6. The distance from the camera to an object was about 1.2 m, while the object size was less than 5 cm. We captured polarization images for eight light directions. For each light direction, we captured four polarization images at $0^o, 54^o, 90^o$ and 144°. The dataset captured with this camera is referred to as the *Lumenera* dataset; this dataset was used in Ngo et al. (2015).

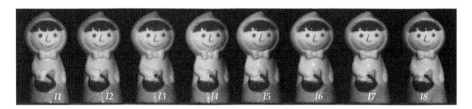

Figure 10.6. *Porcelain doll in our experiment. Images shown have a zero polarizer angle for eight light directions*

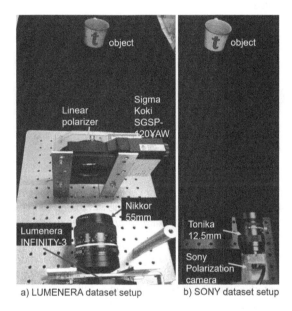

a) LUMENERA dataset setup   b) SONY dataset setup

Figure 10.7. *Experimental setup with a) Lumenera and b) SONY cameras. For a color version of this figure, see www.iste.co.uk/funatomi/computational.zip*

For the second experimental setup, we used a grayscale SONY polarization sensor (Maruyama et al. 2018). In this camera, a micro polarization filter array with fixed angles was integrated on the surface of photodiodes of the image sensor, which resembles the Bayer mosaic of a color filter array. The resolution of the image sensor was $2,064 \times 1,552$ and the pixel depth was 12 bits. We also used eight light sources more than 3 m away from the target object in directions similar to those in the above setup. The intensities of light sources were calibrated using a Lambertian sphere, and the directions of the light sources are visualized in Figure 10.8. The distance from the camera to the target object was about 0.8 m. After decoding the sensor image, we obtained four polarization images at $0°, 45°, 90°$ and $135°$ for each light direction. The experimental setup with the SONY camera is shown in Figure 10.7b. The dataset captured with this camera is referred to as the *SONY* dataset.

| Doll | Cup1 | Cat | Cup2 | Bunny | Bear1 | Squirrel | Bear2 | Egg | Cup3 | Light Directions |

Figure 10.8. *Ten porcelain objects and scatter plot of all eight light directions of the SONY dataset. The shown images were taken with the first light source and a polarizer angle of $0°$. Light directions are visualized with respect to the camera direction using their (x, y) coordinates*

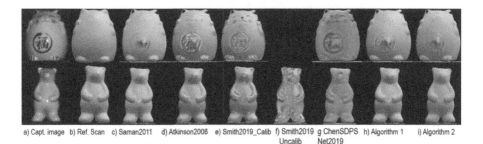

a) Capt. image b) Ref. Scan c) Saman2011 d) Atkinson2006 e) Smith2019_Calib f) Smith2019 g ChenSDPS h) Algorithm 1 i) Algorithm 2
_Uncalib Net2019

Figure 10.9. *SONY dataset: examples of surface normal reconstruction using the proposed and benchmark methods. For a color version of this figure, see www.iste.co.uk/funatomi/computational.zip*

The advantage of the first camera is that the image pixels at different polarizer angles are of the same object points. However, we need to capture several images at different polarizer angles, which takes longer than a conventional photometric stereo method does. Meanwhile, for the second camera, we only need to take one image for each light direction, similar to the case of a conventional photometric stereo method does. However, the disadvantage is that the polarization images of different polarizer angles are of nearby object points.

The shape of objects was captured by a 3Shape E1 3D depth scanner (© 3Shape 2021). The scanner provided object surface meshes computed from depth. The accuracy of the depth sensing was 10 μm. We used the scanned object shapes as reference for evaluating the estimated shapes. We adopted gradient descent optimization to find an optimal reprojection of the reference shape onto the estimated normal map and thus compute the mean absolute difference of an estimated normal map relative to the reference shape. The mean absolute difference was then used to evaluate the performances of the estimation methods. To facilitate the optimization, the initial pose of the 3D scan of an object is manually fixed. This alignment is inspired by the concept of a bundle adjustment (Hartley and Zisserman 2003).

In the following sections, we present experiments conducted for two datasets. For each dataset, we evaluate the results obtained with *Algorithm 1* using ground-truth light directions and then evaluate the results of *Algorithm 2* obtained without knowing the light directions.

SONY dataset

All objects of the dataset are shown in Figure 10.8. There are large variations in shape, surface smoothness and texture. The objects are not always convex, and there is strong inter-reflection and clearly visible specular highlighting.

The reconstruction examples of the proposed and benchmark methods are illustrated in Figure 10.9 for *Cat* and *Bear2*. In the figure, the surface normal $n(p)$ is encoded using *RGB* color with $(n_x(p)+1, n_y(p)+1, n_z(p)+1)$. We also conducted a quantitative evaluation of the dataset by matching the reconstructed shapes with their 3D scanned data. The evaluation results are given in Table 10.1.

Overall, *Atkinson2006* performs worst because it relies only on the polarization cue, which is relatively weak and noisy. Moreover, for the SONY camera, polarization images of different polarizer angles are of nearby object points as mentioned earlier, and the mathematical formulation of the polarization cue may not hold. The formulation only holds when the object is smooth and homogeneous in terms of the refractive index and surface albedo. We see that *Atkinson2006* is more sensitive to surface texture than the other methods. However, if an object is smooth, homogeneous and without inter-reflection, such as *Egg*, *Atkinson2006* delivers the best performance. In contrast, *Woodham1980* works with images that are unpolarized by averaging polarization images and it produces a smoother surface and loses some detail (e.g. at the arms and legs of the cat and the eyes of the bear in Figure 10.9). Overall, it performs much better than *Atkinson2006*. Although *Smith2019_Calib* and *Smith2019_Uncalib* handle polarization like *Atkinson2006*, the performances of these methods are not as good. This is because they disambiguate the surface normal using only a single light source in comparison with three light sources used by *Atkinson2006*. In the case of *Smith2019_Uncalib*, the accuracy is even worse when the estimation of the light source direction is inaccurate. Meanwhile, *Algorithm 1* takes advantage of both the polarization cue (demonstrated in *Atkinson2006*) and shading cue (demonstrated in *Woodham1980*) and delivers the best performance overall. Meanwhile, *Algorithm 2* has to recover light directions simultaneously with shape and thus slightly sacrifices shape accuracy.

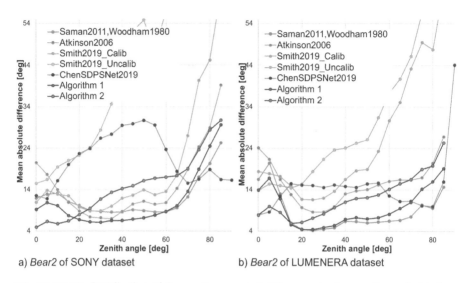

a) *Bear2* of SONY dataset   b) *Bear2* of LUMENERA dataset

Figure 10.10. *Distribution of the surface normal difference by the zenith angle bin for Bear2 of a) SONY dataset and b) Lumenera dataset. The bin size is 5 degrees. For a color version of this figure, see www.iste.co.uk/funatomi/computational.zip*

Methods	Doll	Cup1	Cat	Cup2	Bear1	Squirrel	Bear2	Egg	Cup3
Woodham1980, Saman2011	13.18	9.22	11.06	10.89	**6.88**	10.66	9.68	12.15	7.56
Atkinson2006	18.51	10.77	12.61	19.65	13.41	13.87	10.83	**7.68**	8.88
Smith2019_Calib	36.27	11.44	15.26	33.11	15.89	16.03	12.7	8.34	9.48
Smith2019_Uncalib	NA	NA	NA	NA	23.02	19.69	38.55	24.01	16.30
ChenSDPSNet2019	23.84	19.56	24.47	35.36	25.68	19.00	24.57	21.17	17.06
Algorithm 1	**12.83**	**5.67**	**8.08**	**9.51**	8.42	**9.74**	**7.85**	8.31	**4.53**
Algorithm 2	14.38	6.36	10.94	15.24	14.21	11.23	12.83	10.6	5.27

Table 10.1. *SONY dataset: mean absolute difference (in degrees) of an object's normal in comparison with its 3D scanned data. The best results are shown in bold*

Only *ChenSDPSNet2019*, *Smith2019_Uncalib* can recover the light directions among the benchmark methods. The accuracy for light direction is given in Table 10.2. Overall, the accuracy of the light direction is better than that of the normal. This is because the light direction is estimated from many object points. The estimation accuracy of *Algorithm 2* is much better than those of *Smith2019_Uncalib* and *ChenSDPSNet2019*. For *Smith2019_Uncalib*, this is mainly because *Smith2019_Uncalib* assumes Lambertian reflectance and surface normal estimation using only polarization, which is noisy and ambiguous. In addition, *Smith2019_Uncalib* can only work with objects of uniform albedo.

Methods	Doll	Cup1	Cat	Cup2	Bunny	Bear1	Squirrel	Bear2	Egg	Cup3
Smith2019_Uncalib	NA	NA	NA	NA	NA	11.61	9.93	15.52	11.92	16.76
ChenSDPSNet2019	15.30	13.90	15.36	28.92	**7.20**	25.06	10.33	17.37	10.49	6.60
Algorithm 2	**6.58**	**6.81**	**5.21**	**5.36**	8.08	**6.31**	**6.62**	**4.94**	**4.82**	**5.95**

Table 10.2. *SONY dataset: light direction estimation error (degrees). The best results are shown in bold*

Examples in Figure 10.9 show that the performances of all the methods, except *ChenSDPSNet2019*, were affected by specularity. Although there are not many specularity points, they still affect the final average accuracy of the estimation. These methods, other than *Smith2019_Calib* and *Smith2019_Uncalib*, do not handle specularity explicitly, such that specularity points act as outliers. In contrast, *Smith2019_Calib* and *Smith2019_Uncalib* cope with the specularity by first detecting the specularity points and handling them separately using specular polarization. However, our dataset was captured with a narrow dynamic range and specularity was thus not handled. Meanwhile, the normal estimation network of *ChenSDPSNet2019* uses max pooling to compute a global feature from all the input images and it may be capable of avoiding specularity. Moreover, the smoothness of neighboring pixels is implicitly handled by the convolution kernels. As a result, the estimated normal maps of *ChenSDPSNet2019* are smooth and unaffected by specularity.

Finally, the breakdown of the surface normal error by the zenith angle for all the methods is illustrated for one of the objects shown in Figure 10.9, namely *Bear2*. The distribution of the mean absolute difference from the reference 3D scan by the zenith angle is shown in Figure 10.10a. We see that all methods have low performance for points of a large zenith angle (i.e. larger than 60 degrees) where the image intensity is too dark and noisy. For points of small zenith angle (i.e. less than 20 degrees), where specularity is visible, the performances of all the methods decline. However, the situation is worst for a purely polarization-based method, like *Atkinson2006*, because of weak polarization. For points in the middle range of the zenith angle (i.e. between 20 and 60 degrees), all methods except *Smith2019_Uncalib* perform well. Overall, for this object, *Algorithm 1* performed best because it takes advantage of both polarization and shading.

Lumenera dataset

The dataset includes six objects as shown in Figure 10.11. The reconstruction examples of all the methods are illustrated in Figure 10.12 for *Doll* and *Bear2*. The surface normal reconstruction is evaluated in Table 10.3. The breakdown of surface normal error is illustrated in Figure 10.10b with the same object (*Bear2*) of experiment for the *SONY* dataset. It is noted that there are no 3D scanned data of three objects (*Lemon*, *Bunny* and *Squirrel*) in this dataset and the shape evaluation is not available.

Overall, we obtain results similar to those obtained with the *SONY* dataset. *Algorithm 1* performs best in terms of normal estimation.

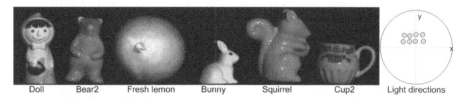

| Doll | Bear2 | Fresh lemon | Bunny | Squirrel | Cup2 | Light directions |

Figure 10.11. *Six objects and scatter plot of all eight light directions of the Lumenera dataset. The shown images were taken with the first light source and a polarizer angle of $0°$. Light directions are visualized with respect to the camera direction using their (x, y) coordinates*

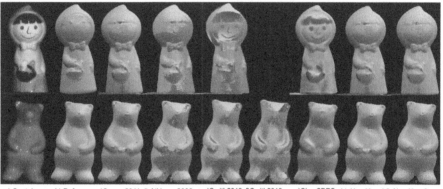

a) Capt. image b) Reference c)Saman2011 d) Atkinson2006   e)Smith2019 f)Smith2019_   g)ChenSDPS   h) Algorithm 1 i) Algorithm 2
_Calib   Uncalib   Net2019

Figure 10.12. *Lumenera dataset: examples (Doll and Bear2) of surface normal reconstruction using the proposed and benchmark methods. For a color version of this figure, see www.iste.co.uk/funatomi/computational.zip*

The light direction accuracies for all objects are given in Table 10.4. The results demonstrate a clear advantage of our proposed method over *Smith2019_Uncalib* and *ChenSDPSNet2019*.

In summary, results obtained for the two datasets show that employing both polarization and shading cues helps the proposed methods recover not only better shape but also light directions. Although we do not have a complete solution for disambiguation of the light directions and surface normals, the results are promising compared with those of Drbohlav and Sara (2001), where two polarizers are controlled to relax the ambiguity of the light directions and surface normals. Their accuracy was about $10°$ even for 14-bit images. This accuracy is shared by an existing uncalibrated photometric stereo method (Chandraker et al. 2005). One limitation of the proposed methods is that they are not robust against specularity and still need some prior with which to disambiguate in estimating the light direction. In the next section, we illustrate the performance of the proposed method when no prior is employed to disambiguate the light directions.

Methods	Doll	Bear2	Cup2
Woodham1980, Saman2011	10.6	**6.47**	10.83
Atkinson2006	16.44	12.72	16.04
Smith2019_Calib	33.49	18.47	31.2
Smith2019_Uncalib	NA	24.35	NA
ChenDSPSNet2019	22.25	14.23	20.90
Algorithm 1	**9.74**	6.88	**10.37**
Algorithm 2	10.2	9.97	12.95

Table 10.3. *Lumenera dataset: mean absolute difference (in degrees) of an object's normal in comparison with the 3D scanned data. The best results are shown in bold*

Method	Doll	Bear2	Lemon	Bunny	Squirrel	Cup2
Smith2019_Uncalib	NA	16.46	11.44	NA	18.84	NA
ChenSDPSNet2019	14.57	6.03	7.05	4.09	5.29	8.55
Algorithm 2	**6.15**	**3.46**	**6.87**	**3.53**	**5.26**	**3.09**

Table 10.4. *Lumenera dataset: light direction estimation error (degree). The best results are shown in bold*

Ambiguity and effect of polarization

In the above experiments conducted with *Algorithm 2*, we employed a simple prior for the light sources in that one light source is on the left and one is on the right of the camera. We conduct an experiment without this prior on the light sources to investigate the ambiguity and effect of polarization.

The results of the ambiguity experiment are shown in Figure 10.13, where most estimated objects clearly differ from their 3D scans. Specifically, we see that the estimated surface normals consistently change direction (e.g. from pointing up to pointing down or from pointing left to pointing right), making the estimated shapes concave. This experiment clearly shows ambiguity without a light source prior.

The next experiment examines the effect of the weight of the polarization or shading constraint. Results of the experiment are shown in Figure 10.14 for *Bear2*, which is unambiguously estimated in the above experiment using the *Lumenera* dataset (see Figure 10.13). Overall, the estimated shape is far from the reference 3D scan when using an extremely high/low weight of polarization/shading. When using an extremely high weight of polarization, there are highly variant estimation results of surface normals in the central area of the object with a small zenith angle where the polarization effect is weak. However, the overall shape is not ambiguous. In contrast, when using an extremely high weight of shading, the estimated shape is ambiguous. The experiment illustrates that the ambiguity problem is more serious with a higher weight of shading.

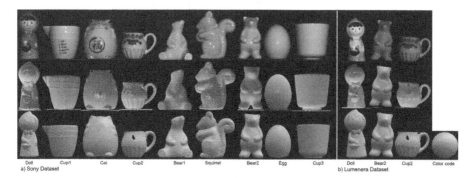

Figure 10.13. *Shapes reconstructed with Algorithm 2 without a prior on light sources for a) SONY and b) Lumenera datasets. Most objects that are actually convex are recovered as concave objects. The captured images of the objects are presented in the first row while the recovered shapes and their 3D scans are, respectively, presented in the second and third rows. The normal color code is on the right of the third row. For a color version of this figure, see www.iste.co.uk/funatomi/computational.zip*

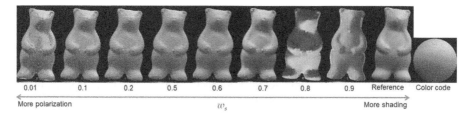

Figure 10.14. *Shapes reconstructed with Algorithm 2 without a prior on light sources for Bear2. The experiment is conducted with a varying weight of the shading constraint. For a color version of this figure, see www.iste.co.uk/funatomi/computational.zip*

In summary, a simple prior on the light sources (i.e. one light source is on the left and one light source is on the right of the camera) can resolve the ambiguity. Using the proposed constraints, we have a heavily constrained equation system for a large number of pixels. The resulting ambiguity is therefore less than the ambiguity of each individual (polarization or shading) constraint. Meanwhile, existing methods for uncalibrated light sources with a number of much more sophisticated priors (e.g. specularity, shadow-coding and employing two polarizers) still face the convex/concave ambiguity; see further discussion in section 10.2. Only methods that assume the shape prior, such as convexity (Okabe et al. 2009) and concavity (Chandraker et al. 2005), fully resolve the ambiguity.

10.8. Conclusion and future works

We presented a shape reconstruction method that integrates shading and polarization to take advantage of and overcome the disadvantages of these two cues. We presented two constraints: one constraint for shading with a pair of light directions and one constraint for polarization with a pair of polarizer angles. As a result, the surface normals, light directions and refractive indexes can be recovered. The proposed algorithms were positively evaluated in both simulations and real-world experiments.

For the estimation of light directions, we currently use a simple method to solve the ambiguity because there is appreciable relaxation when incorporating polarization compared with conventional methods. However, in future work, we would like to improve this solution by employing a stronger technique, such as introducing a smoothness constraint (Belhumeur et al. 1999) or homogeneous refractive index, and thus improve the shape accuracy. A solution to the problem of inter-reflection is another topic left for future work.

10.9. References

3Shape (2021). 3Shape E1 [Online]. Available at: https://www.3shape.com/en/scanners/lab.

Atkinson, G.A. (2017). Polarisation photometric stereo. *Computer Vision and Image Understanding*, 160, 158–167.

Atkinson, G.A. and Hancock, E.R. (2005). Multi-view surface reconstruction using polarization. *IEEE International Conference on Computer Vision (ICCV)*, vol. 1, pp. 309–316.

Atkinson, G.A. and Hancock, E.R. (2006). Recovery of surface orientation from diffuse polarization. *IEEE Transactions on Image Processing*, 15(6), 1653–1664.

Atkinson, G.A. and Hancock, E.R. (2007a). Shape estimation using polarization and shading from two views. *IEEE Transactions on Pattern Analysis and Machine Intelligence*, 29(11), 2001–2017.

Atkinson, G.A. and Hancock, E.R. (2007b). Surface reconstruction using polarization and photometric stereo. *Computer Analysis of Images and Patterns Lecture Notes in Computer Science*, pp. 466–473.

Basri, R., Jacobs, D., Kemelmacher, I. (2006). Photometric stereo with general, unknown lighting. *International Journal of Computer Vision*, 72(3), 239–257.

Belhumeur, P.N., Kriegman, D.J., Yuille, A.L. (1999). The bas-relief ambiguity. *International Journal of Computer Vision*, 35(1), 33–44.

Chandraker, M.K. (2014). What camera motion reveals about shape with unknown BRDF. *IEEE Computer Society Conference on Computer Vision and Pattern Recognition (CVPR)*.

Chandraker, M.K., Kahl, F., Kriegman, D.J. (2005). Reflections on the generalized bas-relief ambiguity. *IEEE Computer Society Conference on Computer Vision and Pattern Recognition (CVPR)*, vol. 1, pp. 788–795.

Chen, L., Zheng, Y., Subpa-asa, A., Sato, I. (2018). Polarimetric three-view geometry. *European Conference on Computer Vision (ECCV)*.

Chen, G., Han, K., Shi, B., Matsushita, Y., Wong, K.-Y.K. (2019). SDPS-Net: Self-calibrating deep photometric stereo networks. *IEEE Conference on Computer Vision and Pattern Recognition*.

Cui, Z., Gu, J., Shi, B., Tan, P., Kautz, J. (2017). Polarimetric multi-view stereo. *IEEE Conference on Computer Vision and Pattern Recognition (CVPR)*, pp. 369–378.

Drbohlav, O. and Sara, R. (2001). Unambiguous determination of shape from photometric stereo with unknown light sources. *IEEE International Conference on Computer Vision (ICCV)*, pp. 581–586.

Drbohlav, O. and Sara, R. (2002). Specularities reduce ambiguity of uncalibrated photometric stereo. *European Conference on Computer Vision (ECCV)*, pp. 46–60.

Foix, S. and Aleny, G. (2011). Lock-in time-of-flight (ToF) cameras: A survey. *IEEE Sensors Journal*, 11(3), 1–11.

Hartley, R. and Zisserman, A. (2003). *Multiple View Geometry in Computer Vision*, 2nd edition. Cambridge University Press.

Hayakawa, H. (1994). Photometric stereo under a light source with arbitrary motion. *Journal of the Optical Society of America A*, 11(11), 3079.

Hecht, E. (2001). *Optics*, 4th edition. Addison Wesley.

Higo, T., Matsushita, Y., Ikeuchi, K. (2010). Consensus photometric stereo. *IEEE Computer Society Conference on Computer Vision and Pattern Recognition*, pp. 1157–1164.

Huynh, C.P., Robles-Kelly, A., Hancock, E.R. (2013). Shape and refractive index from single-view spectro-polarimetric images. *International Journal of Computer Vision*, 101(1), 64–94.

Inoshita, C., Mukaigawa, Y., Matsushita, Y., Yagi, Y. (2014). Surface normal deconvolution : Photometric stereo for optically thick translucent objects. *European Conference on Computer Vision (ECCV)*, pp. 1–15.

Kadambi, A., Taamazyan, V., Shi, B., Raskar, R. (2015). Polarized 3d: High-quality depth sensing with polarization cues. *IEEE International Conference on Computer Vision (ICCV)*, pp. 3370–3378.

Lu, F., Matsushita, Y., Sato, I., Okabe, T., Sato, Y. (2013). Uncalibrated photometric stereo for unknown isotropic reflectances. *IEEE Conference on Computer Vision and Pattern Recognition*, pp. 1490–1497.

Mahmoud, A.H., El-melegy, M.T., Farag, A.A. (2012). Direct method for shape recovery from polarization and shading. *2012 19th IEEE International Conference on Image Processing (ICIP)*, pp. 1769–1772.

Maruyama, Y., Terada, T., Yamazaki, T., Uesaka, Y., Nakamura, M., Matoba, Y., Komori, K., Ohba, Y., Arakawa, S., Hirasawa, Y. et al. (2018). 3.2-mp back-illuminated polarization image sensor with four-directional air-gap wire grid and 2.5-μm pixels. *IEEE Transactions on Electron Devices*, 65(6), 2544–2551.

Miyazaki, D. and Ikeuchi, K. (2010). Photometric stereo under unknown light sources using robust SVD with missing data. *IEEE International Conference on Image Processing*, pp. 4057–4060.

Miyazaki, D., Saito, M., Sato, Y., Ikeuchi, K. (2002). Determining surface orientations of transparent objects based on polarization. *The Journal of the Optical Society of America A*, 19(4), 687–694.

Miyazaki, D., Kagesawa, M., Ikeuchi, K. (2004). Transparent surface modeling from a pair of polarization images. *IEEE Transactions on Pattern Analysis and Machine Intelligence*, 26(1), 73–82.

Nayar, S.K., Fang, X.-S., Boult, T. (1997). Separation of reflection components using color and polarization. *International Journal of Computer Vision*, 21(3), 163–186.

Ngan, A., Durand, F., Matusik, W. (2005). Experimental analysis of BRDF models. *Eurographics Conference on Rendering Techniques, EGSR '05*. Eurographics Association, Aire-la-Ville, pp. 117–126.

Ngo, T.T., Nagahara, H., Taniguchi, R. (2015). Shape and light directions from shading and polarization. *IEEE Conference on Computer Vision and Pattern Recognition (CVPR)*, pp. 2310–2318.

Ngo, T.T., Nagahara, H., Nishino, K., Taniguchi, R., Yagi, Y. (2019). Reflectance and shape estimation with a light field camera under natural illumination. *International Journal of Computer Vision*, 127(11), 1707–1722.

Ngo, T.T., Nagahara, H., Taniguchi, R. (2021a). Polarization and shading datasets [Online]. Available at: https://ngottx.github.io/Projects/ShapeReflectance/Polshading_shape.html.

Ngo, T.T., Nagahara, H., Taniguchi, R. (2021b). Surface normals and light directions from shading and polarization. *IEEE Transactions on Pattern Analysis and Machine Intelligence*, 99, 1–11.

Okabe, T., Sato, I., Sato, Y. (2009). Attached shadow coding: Estimating surface normals from shadows under unknown reflectance and lighting conditions. *IEEE International Conference on Computer Vision*, pp. 1693–1700.

Papadhimitri, T. and Favaro, P. (2013). A new perspective on uncalibrated photometric stereo. *IEEE Conference on Computer Vision and Pattern Recognition*, pp. 1474–1481.

Papadhimitri, T. and Favaro, P. (2014). Uncalibrated near-light photometric stereo. *British Machine Vision Conference (BMVC)*, pp. 1–12.

Rahmann, S. (1999). Inferring 3D scene structure from a single polarization image. *Industrial Lasers and Inspection*, Munich.

Saito, M., Sato, Y., Ikeuchi, K., Kashiwagi, H. (1999). Measurement of surface orientations of transparent objects using polarization in highlight. *IEEE Computer Society Conference on Computer Vision and Pattern Recognition (CVPR)*, vol. 1, pp. 381–386.

Saman, G., Hancock, E., Science, C. (2011). Refractive index estimation using photometric stereo. *IEEE International Conference on Image Processing (ICIP)*, pp. 1925–1928.

Santo, H., Samejima, M., Sugano, Y., Shi, B., Matsushita, Y. (2017). Deep photometric stereo network. *2017 IEEE International Conference on Computer Vision Workshops (ICCVW)*, pp. 501–509.

Shi, B., Matsushita, Y., Wei, Y., Xu, C., Tan, P. (2010). Self-calibrating photometric stereo. *IEEE Computer Society Conference on Computer Vision and Pattern Recognition (CVPR)*, pp. 1118–1125.

Smith, W.A.P., Ramamoorthi, R., Tozza, S. (2016). Linear depth estimation from an uncalibrated, monocular polarisation image. *European Conference on Computer Vision (ECCV)*, vol. 9912, pp. 109–125.

Smith, W.A.P., Ramamoorthi, R., Tozza, S. (2019). Height-from-polarisation with unknown lighting or albedo. *IEEE Transactions on Pattern Analysis and Machine Intelligence*, 41(12), 2875–2888.

Sunkavalli, K., Zickler, T., Pfister, H. (2010). Visibility subspaces: Uncalibrated photometric stereo with shadows. In *European Conference on Computer Vision (ECCV)*, Daniilidis, K., Maragos, P., Paragios, N. (eds). Springer, Berlin, Heidelberg.

Wolff, L.B. and Boult, T. (1991). Constraining object features using a polarization reflectance model. *IEEE Transactions on Pattern Analysis and Machine Intelligence*, 13(7), 635–657.

Wolff, L.B., York, N., Oren, M. (1998). Improved diffuse reflection models for computer vision. *International Journal of Computer Vision*, 30(1), 55–71.

Woodham, R.J. (1980). Photometric method for determining surface orientation from multiple images. *Optical Engineering*, 19(1).

Yang, L., Tan, F., Li, A., Cui, Z., Furukawa, Y., Tan, P. (2018). Polarimetric dense monocular slam. *IEEE/CVF Conference on Computer Vision and Pattern Recognition*, pp. 3857–3866.

Zhang, L. and Hancock, E.R. (2013). Robust estimation of shape and polarisation using blind source separation. *Pattern Recognition Letters*, 34(8), 856–862.

Zhu, D. and Smith, W.A. (2019). Depth from a polarisation + RGB stereo pair. *IEEE/CVF Conference on Computer Vision and Pattern Recognition (CVPR)*.

Zickler, T.E., Belhumeur, P.N., Kriegman, D.J. (2002). Helmholtz stereopsis: Exploiting reciprocity for surface reconstruction. *International Journal of Computer Vision*, 49(2), 215–227.

11

Polarization Imaging in the Wild Beyond the Unpolarized World Assumption

Jérémy Maxime RIVIERE

Realistic Graphics and Imagining, Imperial College London, UK

11.1. Introduction

Polarization imaging has been widely studied in the past for material classification (Wolff 1990), shape estimation (Wolff 1989; Miyazaki and Ikeuchi 2005; Guarnera et al. 2012; Kadambi et al. 2015; Smith et al. 2016) and reflectometry (Miyazaki et al. 2003). All of these methods make the "unpolarized world" assumption, i.e. they assume the incident illumination is completely unpolarized. The common guiding thread of all these methods is to measure the polarization induced by reflection off the material's surface, by taking multiple photographs (at least three) of the object through a rotating linear polarizer placed in front of the camera sensor. In this chapter, we will first cover some background on polarization from specular reflection and the necessary mathematical tools to handle polarized light. We continue to follow the assumption that diffuse reflection completely depolarizes the incident illumination. While this has been shown not to hold for 3D objects near occluding contours (Atkinson and Hancock 2006), it greatly simplifies the foundational theories for the case covered in this chapter: capturing the appearance of spatially varying, planar surfaces under unconstrained, outdoor illumination.

Computational Imaging for Scene Understanding,
coordinated by Takuya FUNATOMI and Takahiro OKABE.
© ISTE Ltd 2024.

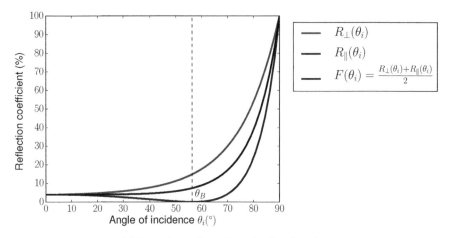

(a) Fresnel equations at an air–glass interface

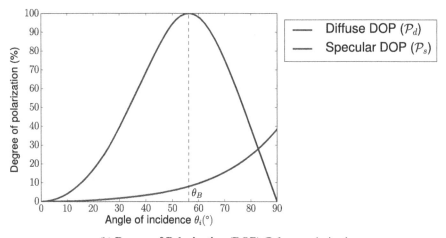

(b) **Degree of Polarization (DOP)** $\mathcal{P}$ for unpolarized
incident illumination reflected at an air–glass interface

Figure 11.1. *Plots of the Fresnel equations and degree of polarization
at an air–dielectric interface. For a color version of this figure, see
www.iste.co.uk/funatomi/computational.zip*

   Polarization refers to the intrinsic property of a wave to oscillate at a preferred
orientation, orthogonal to the direction of propagation. For light waves, polarization
refers to the oscillation of the electric field ($\vec{E}$-vector) along the direction of
propagation $\vec{\omega}$. When at any point in time, no particular direction of oscillation can
be determined, the ray of light is said to be unpolarized. Light bulbs are an example
of light sources emitting unpolarized light.

One way to obtain polarized light from an unpolarized beam of light is by reflection off a dielectric surface: according to Fresnel equations, a portion of the light gets reflected perpendicularly to the plane of incidence (plane containing the normal to the surface $\vec{n}$ and ray of light incident at $\vec{\omega}_i$) while the rest is reflected parallel to the plane of incidence (Figure 11.1a). At a particular angle of incidence θ_B, the component of light parallel to the plane of incidence gets completely transmitted, thus reflecting purely linearly polarized light perpendicular to the plane of incidence. At any other angle of incidence, the reflected ray is partially polarized, i.e. a mixture of unpolarized and linearly polarized light where the degree of polarization (DOP) depends on the angle of incidence (Figure 11.1b, blue curve). A similar, weaker effect can also be observed on reflection on metals, where the main difference is that the p-polarized component never goes to 0 and hence complete polarization is never attained.

In optics, there exist two common mathematical frameworks to describe the polarization state of light and its changes through polarizing elements: Jones calculus, developed by Robert Clarke Jones in 1941 and Mueller calculus, developed by Hans Mueller in 1943 to model the interactions of Stokes vectors with polarizing optical elements (section 11.2). In this chapter, as in related literature, we consider the latter framework, as Jones calculus can only account for fully polarized light. As such, Mueller calculus can be seen as a generalization of Jones calculus to partial polarization, which covers unpolarized, partially polarized and fully polarized light.

11.2. Mueller calculus

Mueller calculus is a mathematical framework used to manipulate Stokes vectors, where the effects of a particular optical element are represented by a 4x4 real-valued matrix called the Mueller matrix. The Stokes parameters, developed by George Gabriel Stokes in 1852, are a set values that describe the polarization of light in terms of its total intensity ($L(\vec{\omega})$), DOP ($\mathcal{P}$) and the parameters of the polarization ellipse which is the geometric figure traced by the tip of the electric field as it oscillates, when looking down the direction of propagation (Figure 11.2a). For mathematical convenience, Stokes parameters are represented as a four-dimensional vector (Collett 2005) as follows:

$$\vec{s} = \begin{bmatrix} s_0 \\ s_1 \\ s_2 \\ s_3 \end{bmatrix} = \begin{bmatrix} L(\vec{\omega}) \\ L(\vec{\omega})\mathcal{P}\cos 2\psi \cos 2\chi \\ L(\vec{\omega})\mathcal{P}\sin 2\psi \cos 2\chi \\ L(\vec{\omega})\mathcal{P}\sin 2\chi \end{bmatrix} \qquad [11.1]$$

where s_0 is the total intensity of the incident beam, s_1 and s_2, respectively, are the intensities of $0°$ and $+45°$ linear polarization and s_3 is the intensity of right circular polarization. It is common to look at Stokes vectors in a normalized space (i.e. $s_0 = 1$) by dividing each component by s_0. The Poincaré sphere (Figure 11.2b) then serves as a convenient visualization tool where the last three normalized Stokes parameters are parametrized in spherical coordinates. In this parametrization, the Stokes vectors form an orthonormal basis spanning the space of unpolarized ($\mathcal{P} = 0$), partially polarized ($0 \leq \mathcal{P} \leq 1$) and fully polarized ($\mathcal{P} = 1$) light.

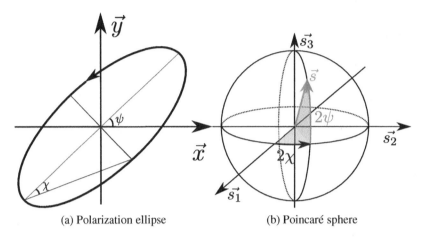

(a) Polarization ellipse   (b) Poincaré sphere

Figure 11.2. *Visualizing polarization: When looking down the propagation direction of a light wave, the tip of its electric field traces an ellipse as it oscillates (a). Stokes parameters are related to the parameters of the polarization ellipse as per equation [11.1]. They span a 3D space represented in spherical coordinates on the Poincaré sphere (b). For a color version of this figure, see www.iste.co.uk/funatomi/ computational.zip*

When a beam of light interacts with an optical element, the resulting polarization state is given by a simple matrix-vector multiplication in Mueller calculus:

$$\vec{s_o} = M\vec{s_i} \qquad\qquad [11.2]$$

where $\vec{s_i}$ is the polarization state of the beam of light incident on the optical element, M encapsulates the optical properties of the polarizing element and $\vec{s_o}$ is the resulting Stokes vector.

The combined effect of light interacting with multiple optical elements is simply modeled by stacking the Mueller matrices of each element by the right multiplication rule. Given N polarizing elements, each associated with a Mueller matrix $M_k, \forall k \in [1, N]$, their combined effect on the input Stokes vector $\vec{s_i}$ is defined as:

$$\vec{s_o} = M_N M_{N-1}...M_k...M_1\vec{s_i} \qquad\qquad [11.3]$$

For the purpose of this chapter, we will be interested particularly in two types of optical elements: polarizing filters (section 11.3) and reflectors (section 11.3.2), which are the building blocks for polarization imaging.

11.3. Polarizing filters

Polarizing filters are optical elements designed to selectively transmit only light in a particular polarization state. They exist in two flavors: circular and linear polarizers. For brevity, we will only present the latter as the application presented in section 11.6 does not involve measuring the effects of circular polarization. Refer to Chapter 5 of Riviere (2017) and related articles for details on circular polarizers.

11.3.1. *Linear polarizers*

Linear polarizers are extremely popular with outdoors hobbyists, especially for water-sports, as they are well suited to reduce glare caused by the sun reflecting at a glancing angle off the surface of water, where the unpolarized light from the sun becomes largely horizontally polarized on reflection, due to Fresnel effects. Polarized sunglasses are therefore vertically polarized in order to suppress the unpleasant glare. In graphics and vision, linear polarization is often used for diffuse-specular separation, where linear polarizers are affixed to both lights and cameras.

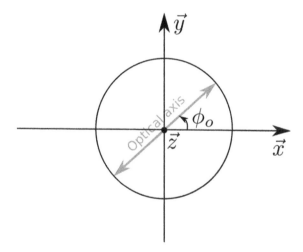

Figure 11.3. *Geometry of a general linear polarizer rotated at an angle ϕ_o from its local coordinate system. For a color version of this figure, see www.iste.co.uk/funatomi/computational.zip*

In Mueller formalism, a linear polarizer is defined with respect to a local Cartesian coordinate $(\vec{x}, \vec{y}, \vec{z})$, where $\vec{x}$ is the horizontal direction, $\vec{y}$ the vertical direction and

$\vec{z}$ the direction of transmission (Figure 11.3). A linear polarizer with its optical axis rotated at an angle ϕ_o to the horizon is thus defined as:

$$M_{pol}(\phi_o) = M_{rot}(-\phi_o)M_{horiz\_pol}M_{rot}(\phi_o)$$

$$= \frac{1}{2}\begin{bmatrix} 1 & \cos 2\phi_o & \sin 2\phi_o & 0 \\ \cos 2\phi_o & \cos^2 2\phi_o & \cos 2\phi_o \sin 2\phi_o & 0 \\ \sin 2\phi_o \cos 2\phi_o \sin 2\phi_o & \sin^2 2\phi_o & 0 \\ 0 & 0 & 0 & 0 \end{bmatrix} \quad [11.4]$$

where M_{horiz_pol} is the Mueller matrix for an ideal horizontal polarizer:

$$M_{horiz\_pol} = \frac{1}{2}\begin{bmatrix} 1 & 1 & 0 & 0 \\ 1 & 1 & 0 & 0 \\ 0 & 0 & 0 & 0 \\ 0 & 0 & 0 & 0 \end{bmatrix} \quad [11.5]$$

and $M_{rot}(\alpha)$ is the Mueller matrix of a rotator that transforms local into global coordinates:

$$M_{rot}(\alpha) = \begin{bmatrix} 1 & 0 & 0 & 0 \\ 0 & \cos 2\alpha & -\sin 2\alpha & 0 \\ 0 & \sin 2\alpha & \cos 2\alpha & 0 \\ 0 & 0 & 0 & 1 \end{bmatrix} \quad [11.6]$$

11.3.2. Reflectors

As previously mentioned, unpolarized light becomes partially linearly polarized on reflection off a surface, due to Fresnel effects. Such effects are represented, in Mueller calculus, by the combined effects of a linear diattenuator and linear retarder of phase δ:

$$M_r = \begin{bmatrix} \dfrac{R_\perp + R_\parallel}{2} & \dfrac{R_\perp - R_\parallel}{2} & 0 & 0 \\ \dfrac{R_\perp - R_\parallel}{2} & \dfrac{R_\perp + R_\parallel}{2} & 0 & 0 \\ 0 & 0 & \sqrt{R_\parallel R_\perp}\cos\delta & \sqrt{R_\parallel R_\perp}\sin\delta \\ 0 & 0 & -\sqrt{R_\parallel R_\perp}\sin\delta & \sqrt{R_\parallel R_\perp}\cos\delta \end{bmatrix} \quad [11.7]$$

where $R_\parallel$ and $R_\perp$ are the parallel (respectively perpendicular) reflectance coefficients as predicted by Fresnel equations and δ is the relative phase between the parallel and perpendicular polarized components. For dielectrics, δ is a simple step-edge function of the angle of incidence: $\delta = \pi$ for any angle of incidence before Brewster angle (θ_B)

and 0 otherwise. For conductors, it is a complex function of the angle of incidence (Chen and Wolff 1998). Here, the input and output coordinate systems are defined with respect to the plane of incidence containing the ray of light incident at $\vec{\omega}_i$, surface normal $\vec{n}$ and direction of perfect reflection $\vec{\omega}_r$.

As with linear polarizers, reflectors can be rotated with respect to their local coordinate system:

$$M'_r(\phi) = M_{rot}(\phi) M_r M_{rot}(-\phi) \tag{11.8}$$

where ϕ is the inclination angle of the plane of incidence with respect to the local up vector (Figure 11.4).

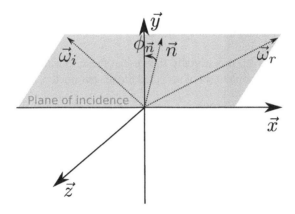

Figure 11.4. *Geometry of a general reflecting optical element rotated by an angle $\phi_{\vec{n}}$ from its local coordinate system. For a color version of this figure, see www.iste.co.uk/funatomi/computational.zip*

11.4. Polarization imaging

The goal of polarization imaging is to recover the optical properties of a surface's material from observations of its reflected Stokes parameters. In particular, polarized light from reflection provides useful cues for shape estimation. A common technique to measure the complete set of parameters of the reflected Stokes ($\vec{s}_r$) is to take three observations of the sample under study with a linear polarizer rotated at three different orientations in front of a camera and an additional measurement with a circular polarizer (assumed right-handed). In the interest of brevity, we will restrict our discussion to linear Stokes measurement only. By direct applications of the theory presented in section 11.2, we explain why $0°$, $45°$ and $90°$ are commonly used as the orientations of the linear polarizer when measuring linear reflected Stokes parameters (Ghosh et al. 2010; Guarnera et al. 2012; Kadambi et al. 2015):

$$\vec{s}_o(0°) = \left[\frac{s_{r,0} + s_{r,1}}{2}, \frac{s_{r,0} + s_{r,1}}{2}, 0, 0 \right]^T$$

$$\vec{s}_o(45°) = \left[\frac{s_{r,0} + s_{r,2}}{2}, 0, \frac{s_{r,0} + s_{r,2}}{2}, 0 \right]^T$$

$$\vec{s}_o(90°) = \left[\frac{s_{r,0} - s_{r,1}}{2}, \frac{s_{r,1} + s_{r,0}}{2}, 0, 0 \right]^T \qquad [11.9]$$

where $s_o(\vec{\phi}_o)$ indicates the output Stokes vector as observed through a linear polarizer rotated at angle ϕ_o with the camera's coordinate system as the frame of reference. The superscript T indicates matrix transposition.

From equation [11.9], it is trivially shown that these four measurements are well suited to recover the complete set of reflected Stokes parameters as:

$$s_{r,0} = s_{o,0}(0°) + s_{0,0}(90°)$$

$$s_{r,1} = s_{o,0}(0°) - s_{0,0}(90°)$$

$$s_{r,2} = 2s_{o,0}(45°) - s_{r,0} \qquad [11.10]$$

It is worth noting that the first parameter of the reflected Stokes vector ($s_{r,0}$) can be obtained from any combination of measurements taken at orthogonal directions of the linear polarizer. If we denote ϕ_o an arbitrary rotation of the polarizer, we obtain:

$$s_{o,0}(\phi_o) + s_{o,0}\left(\phi_o + \frac{\pi}{2}\right) = \frac{s_{r,0} + s_{r,1}\cos(2\phi_o) + s_{r,2}\sin(2\phi_o)}{2}$$

$$+ \frac{s_{r,0} + s_{r,1}\cos\left(2\left(\phi_o + \frac{\pi}{2}\right)\right) + s_{r,2}\sin\left(2\left(\phi_o + \frac{\pi}{2}\right)\right)}{2}$$

$$= s_{r,0} + \frac{s_{r,1}\cos(2\phi_o) + s_{r,2}\sin(2\phi_o)}{2}$$

$$- \frac{s_{r,1}\cos(2\phi_o) + s_{r,2}\sin(2\phi_o)}{2}$$

$$= s_{r,0} \qquad [11.11]$$

The reflected Stokes vector $\vec{s}_r$ takes a different form depending on the polarization state of the incident illumination. Most prior work on polarization has focused on resolving shape from polarization (SfP) under unpolarized incident illumination, by measuring the first three components of the reflected Stokes vector as per the protocol described above. In the following, we present the theoretical foundations for polarization imaging under partially linearly polarized incident illumination (section 11.5.1), first studied in Riviere et al. (2017), who showed it to be a generalization of the well-studied problem of polarization imaging under the unpolarized world assumption. These derivations constitute the theoretical basis for polarization imaging reflectometry under natural illumination (section 11.6), known to be partially linearly polarized (in general) due to single-scattering of light by molecules in the atmosphere (Strutt 1871a, 1871b).

11.5. Image formation model

11.5.1. *Partially linearly polarized incident illumination*

In Stokes formalism, partially linearly polarized light is obtained from equation [11.1] when the fourth component of the Stokes vector is 0. This corresponds to $\chi = k\frac{\pi}{2}, \forall k \in \mathbb{Z}$. The incident Stokes vector for partially linearly polarized illumination is thus defined as:

$$\vec{s}_i = \begin{bmatrix} L_i(\vec{\omega}_i) \\ \pm L_i(\vec{\omega}_i)\mathcal{P}_i \cos(2\psi_i) \\ \pm L_i(\vec{\omega}_i)\mathcal{P}_i \sin(2\psi_i) \\ 0 \end{bmatrix} \tag{11.12}$$

where the sign of the second and third parameters is the same and depends on the sign of $\cos(2\chi)$. The angle of polarization (ψ_i) is defined with respect to the right-handed local coordinate system spanned by $\vec{\omega}_i$ the direction of incidence, $\vec{E}_\perp$ the direction orthogonal to the plane of incidence and $\vec{E}_\parallel$ the direction parallel to the plane of incidence (Figure 11.5).

Upon reflection off a surface, the polarization state of an incident beam of light is changed differently by diffuse and specular reflection. It is commonly assumed that diffuse reflection depolarizes light, which is the assumption adopted by Riviere et al. (2017) and hence in this chapter. Therefore, by application of equation [11.4], the intensity observed at any orientation of the linear polarizer for diffuse reflection (assuming Lambertian reflection) is:

$$I_d(\phi_o) = \frac{1}{2}\frac{\rho_d}{\pi}L_i(\vec{\omega}_i) \tag{11.13}$$

where ρ_d is the material's diffuse albedo.

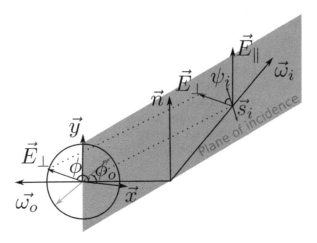

Figure 11.5. *Polarization imaging geometry: The angle of polarization ψ_i of the incident beam of light is relative to the local coordinate system spanned by $[\vec{E}_\perp, \vec{E}_\parallel, \vec{\omega}_i]$. The outgoing coordinate system is defined by $[\vec{x}, \vec{y}, \vec{\omega}_o]$, the local coordinate system of the camera. ϕ_o is the angle of rotation of the linear polarizer in front of the camera, and ϕ is the angle between the camera's $\vec{x}$-axis and the direction perpendicular to the plane of incidence ($\vec{E}_\perp$). For a color version of this figure, see www.iste.co.uk/funatomi/computational.zip*

On the other hand, specular reflection changes the polarization state of incident light according to equation [11.8]:

$$\vec{s}_r = M_{rot}(\phi) M_r \vec{s}_i$$

$$= L_i(\vec{\omega}_i) \begin{bmatrix} \dfrac{R_\perp + R_\parallel}{2} + \mathcal{P}_i \dfrac{R_\perp - R_\parallel}{2} \cos 2\psi_i \\[2ex] \dfrac{R_\perp - R_\parallel}{2} \cos 2\phi + \\[1ex] \mathcal{P}_i \left(\dfrac{R_\perp + R_\parallel}{2} \cos 2\phi \cos 2\psi_i - \sqrt{R_\perp R_\parallel} \sin 2\phi \sin 2\psi_i \cos \delta \right) \\[2ex] \dfrac{R_\perp - R_\parallel}{2} \sin 2\phi + \\[1ex] \mathcal{P}_i \left(\dfrac{R_\perp + R_\parallel}{2} \sin 2\phi \cos 2\psi_i + \sqrt{R_\perp R_\parallel} \cos 2\phi \sin 2\psi_i \cos \delta \right) \\[2ex] -\mathcal{P}_i \sqrt{R_\perp R_\parallel} \sin 2\psi_i \sin \delta \end{bmatrix}$$

$$[11.14]$$

Note that the right-most rotation matrix ($M_{rot}(-\phi)$) was dropped here because the incident Stokes vector is directly expressed in the rotated local coordinate system spanned by $[\vec{E}_\perp, \vec{E}_\parallel, \vec{\omega}_i]$ (Figure 11.5).

Finally, the intensity profile of $\vec{s}_r$ observed through a rotated linear polarizer is obtained by the inner product of the first row of equation [11.4] with equation [11.14]:

$$I_s(\phi_o) = L_i(\vec{\omega}_i) \left(\frac{R_\perp + R_\parallel}{4} + \frac{R_\perp - R_\parallel}{4} \cos\left(2(\phi_o - \phi)\right) \right)$$

$$+ L_i(\vec{\omega}_i) \left(\frac{R_\perp - R_\parallel}{4} + \frac{R_\perp + R_\parallel}{4} \cos\left(2(\phi_o - \phi)\right) \right) \mathcal{P}_i \cos\left(2\psi_i\right)$$

$$+ L_i(\vec{\omega}_i) \frac{\sqrt{R_\perp R_\parallel}}{2} \sin\left(2(\phi_o - \phi)\right) \mathcal{P}_i \sin\left(2\psi_i\right) \qquad [11.15]$$

The complete image formation model under partially linearly polarized incident illumination is obtained by the addition of equation [11.13] and equation [11.15]:

$$I(\phi_o) = I_d(\phi_o) + I_s(\phi_o)$$

$$= \frac{\rho_d L_i(\vec{\omega}_i)}{2\pi} + L_i(\vec{\omega}_i) \left(\frac{R_\perp + R_\parallel}{4} + \frac{R_\perp - R_\parallel}{4} \cos\left(2(\phi_o - \phi)\right) \right)$$

$$+ L_i(\vec{\omega}_i) \left(\frac{R_\perp - R_\parallel}{4} + \frac{R_\perp + R_\parallel}{4} \cos\left(2(\phi_o - \phi)\right) \right) \mathcal{P}_i \cos\left(2\psi_i\right)$$

$$+ L_i(\vec{\omega}_i) \frac{\mathcal{P}_i \sin\left(2\psi\right) \sin\left(2(\phi_o - \phi)\right) \sqrt{R_\perp R_\parallel}}{2} \qquad [11.16]$$

11.5.2. *Unpolarized incident illumination*

By following a similar approach to section 11.5.1, it is trivial to obtain the counterparts to equation [11.13] and equation [11.16] under unpolarized incident illumination where the input Stokes vector is instead $\vec{s}_i = [L_i(\vec{\omega}_i), 0, 0, 0]^T$. Equation [11.13] and equation [11.16] remain identical while $\mathcal{P}_i$ is now 0 in equation [11.12] and equation [11.15], such that the expression for polarization from specular reflection simplifies to:

$$\vec{s}_r = L_i(\vec{\omega}_i) \begin{bmatrix} \dfrac{R_\perp + R_\parallel}{2} \\ \dfrac{R_\perp - R_\parallel}{2} \cos\left(2(\phi)\right) \\ \dfrac{R_\perp - R_\parallel}{2} \sin\left(2(\phi)\right) \\ 0 \end{bmatrix} \qquad [11.17]$$

and the intensity through a rotated linear polarizer (referred to as the transmitted radiance sinusoid (TRS)) simplifies to (with diffuse reflection term):

$$I_s(\phi_o) = \frac{I_{max} + I_{min}}{2} + \frac{I_{max} - I_{min}}{2} \cos\left(2(\phi_o - \phi)\right)$$

where

$$I_{max} = \frac{\rho_d L_i(\vec{\omega}_i)}{2\pi} + \frac{L_i(\vec{\omega}_i) R_\perp}{2}$$

$$I_{min} = \frac{\rho_d L_i(\vec{\omega}_i)}{2\pi} + \frac{L_i(\vec{\omega}_i) R_\parallel}{2} \qquad [11.18]$$

11.5.3. Discussion

Upon comparison of the expressions obtained under partially polarized incident illumination (equation [11.12] and equation [11.15]) and their counterparts under unpolarized incident illumination (equation [11.17] and equation [11.18]), it is easy to understand why most prior work on SfP (Wolff 1989; Guarnera et al. 2012; Miyazaki et al. 2012; Kadambi et al. 2015) has been carried out under the unpolarized world assumption. In these conditions, the angle ϕ is easily obtained either from equation [11.17] (Guarnera et al. 2012) as:

$$\phi = \frac{1}{2} \arctan \frac{s_{r,2}}{s_{r,1}} \qquad [11.19]$$

or by fitting observations through a linear polarizer oriented to three or more positions to equation [11.18] (Wolff 1989; Huynh et al. 2010; Miyazaki et al. 2012; Kadambi et al. 2015). This in turn provides cues for SfP.

Under partially polarized incident illumination, however, it is not clear in general how the angle ϕ can be easily obtained, as the reflected Stokes vector now also depends on the polarization state of the incident illumination, to the extent that some researchers (Atkinson and Hancock 2006) have stated incident linear polarization as a fundamental limitation of SfP.

On close inspection of equation [11.12], however, we argue that those limitations can be relaxed under certain assumptions. First, for dielectric materials, the component of reflection parallel to the plane of incidence is completely transmitted (i.e. $R_\parallel = 0$) at a particular angle of incidence called Brewster angle (θ_B). Under this assumption, equation [11.16] simplifies to:

$$I_{\theta_B}(\phi_o) = \frac{\rho_d L_i(\vec{\omega}_i)}{2\pi} + \frac{L_i(\vec{\omega}_i) R_\perp}{4} \cos\left(2(\phi_o - \phi)\right)$$

$$+ \frac{L_i(\vec{\omega}_i) R_\perp \mathcal{P}_i \cos(2\psi_i)}{4} \cos\left(2(\phi_o - \phi)\right) \qquad [11.20]$$

which can be rearranged in the common form of the TRS (equation [11.18] (plus diffuse term)):

$$I_{\theta_B}(\phi_o) = \frac{I_{max} + I_{min}}{2} + \frac{I_{max} - I_{min}}{2}\cos\left(2(\phi_o - \phi)\right)$$

where

$$I_{max} = \frac{\rho_d L_i(\vec{\omega}_i)}{2\pi} + L_i(\vec{\omega}_i)\frac{\left(1 + \mathcal{P}_i \cos\left(2\psi_i\right)\right) R_\perp}{2}$$

$$I_{min} = \frac{\rho_d L_i(\vec{\omega}_i)}{2\pi} \tag{11.21}$$

We further observe that equation [11.16] simplifies greatly when $\sin\left(2\psi_i\right) = 0$, which corresponds to the incident illumination being either horizontal (i.e. $\psi_i = 0$) or vertical (i.e. $\psi_i = \frac{\pi}{2}$). In these cases, equation [11.16] can be written in a similar way as equation [11.18] (with diffuse reflection term):

$$I(\phi_o) = \frac{I_\perp + I_\perp}{2} + \frac{I_\perp - I_\parallel}{2}\cos\left(2(\phi_o - \phi)\right)$$

where

$$I_\perp = \frac{\rho_d L_i(\vec{\omega}_i)}{2\pi} + L_i(\vec{\omega}_i)\frac{\left(1 + \mathcal{P}_i \cos\left(2\psi_i\right)\right)R_\perp}{2}$$

$$I_\parallel = \frac{\rho_d L_i(\vec{\omega}_i)}{2\pi} + L_i(\vec{\omega}_i)\frac{\left(1 - \mathcal{P}_i \cos\left(2\psi_i\right)\right)R_\parallel}{2} \tag{11.22}$$

Note that here, the TRS is defined in terms of $I_\perp$ and $I_\parallel$ instead of I_{max} and I_{min} since the way $I_\perp$ and $I_\parallel$ compare to each other depends on the sign of $\cos\left(2\psi_i\right)$.

The expressions in equation [11.21] and equation [11.22] mean that, in principle, we should recover the incident polarization in order to recover $R_\perp$ and $R_\parallel$. However, in section 11.8.3, we will discuss the challenges in recovering the incident polarization, as well as practical guidelines to overcome those challenges. Also note that equation [11.21] and equation [11.22] are generalizations of equation [11.18] where the latter is obtained when $\mathcal{P}_i = 0$.

Finally, the derivations in this chapter were made under the assumption of perfect mirror reflection and point light illumination. To account for reflection from rough surfaces under spherical illumination (as is the case under natural illumination), specular reflections should be modeled according to the microfacet theory (Torrance and Sparrow 1967). Equation [11.13] and equation [11.15] thus become:

$$I'_d(\phi_o) = \int_{\Omega^+} I_d(\phi_o)(\vec{n}.\vec{\omega}_i)d\vec{\omega}_i$$

$$I'_s(\phi_o) = \int_{\Omega^+} I_s(\phi_o)f_s(\vec{\omega}_i,\vec{\omega}_o)(\vec{n}.\vec{\omega}_i)d\vec{\omega}_i \qquad [11.23]$$

where $f_s(\vec{\omega}_i,\vec{\omega}_o)$ is a Cook-Torrance microfacet bidirectional reflectance distribution function (BRDF) with a GGX (Walter et al. 2007) distribution term, forming a narrow lobe around the reflection vector. Within the extent of the specular lobe, the incident polarization can be assumed constant, as the polarization field typically varies smoothly over the sky (Können 1985). The derivations in this chapter serve as a theoretical basis to inform the design of a measurement protocol for reflectometry in general outdoors conditions as proposed by Riviere et al. (2017) and which we detail in section 11.6.

11.6. Polarization imaging reflectometry in the wild

Based on the theoretical derivations presented earlier for polarization imaging under partially linearly polarized incident illumination, we (Riviere et al. 2017) derive a measurement protocol for reflectometry "in the wild" to recover high-resolution, spatially varying reflectance maps for planar, isotropic dielectric surfaces based on commodity hardware.

In particular, we propose to image a sample at a minimum of three vantage points: one observation should be made close to normal incidence in order to have a canonical frame of reference for data registration, while two more observations should be taken close to Brewster angle of incidence, roughly orthogonal to each other, in order for equation [11.20] to apply. This measurement protocol is further supported by the work of Nielsen et al. (2015) who show measurements near Brewster angle (around 60° for dielectrics) to be nearly optimal for reflectance estimation restricted to a single measurement.

The main reason for limiting the number of measurements to only three viewpoints stems from a practical point of view: some of the acquired samples (Figures 11.11 and 11.12, first row) were captured on busy walking paths where the capture process was interrupted multiple times by passers-by.

11.7. Digital single-lens reflex (DSLR) setup

11.7.1. *Data acquisition*

The data acquisition setup is designed around commodity photography equipment, often used for image-based lighting applications (Debevec 1998):

– **a DSLR camera**: in the experiments presented thereafter, a 17.9 MegaPixel (MP) Canon EOS 650D camera with an 18–55 mm lens was used, to which a 58 mm rotatable linear polarizer from Edmund Optics was mounted. Throughout the capture process, the camera was mounted on a heavy duty tripod for stability.

– **a calibration target:** an X-Rite ColorChecker® chart placed next to the sample for radiometric calibration.

– **a chrome ball:** to record the incident illumination.

Figure 11.6 shows a typical arrangement of the capture setup for outdoor measurements, where the linear polarizer was manually rotated in front of the camera at three marked orientations $(0°, 45°, 90°$, marked on the polarizer) to image the polarization of light reflected off the sample, as per equation [11.9].

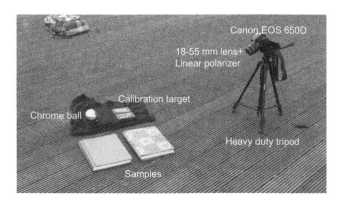

Figure 11.6. *Principal polarization imaging setup: Commodity photography equipment often used for image-based lighting applications (Debevec 1998) can be used for data acquisition. For a color version of this figure, see www.iste.co.uk/funatomi/computational.zip*

The measurement protocol (Figure 11.7) proceeds as follows: first, image the s_0 component of the reflected Stokes field of the sample close to normal incidence, to provide a canonical frame of reference for data registration. While this could be done by photographing the sample without polarizer, it was found to be more practical to leave the polarizer on at all times during capture and simply capture the $0°$ and $90°$ orientations only at normal incidence. Then, proceed to the complete measurement of the linear Stokes parameters for two roughly orthogonal views close to Brewster angle of incidence. While in principle, equation [11.20] and equation [11.21] are only true at the exact Brewster angle of incidence, it was found that being in a ±15° window around Brewster angle is sufficient, in practice, for good qualitative measurements. This will be discussed in more detail in section 11.9.2. Typical measurements took around 5 minutes per sample when uninterrupted and up to 20 minutes when interrupted by passers-by, as was the case for the "drain cover" sample (Figures 11.11 and 11.12, first row).

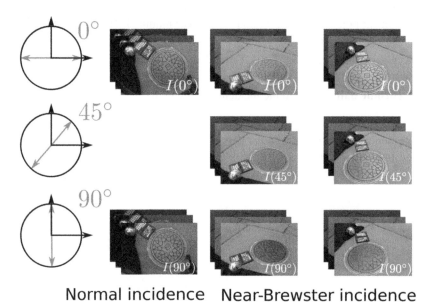

Normal incidence   Near-Brewster incidence

Figure 11.7. *Polarization imaging reflectometry in the wild: A set of high dynamic range (HDR) sequences near normal incidence and close to Brewster angle of incidence is acquired. For a color version of this figure, see www.iste.co.uk/funatomi/ computational.zip*

A simple practical guideline for finding the best near-Brewster measurement in practice proceeds as follows: with the polarizer oriented at 90° (i.e. vertical) with respect to the camera's $\vec{x}$-axis pointed at the sample, adjust the camera's height until minimum transmission through the polarizer is achieved. In this configuration, the camera is seeing the sample at a near-Brewster angle defined from the sample's mean up-vector. The rationale behind this idea is that given the planar nature of the samples, the plane of incidence is mainly vertical and the horizontally reflected light at Brewster angle is therefore cross-polarized with the vertical polarizer on the camera.

Each near-Brewster measurement consists of multiple frames per orientation of the polarizer, where each frame is taken at a different exposure level in order to recover HDR maps of the reflected Stokes parameters.

For each view, the HDR sequences captured at $0°$, $45°$ and $90°$ are then combined into the linear sRGB HDR radiance maps $I(0°)$, $I(45°)$ and $I(90°)$, respectively, using pfstools (Mantiuk et al. 2007), although any other suitable method to obtain HDR imagery can be used instead.

11.7.2. *Calibration*

Radiometric calibration

As for any data-driven approach for reflectometry, calibration of the acquired data is essential before any further processing. The X-Rite ColorChecker® next to the sample is thus used for radiometric calibration of both the sample's response and light probe captured by the chrome ball. After assembling the HDR frames ($I(0°)$, $I(45°)$ and $I(90°)$), their overall brightness is scaled such that the white point of the calibration target reads an intensity of $[0.45, 0.45, 0.45]$. Note that the linear sRGB value for this patch is $[0.9, 0.9, 0.9]$ but equation [11.13] predicts that the observed intensity through a linear polarizer for diffusely reflected light is halved.

Geometric calibration

For surface normal estimation (section 11.8.1) and roughness estimation through inverse rendering (section 11.8.3), camera pose estimations are further required. For this purpose, Wu's VisualSfM (Wu 2011), a GUI-based structure from motion (SfM) software package, is used to obtain camera poses and tracks (set of correspondences between 3D world coordinates and 2D projected points on the image plane), which are refined through bundle adjustment (Wu et al. 2011). From those tracks, it is possible to compute homography matrices to register the acquired data to the canonical view captured at normal incidence and perform all subsequent calculations in image space of the canonical view.

11.7.3. *Polarization processing pipeline*

Given a set of HDR observations through a linear polarizer oriented at $\phi_o = 0°, 45°, 90°$ for two roughly orthogonal observations near the Brewster angle of incidence, start by fitting equation [11.21] per pixel (in the least-squares sense) to each set of observations. While, in principle, any nonlinear optimization package could be used to perform this task, it was found that such approach is inefficient in practice, with fitting times in the order of two hours at the camera's resolution, when fitting with the curve fitting routine from SciPy's optimization package (Jones et al. 2001).

Instead, note that by rearranging the terms in equation [11.21] as in equation [11.24], we obtain a linear problem of the form $Ax = b$ which can be solved for very efficiently through singular value decomposition (SVD), bringing the fitting times under 10 seconds:

$$\underbrace{I_{\theta_B}(\phi_o; x, y)}_{b} = \underbrace{\begin{bmatrix} 1 & \cos 2\phi_o & \sin 2\phi_o \end{bmatrix}}_{A} \underbrace{\begin{bmatrix} \dfrac{I_{max} + I_{min}}{2} \\ \dfrac{I_{max} - I_{min}}{2} \cos 2\phi \\ \dfrac{I_{max} - I_{min}}{2} \sin 2\phi \end{bmatrix}}_{x}$$ [11.24]

The per-pixel parameters of the TRS can then be obtained from the intermediate result of this linear formulation ($\hat{x} = \begin{bmatrix} x_1 & x_2 & x_3 \end{bmatrix}^T$) as:

$$\hat{I}_{max} = x_1 + \sqrt{x_2^2 + x_3^2}$$

$$\hat{I}_{min} = x_1 - \sqrt{x_2^2 + x_3^2}$$

$$\hat{\phi} = \frac{1}{2} \arctan \frac{x_3}{x_2}$$ [11.25]

Figure 11.8 presents TRS parameter maps for a "drain cover" sample, obtained from the fitting protocol described above. These maps constitute the building blocks from which the reflectance estimation pipeline is derived (section 11.8).

(a) $\hat{I}_{max}(x, y)$ (b) $\hat{I}_{min}(x, y)$ (c) $\hat{\phi}(x, y)$

Figure 11.8. *TRS fitting: For each near-Brewster view, a per-pixel fit of equation [11.21] to the acquired data is computed. For a color version of this figure, see www.iste.co.uk/funatomi/computational.zip*

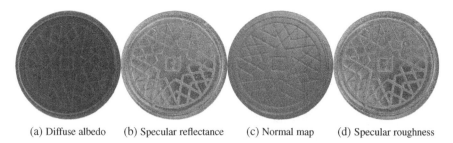

(a) Diffuse albedo   (b) Specular reflectance   (c) Normal map   (d) Specular roughness

Figure 11.9. *Polarization imaging reflectometry: Example reflectance maps recovered for a permanent on-site specular "drain cover", captured on a busy sidewalk close to Imperial College's campus in South Kensington. For a color version of this figure, see www.iste.co.uk/funatomi/computational.zip*

11.8. Reflectance recovery

11.8.1. *Surface normal estimation*

Normal estimation is formulated in a multiview SfP framework, akin to that of Wolff (1989) and Miyazaki et al. (2012). Consider the projection of $\vec{E}_\perp$ on the image plane. From Figure 11.5, it is clear that this vector, denoted $\vec{b}$, is defined as:

$$\vec{b} = \begin{bmatrix} \cos\hat{\phi} & sin\hat{\phi} & 0 \end{bmatrix}^T \tag{11.26}$$

where $\hat{\phi}$ is obtained from [equation (11.24)]. The vector $\vec{b}$ is, by definition, orthogonal to the surface normal $\vec{n}$; therefore, the following expression always holds true:

$$(\vec{b}.\vec{n}) = 0 \tag{11.27}$$

At each viewpoint, knowing $\vec{b}$ essentially constrains the surface normal to lie in the plane of incidence. Combining at least two observations of the TRS phase angle $\hat{\phi}$ from different viewpoints therefore provides enough constraints to fully recover $\vec{n}$.

Given two observations of the TRS phase angle ($\hat{\phi}_0$ and $\hat{\phi}_1$ at two viewpoints close to Brewster angle of incidence (Figure 11.8), with their respective camera rotation matrices in world coordinates (R_0 and R_1) obtained from VisualSfM, the surface normal is recovered by solving the following linear problem:

$$\underbrace{\begin{bmatrix} R_0^T \vec{b}_0 \\ R_1^T \vec{b}_1 \end{bmatrix}}_{2\times 3} \begin{bmatrix} n_x \\ n_y \\ n_z \end{bmatrix} = \begin{bmatrix} 0 \\ 0 \\ 0 \end{bmatrix} \tag{11.28}$$

Similarly to equation [11.24], the above problem can be solved for efficiently using SVD. Figure 11.9c shows the normal map recovered for the "drain cover" dataset.

11.8.2. *Diffuse albedo estimation*

In the framework presented in section 11.5.1, diffuse albedo estimation is straightforward from the observation of $\hat{I}_{min}$. From equation [11.21] and equation [11.23], we have:

$$\hat{I}_{min} = \frac{\rho_d}{2\pi} \underbrace{\int_{\Omega^+} (\vec{n}.\vec{\omega}_i) L_i(\vec{\omega}_i) d\vec{\omega}_i}_{\pi}$$

$$= \frac{\rho_d}{2} \qquad\qquad\qquad [11.29]$$

where the integral part is simplified because of the radiometric calibration step presented in section 11.7.2. The diffuse albedo is thus simply obtained as $\rho_d = 2\hat{I}_{min}$. In principle, any of the two views close to Brewster incidence can be used to estimate the diffuse albedo. However, in practice, it was found that despite best efforts to be as close as possible to Brewster angle as per the protocol described in section 11.7.1, one of the two views may, in some cases, show slightly better specular cancellation (see Figure 11.13 for an example). A better solution is thus to compute $2\hat{I}_{min}$ for each view and set the diffuse albedo (see Figure 11.9a) as the minimum of the two.

11.8.3. *Specular component estimation*

11.8.3.1. *Influence of incident polarization*

As pointed out at the end of section 11.4, the measured reflected intensities ($s_{r,0}$) depend on the state of polarization of the incident illumination which, in principle, should be recovered in order to solve for equation [11.30] to equation [11.32] described thereafter for specular reflectance and roughness estimation. However, in practice, recovering the incident Stokes field from a mirror ball is challenging as reflection from metals becomes elliptical, thus requiring measurements with both a linear and circular polarizer.

However, constantly swapping a linear and circular polarizer in front of the camera proved cumbersome for outdoors measurements and defeated the purpose of simplicity of the presented method. However, using a black shiny dielectric sphere instead to solve for the issue of reflected elliptical illumination did not work either, as recovering the incident Stokes field requires inverting the 4x4 Mueller matrix of the dielectric sphere, which becomes singular at the Brewster angle due to the total transmission of the p-polarized component.

Instead, a practical solution is based on the observation that the incident radiance recorded using the mirror ball already encodes the modulation of intensity in the $s_{r,0}$ component due to the incident partial linear polarization (Figure 11.10). Hence, light probes captured outdoors exhibit darker and brighter sections in the sky due to polarization effects. Figure 11.10 shows that this change in intensity of reflected light observed on a stainless-steel mirror ball is very similar to that on a dielectric around Brewster angle, allowing for a first-order approximation to solve for specular reflectance and roughness.

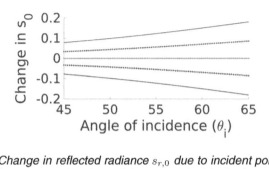

Figure 11.10. *Change in reflected radiance $s_{r,0}$ due to incident polarized illumination (DOP = 80%) is very similar for stainless steel (solid plots) and a dielectric ($\eta = 1.5$, dotted plots) around Brewster angle of incidence. The colors indicate three different angles of polarization with respect to the plane of incidence (Red: $\psi_i = 0°$, Green: $\psi_i = 45°$, Blue: $\psi_i = 90°$). For a color version of this figure, see www.iste.co.uk/funatomi/computational.zip*

11.8.3.2. *Specular reflectance*

From the same view as that used for diffuse albedo estimation, a specular only observation is obtained by subtracting half the diffuse albedo from $\hat{I}_{max}$:

$$\hat{I}_{s,max} = \hat{I}_{max} - \frac{\rho_d}{2}$$

$$= \int_{\Omega^+} \frac{R_\perp(\mathcal{P}_i \cos 2\psi_i + 1)}{2} f_s(\vec{\omega}_i, \vec{\omega}_o) L_i(\vec{\omega}_i)(\vec{n}.\vec{\omega}_i)d\vec{\omega}_i \qquad [11.30]$$

This diffuse-free image encodes $R_\perp$ up to a scale factor, and knowledge of $R_\perp$ is sufficient to recover the material's index of refraction as demonstrated by Ghosh et al. (2010), and subsequently its reflectance at normal incidence ($F(0°)$, see Figure 11.9b) for use with Schlick's approximation (Schlick 1994) to model Fresnel reflectance:

$$\eta = \sqrt{\frac{1 + \sqrt{R_\perp}}{1 - \sqrt{R_\perp}}}$$

$$F(0°) = \frac{(\eta - 1)^2}{(\eta + 1)^2} \qquad [11.31]$$

In practice, recovering $R_\perp$ is challenging as it depends both on the surface's specular roughness and on the polarization state of incident illumination, which is not easily obtained, even when capturing a light probe for the reasons stated earlier.

Instead, a practical solution is to use a template-based approximation where the scale factor is recovered from the observation of the diffuse-free signal on the plastic casing surrounding the calibration target. The latter is made of plastic for which the index of refraction ($\eta_{chart} = 1.46$) can be looked up from online sources (CGSociety 2007; Pixel and Poly n.d.). From this, specular reflectance recovery proceeds as follows:

– first, compute the chart's perpendicular reflection coefficient at Brewster angle under uniform spherical illumination, $R_{\perp,chart}$, and compute the scale factor between the chart's measured diffuse-free intensity and $R_{\perp,chart}$;

– this scale factor is then applied to the sample's diffuse-free intensity image, to obtain an estimate of $R_{\perp}$.

The idea is that, since the samples are planar, they are roughly subject to the same incident illumination as the calibration target placed flat close to the sample.

11.8.3.3. *Specular roughness*

Finally, roughness estimation is formulated as an inverse rendering problem where the goal is to find the value σ that best explains the acquired data by solving the following least-squares problem:

$$\min_{\sigma} \frac{1}{2} \sum_{i} ||I_{r,i} - \hat{I}_{r,i}(\sigma)||_2 \qquad [11.32]$$

where $I_{r,i}$ is the measured reflected intensity at each near-Brewster views and $\hat{I}_{r,i}$ is rendered at each stage of the optimization, given estimates of the diffuse and specular reflection components and normals from the previous steps, and camera poses with their respective light probes. The latter encodes the modulation of intensity in I_r due to incident partial linear polarization. This first-order approximation of incident polarization with view-dependent light probes was found to give satisfying results in practice (see Figures 11.11 and 11.12). The values of $I_{r,i}$ are never measured directly, but recalling equation [11.11], it was shown that the sum of observations from any two orthogonal orientations of the polarizer yields $s_{r,0}$, so in particular:

$$I_r = s_{r,0} = \hat{I}_{min} + \hat{I}_{max} \qquad [11.33]$$

The rendered intensities ($\hat{I}_r(\sigma)$) are computed as follows:

$$\hat{I}_r(\sigma) = \int_{\Omega^+} \left(\frac{\rho_d}{\pi} + \frac{F(\theta)G(\vec{\omega}_i, \vec{\omega}_o, \vec{n})D(\vec{\omega}_i, \vec{\omega}_o, \vec{n}, \sigma)}{4|\vec{n}.\vec{\omega}_i||\vec{n}.\vec{\omega}_o|} \right) L_i(\vec{\omega}_i)(\vec{n}.\vec{\omega}_i)d\vec{\omega}_i$$

$$[11.34]$$

where $L_i(\vec{\omega}_i)$ is obtained from each view's light probe, ρ_d is the diffuse albedo estimated as in section 11.8.2, $\vec{n}$ is the surface normal estimated as described in section 11.8.1, $F(\theta)$ is computed by Schlick's approximation (Schlick 1994) given an estimate of $F(0°)$ (equation [11.31]) and $D(\vec{\omega}_i, \vec{\omega}_o, \vec{n}, \sigma)$ is a GGX distribution term (Walter et al. 2007).

11.9. Results and analysis

11.9.1. *Results*

To assess the validity of the presented method, a series of outdoor measurements were conducted for planar surfaces under varying illumination conditions. Note that most of the samples are permanent on-site structures that cannot be brought in a laboratory for measurement. Figure 11.11 shows reflectance maps (a–d) that were estimated for some datasets following the steps outlined in section 11.8, each measured under different illumination conditions (e). Photo-rendering comparisons are provided in Figure 11.12(a,b), as well as renderings in novel light environments (c). For more results, see Chapter 6 of Riviere (2017).

These exemplar datasets were chosen as they exhibit very different reflectance properties. The "drain cover" (Figures 11.11 and 11.12, first row) was captured on a sidewalk near Queen's Gate in South Kensington, London, which is a busy area due to the close proximity of the Natural History Museum, Victoria & Albert Museum and Hyde Park. As can be seen on the light probe (Figure 11.11, first column, e), it was captured in an environment with trees and buildings all around, showing robustness of the method to clutter in the incident illumination.

Furthermore, it is an interesting sample as it is not strictly speaking a dielectric. Indeed, drain covers are generally made of cast iron, a composite of iron (metal) and carbon (dielectric). Nonetheless, good qualitative results were obtained for this dataset, which can be attributed to two main factors:

1) dielectric behavior tends to dominate in metal–dielectric composites;

2) outdoor metallic surfaces are subject to weathering effects causing oxidation, which adds to the dielectric-like behavior.

The first observation is further validated by the "red book" sample (Figures 11.11 and 11.12, second row), which is covered with a thin layer of metal–dielectric composite paint. It was captured in an open environment with partial cloud coverage and presents a complex texture pattern which is faithfully captured in the reflectance maps.

The "red bricks" and "garden pavement" samples (Figures 11.11 and 11.12, third and fourth rows) are both diffuse-dominated samples captured in very different lighting environments: the first one was captured under a fairly uniform blue sky with a mild cloud coverage, while the second was captured under full overcast conditions.

Note that the latter is an ideal condition in the sense that it corresponds exactly to the well-studied case of polarization imaging under the unpolarized world assumption. It can be seen in Figure 11.11, fourth row, that the surface normals are slightly noisy in this case. This is to be expected as the sample is mostly diffuse thus giving a low signal-to-noise ratio (SNR) in its specular polarization signal. Even then, the proposed method is still able to recover reflectance maps that go a long way towards realism (Figure 11.12, fourth row).

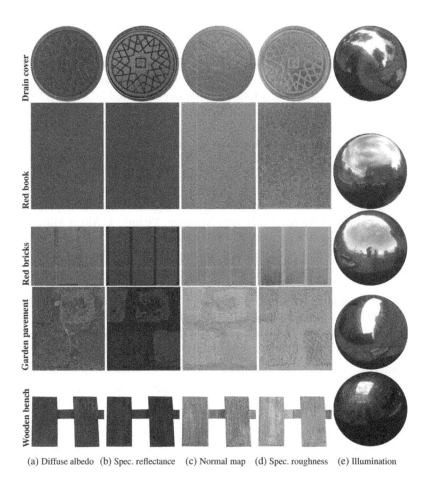

(a) Diffuse albedo   (b) Spec. reflectance   (c) Normal map   (d) Spec. roughness   (e) Illumination

Figure 11.11. *Reflectance maps ((a)–(d)) estimated from two views of the sample close to Brewster angle of incidence, under natural outdoors illumination (e). The proposed method is agnostic to the incident illumination and robust to changes in illumination during capture. For a color version of this figure, see www.iste.co.uk/ funatomi/computational.zip*

Finally, the "wooden bench" sample (Figures 11.11 and 11.12, last row), captured in an interior courtyard surrounded by tall buildings, shows significant variations in specularity due to varying levels of wear and tear to the layer of varnish. Some artifacts can be seen on the support panel in the middle, due to ambient occlusion which the proposed method does not model.

Overall, the proposed method captures reflectance maps with rich details, for samples ranging from diffuse-dominated (Figures 11.11 and 11.12, third and fourth rows) to highly specular (Figures 11.11 and 11.12, first and second rows), which provide good qualitative results for photo-realistic rendering.

11.9.2. *Discussion and error analysis*

The main assumption for the proposed method to work is that measurements be made close to Brewster angle. The following provides a theoretical analysis of the Brewster measurement assumption and in particular, how close is "close to Brewster angle".

11.9.2.1. *Surface normal accuracy*

We rely on the detection of the maximum intensity of the TRS at two different views, to provide cues to reliably estimate surface normals. We thus conducted a theoretical analysis of the changes in transmitted radiance through a linear polarizer both under unpolarized illumination (Figure 11.13, first row) and partially linearly polarized illumination with a DOP of 80%, which is the maximum predicted by the Rayleigh sky model (Figure 11.13, second row). The latter is essentially the worst-case scenario for passive outdoors reflectometry. The TRS was simulated at different angles of incidence, for a flat surface pointing at the vertical (i.e. its azimuth is at 90° from the horizon).

As can be seen in Figure 11.13 (first row), the maximum of the sinusoid is at 0° for any angle of incidence, which is exactly what is expected: the phase of the sinusoid does not depend on the angle of incidence under the unpolarized world assumption and multiview surface normal estimation is reliable even for curved surfaces as has been demonstrated multiple times in previous work (Wolff 1989; Miyazaki et al. 2012). Under partial linear polarization, however (Figure 11.13, second row), the phase of the sinusoid varies with the angle of polarization of the incident illumination (colored curves). A behavior similar to that of the unpolarized world assumption can be observed again at Brewster angle ($\theta_i = \theta_B$). Interestingly, this still holds to some extent in a 15° window around Brewster angle where the estimation error on ϕ was found to be less than $4°$, which is acceptable for rendering applications. Furthermore, while Brewster angle can vary over large surfaces, only a 5–6° variation was found over the largest measured sample, which is still within the acceptable range.

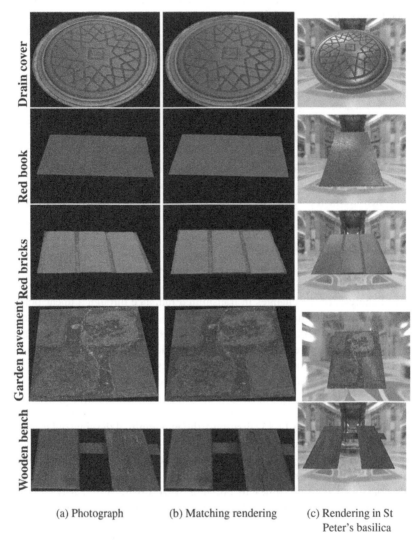

(a) Photograph   (b) Matching rendering   (c) Rendering in St
Peter's basilica

Figure 11.12. *Comparisons of sample photographs (a) to matching renderings under the same incident illumination (b), as well as renderings in novel lighting environment (c). For a color version of this figure, see www.iste.co.uk/funatomi/computational.zip*

Surf. normals	Riviere et al. (2017)		Ghosh et al. (2009)		Aittala et al. (2015)	
Time of Day	Std. dev.	Mean	Std. dev.	Mean	Std. dev.	Mean
Cloudy (midday)	3.80°					
Sunny (10-10:30 am)	8.91°	[0, 0, 1]	5.32°	[0, 0, 1]	5.13°	[0, 0, 1]
Sunny (3-3:30 pm)	5.97°					
Sunny (6-6:30 pm)	8.97°					

Table 11.1. *Statistical variation in surface normals of "red book" under different lighting conditions (left column), compared to two measurement methods using controlled illumination*

Furthermore, Table 11.1 shows a statistical analysis of surface normals estimated from the presented method for the "red book" (Figures 11.11 and 11.12, second row) at different times of day under an open sky, compared to the recent work of Ghosh et al. (2009) and Aittala et al. (2015) on surface reflectometry using commodity hardware.

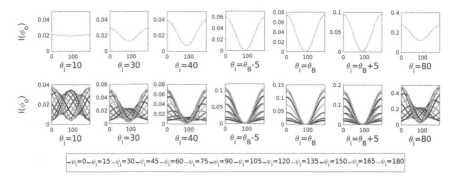

Figure 11.13. *Brewster angle measurement validation: Simulated TRS for a glass material (index of refraction $\eta = 1.5$) oriented at an azimuth $\phi_{\vec{n}} = 90°$. First row: Simulation under unpolarized incident illumination – the maximum of the TRS is found at $\phi_o = 0°$, as expected, for any angle of incidence θ_i. Second row: Simulation under partially linearly polarized illumination with a DOP of 80%. The different colors represent different angles of polarization ψ_i. Unlike under unpolarized incident illumination, the phase of the TRS depends on the angle of polarization of the incident illumination. However, behavior similar to that obtained under unpolarized illumination can be observed again at and around Brewster angle (i.e. when $\theta_i = \theta_B$). For a color version of this figure, see www.iste.co.uk/funatomi/computational.zip*

The different times of day were chosen to reflect changes in polarization of the skylight as predicted by the Rayleigh sky model: at sunset and sunrise, the sky is strongly vertically polarized at the zenith in the north–south direction, while it is mostly horizontally polarized at midday. From the figures in Table 11.1, the following observations can be made: the cloudy measurements (Table 11.1, first column) and sunny measurements at 3 pm (Table 11.1) seem to be the best conditions for data acquisition. The former makes sense as the unpolarized world assumption (ideal case) prevails. The latter comes from the fact that the measurements were carried close to the midday sun, which was determined to be around 1–1:30 pm on the day of the capture, where the incident illumination was mostly horizontal. From the theoretical analysis in section 11.4, this makes sense as horizontal polarization simplifies the mathematics of the TRS to a form close to that of the unpolarized world assumption. The other two conditions at 10 am (Table 11.1, second row) and 6 pm (Table 11.1, fourth row) correspond to nonideal cases with, respectively, an arbitrary polarization angle and vertical polarization. These effects are however mitigated in part by the fact that the measurements are carried out close to Brewster angle of incidence. The 6 pm measurement is interesting for another reason, however, as measurements at normal incidence provide a good way of separating diffuse and specular reflectance (Figure 11.14). This is because at that time, the band of highest degree of polarization is directly overhead.

(a) Lighting environment   (b) Photograph   (c) Specular-free image   (d) Diffuse-free image

Figure 11.14. *Diffuse-specular separation at normal incidence: At 6 pm, the sky is strongly linearly polarized at the zenith (a), which allows for good diffuse-specular separation ((c),(d)) near normal incidence. For a color version of this figure, see www.iste.co.uk/funatomi/computational.zip*

11.10. References

Aittala, M., Weyrich, T., Lehtinen, J. (2015). Two-shot SVBRDF capture for stationary materials. *ACM Transactions on Graphics*, 34(4), 110.

Atkinson, G. and Hancock, E.R. (2006). Recovery of surface orientation from diffuse polarization. *IEEE Transactions on Image Processing*, 15(6), 1653–1664.

CGSociety (2007). A complete IOR list [Online]. Available at: http://forums. cgsociety.org/showthread.php?t=513458.

Chen, H. and Wolff, L.B. (1998). Polarization phase-based method for material classification in computer vision. *International Journal of Computer Vision*, 28(1), 73–83.

Collett, E. (2005). *Field Guide to Polarization*. SPIE Field Guides, vol. FG05. SPIE.

Debevec, P. (1998). Rendering synthetic objects into real scenes: Bridging traditional and image-based graphics with global illumination and high dynamic range photography. *Proceedings of ACM SIGGRAPH 98*.

Ghosh, A., Chen, T., Peers, P., Wilson, C.A., Debevec, P.E. (2009). Estimating specular roughness and anisotropy from second order spherical gradient illumination. *Computer Graphics Forum*, 28(4), 1161–1170.

Ghosh, A., Chen, T., Peers, P., Wilson, C.A., Debevec, P. (2010). Circularly polarized spherical illumination reflectometry. *ACM Transactions on Graphics (Proceedings of SIGGRAPH Asia)*, 29, 162:1–12.

Guarnera, G.C., Peers, P., Debevec, P.E., Ghosh, A. (2012). Estimating surface normals from spherical stokes reflectance fields. *ECCV Workshop on Color and Photometry in Computer Vision*, pp. 340–349.

Huynh, C.P., Robles-Kelly, A., Hancock, E. (2010). Shape and refractive index recovery from single-view polarisation images. *2010 IEEE Conference on Computer Vision and Pattern Recognition (CVPR)*, pp. 1229–1236.

Jones, E., Oliphant, T., Peterson, P. (2001). SciPy: Open source scientific tools for Python [Online]. Available at: http://www.scipy.org/.

Kadambi, A., Taamazyan, V., Shi, B., Raskar, R. (2015). Polarized 3d: High-quality depth sensing with polarization cues. *Proceedings of the IEEE International Conference on Computer Vision*, pp. 3370–3378.

Können, G. (1985). *Polarized Light in Nature*. Cambridge University Press.

Mantiuk, R., Krawczyk, G., Mantiuk, R., Seidel, H.-P. (2007). High dynamic range imaging pipeline: Perception-motivated representation of visual content. *Electronic Imaging 2007, International Society for Optics and Photonics*, article 649212.

Miyazaki, D. and Ikeuchi, K. (2005). Inverse polarization raytracing: Estimating surface shapes of transparent objects. *CVPR*, pp. 910–917.

Miyazaki, D., Tan, R.T., Hara, K., Ikeuchi, K. (2003). Polarization-based inverse rendering from a single view. *ICCV*, pp. 982–987.

Miyazaki, D., Shigetomi, T., Baba, M., Furukawa, R., Hiura, S., Asada, N. (2012). Polarization-based surface normal estimation of black specular objects from multiple viewpoints. *2012 Second International Conference on 3D Imaging, Modeling, Processing, Visualization and Transmission (3DIMPVT)*, pp. 104–111.

Nielsen, J.B., Jensen, H.W., Ramamoorthi, R. (2015). On optimal, minimal BRDF sampling for reflectance acquisition. *ACM Transactions on Graphics*, 34(6), 186:1–11 [Online]. Available at: http://doi.acm.org/10.1145/2816795.2818085.

Pixel and Poly (n.d.). Pixelandpoly IOR list [Online]. Available at: https://pixelandpoly.com/ior.html.

Riviere, J. (2017). On-site surface reflectometry. PhD Thesis, EThOS.

Riviere, J., Reshetouski, I., Filipi, L., Ghosh, A. (2017). Polarization imaging reflectometry in the wild. *ACM Transaction on Graphics (TOG)*, 36(6), 206.

Schlick, C. (1994). An inexpensive brdf model for physically-based rendering. *Computer Graphics Forum*, 13(3), 233–246.

Smith, W., Ramamoorthi, R., Tozza, S. (2016). Linear depth estimation from an uncalibrated, monocular polarisation image. *Proc. of European Conference on Computer Vision (ECCV)*, pp. 517–526.

Strutt, J.W. (1871a). On the scattering of light by small particles. *The London, Edinburgh, and Dublin Philosophical Magazine and Journal of Science*, 41(275), 447–454.

Strutt, J.W. (1871b). On the light from the sky, its polarization and colour. *The London, Edinburgh, and Dublin Philosophical Magazine and Journal of Science*, 41(271), 107–120.

Torrance, K.E. and Sparrow, E.M. (1967). Theory for off-specular reflection from roughened surfaces. *JOSA*, 57(9), 1105–1114.

Walter, B., Marschner, S.R., Li, H., Torrance, K.E. (2007). Microfacet models for refraction through rough surfaces. *Proceedings of the 18th Eurographics Conference on Rendering Techniques*. Eurographics Association, pp. 195–206.

Wolff, L.B. (1989). Surface orientation from two camera stereo with polarizers. *Proc. SPIE Conf. Optics, Illumination and Image Sensing for Machine Vision IV*, 1194, 287–297.

Wolff, L.B. (1990). Polarization-based material classification from specular reflection. *PAMI*, 12(11), 1059–1071.

Wu, C. (2011). Visualsfm: A visual structure from motion system [Online]. Available at: http://ccwu.me/vsfm/index.html.

Wu, C., Agarwal, S., Curless, B., Seitz, S.M. (2011). Multicore bundle adjustment. *2011 IEEE Conference on Computer Vision and Pattern Recognition (CVPR)*, pp. 3057–3064.

12

Multispectral Polarization Filter Array

Kazuma SHINODA

Graduate School of Regional Development and Creativity,
Utsunomiya University, Japan

12.1. Introduction

Imaging of the spectral or polarization spatial distributions of objects is expected to have various applications in the field of computer vision. Although various methods have been studied for taking multispectral and polarization images, the optical systems employed to date have been too large and the methods require capturing the image several times. Filter arrays that transmit different wavelengths and polarization for each pixel have been considered as a method to capture images with a single exposure. Some representative examples are shown in Figure 12.1. A 5-band multispectral filter array (MSFA) with RGB plus cyan and orange has been developed, and a method to reconstruct multispectral images from single exposure images (i.e. demosaicking) has been proposed by Monno et al. (2015). Jia et al. (2016) proposed a 16-channel MSFA with sinusoidal sensitivity using a Fabry–Perot type filter (Jia et al. 2016), called a Fourier spectral filter array, which can remove aliasing. Gruev et al. (2010) also proposed a filter array method for capturing polarization images, where linear polarizers of 0, 45, 90 and 135° are periodically arranged in each pixel to visualize the polarization intensity and angle (Gruev et al. 2010). Sony's IMX250MYR is a method to achieve RGB and polarization images on the same imager (Sony Semiconductor Solutions Group 2018) using a wire-grid polarization filter array

Computational Imaging for Scene Understanding,
coordinated by Takuya FUNATOMI and Takahiro OKABE.
© ISTE Ltd 2024.

mounted on the surface of a conventional RGB filter array. The imaging principle is simple and can be achieved by a combination of conventional techniques. However, a dedicated semiconductor manufacturing process is required that can fabricate both color resist-based filter arrays and wire grids and mount them on the imager.

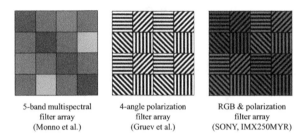

| 5-band multispectral filter array (Monno et al.) | 4-angle polarization filter array (Gruev et al.) | RGB & polarization filter array (SONY, IMX250MYR) |

Figure 12.1. *Multispectral and polarization filter array examples. For a color version of this figure, see www.iste.co.uk/funatomi/computational.zip*

In recent years, the simultaneous imaging of multispectral and polarization images has been studied and is expected to be applied in various fields such as industry, agriculture and medicine. A multispectral polarization image of a captured object (in this section, a multispectral polarization image means four-dimensional data for two-dimensional space, spectral and linear polarization) is obtained by separating the incident light with a beamsplitter, passing it through a color filter and a polarization filter, and then capturing it with two CCDs (Zhao et al. 2009). Fu et al. (2015) proposed a compressive sensing approach to obtain multispectral polarization images by taking multiple images with random observation bases and estimating them with l_1-l_2 norm minimization (Fu et al. 2015). However, both of these methods require complex optical systems and multiple exposures. The use of filter arrays, which can acquire multispectral polarization images of objects with a small set of equipment and a single exposure, can be considered as an improved method.

Snapshot multispectral polarization imaging can be achieved by implementing a polarization filter array on an MSFA using the same principle as the RGB polarization camera in Figure 12.1. Tu and Pau (2016) theoretically detailed a method to determine the sensitivity of a filter array when stacking an MSFA and a polarization filter array, assuming a device with ideal optical transmission characteristics. Junger et al. (2011) proposed a method for adjusting the polarization transmission characteristics by wire grid orientation and the spectral transmission characteristics based on the sub-wavelength structure on a CMOS (Junger et al. 2011). All these methods are based on the assumption that the linear polarization component is different for each pixel, and the device corresponding to the linear polarizer is tilted in the directions of 0, 45, 90 or 135° for each pixel. The linear polarizer removes polarization components other than the target angle and acts as an all-pass filter in the desired wavelength range. In contrast, the MSFA should ideally adjust only the spectral transmission characteristics and should have no effect on the polarization characteristics of the incident light. Due to these constraints, conventional techniques for taking multispectral polarization images are limited in the observation wavelength range and the difficulty of manufacturing them limits the devices that can be fabricated. Therefore, although it is easy to study the imaging of spectral polarization images on a simulation basis, it is difficult to make a small number of prototypes of multispectral polarization filter arrays with specific transmission characteristics and patterns using a combination of existing technologies, and this can be a bottleneck in the research and development of multispectral polarization imaging.

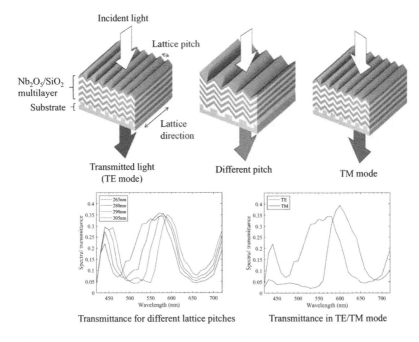

Figure 12.2. *Conceptual diagram of a multilayer photonic crystal. For a color version of this figure, see www.iste.co.uk/funatomi/computational.zip*

In this section, based on the above background, we introduce a method to realize a multispectral polarization filter array based on multilayer photonic crystals and snapshot imaging of multispectral polarization images using this filter array (Shinoda et al. 2018). A conceptual diagram of a multilayer photonic crystal is shown in Figure 12.2. A multilayer photonic crystal is an oxide multilayer with a wave-like structure, and its spectral and polarization transmission characteristics can be adjusted by changing its wave-like structure (lattice pitch and lattice direction), which will be described in detail in the next section. Therefore, it is easy to realize a multispectral polarization filter array with different transmission characteristics for each pixel by changing the lattice pitch and direction for each pixel. A captured grayscale image taken on a monochrome imager equipped with this filter array gives different values of the spectral and polarization properties for each pixel, and the multispectral polarization image can be recovered by demosaicking. It is possible to reconstruct multispectral images at each polarization angle, non-polarized multispectral images, polarized images at each wavelength and even RGB images from multispectral polarization images, thus enabling us to realize conventional multispectral, polarization and RGB cameras simultaneously in a single shot. In this section, we introduce the principle of multilayer photonic crystals, prototype cameras, spectral and polarization transmission characteristics, and explain the demosaicking method of multispectral polarization images.

12.2. Multispectral polarization filter array with a photonic crystal

It is known that periodic structural modulation of optical multilayers can be used to control the transmittance of incident light and to add polarization selection functions to the multilayers. In particular, a method to realize a wavelength-selective filter has been proposed called a photonic crystal type wavelength filter (PhCF), which works by forming a triangular wave-shaped multilayer film consisting of a high refractive index layer and a low refractive index layer and adjusting the film thickness, lattice period and film material (Ohtera et al. 2007, 2011). Since the lattice period influences the transmitted wavelength, this PhCF can be used to produce various wavelength filters, such as that shown in Figure 12.2, even when the lattice period is changed for the same film material and thickness. Furthermore, due to the structural anisotropy of the PhCF, the optical properties of the TE mode (transmittance of polarized light parallel to the lattice direction) and the TM mode (transmittance of polarized light perpendicular to the lattice direction) are different. When linearly polarized light enters the PhCF, the output spectrum is due to the TE mode if the polarization angle is parallel to the lattice, and due to the TM mode if it is perpendicular to the lattice. When non-polarized light enters the PhCF, the transmittance is the sum of the TE and TM modes.

Based on the above characteristics, multilayer photonic crystals can be used to realize filters for various wavelengths by adjusting their microstructure, and can act as polarization filters at the same time. Taking advantage of this feature, a multispectral polarization filter array can be realized by depositing multilayer photonic crystals with different microstructures (lattice pitch and lattice angle) in each pixel region of the imager. However, it is important that the same film material is used for all pixels and that the film thickness is not changed for each pixel. By changing the lattice pitch and direction for each imager pixel size, the spectral polarization transmission characteristics of each pixel can be changed, which has the advantage that a prototype filter array can be manufactured stably using the same process.

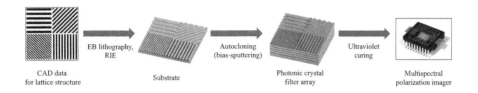

Figure 12.3. *Fabrication process for photonic crystal filter array. For a color version of this figure, see www.iste.co.uk/funatomi/computational.zip*

A more specific procedure for manufacturing a filter array is shown in Figure 12.3. First, a rectangular region is defined based on the pixel size of an imager, and a lattice pattern with different intervals and angles for each region is created by CAD or using other data. The following approach uses the same material and manufacturing process for any lattice pattern. Initially, the microstructure of the pattern data is transferred to a quartz substrate by electron beam (EB) lithography according to the intended pattern, and then the microstructure of the quartz substrate is formed by reactive ion etching (RIE). Next, an auto-cloning method based on bias sputtering (Kawashima et al. 2000) is used to deposit a predetermined number of high and low refractive index materials in the desired film configuration. Because it is difficult to deposit the film directly on the imager due to the semiconductor manufacturing process, the filter arrays are fabricated separately as photonic crystal filter arrays and mounted on the monochrome imager by UV curing. During bonding, the imager and filter array are aligned by a micromanipulator while observing the images taken by the monochrome imager.

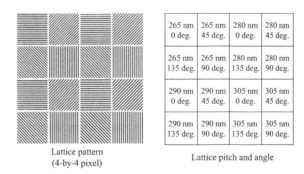

265 nm 0 deg.	265 nm 45 deg.	280 nm 0 deg.	280 nm 45 deg.
265 nm 135 deg.	265 nm 90 deg.	280 nm 135 deg.	280 nm 90 deg.
290 nm 0 deg.	290 nm 45 deg.	305 nm 0 deg.	305 nm 45 deg.
290 nm 135 deg.	290 nm 90 deg.	305 nm 135 deg.	305 nm 90 deg.

Lattice pattern
(4-by-4 pixel)

Lattice pitch and angle

Figure 12.4. *Design overview for prototype photonic crystal multispectral filter array*

As an example, an overview of the design of our prototype photonic crystal-type multispectral polarization filter array is shown in Figure 12.4. The size of the wave-like structure is set to $4,650\times4,650$ nm per pixel to match the imager's pixel size. One block of the filter array is 4×4 pixels, and four PhCFs with different lattice pitches (265, 280, 290, 305 nm, referred to as Filters 1–4) are arranged at four different angles to create a multispectral polarization filter array with a total of 16 different filters. The pattern is periodically repeated vertically and horizontally for 100×100 pixels. The deposition procedure, film material, film thickness and other deposition parameters were determined according to the literature (Ohtera et al. 2007), with the high refractive index layer being Nb_2O_5, the low refractive index layer being SiO_2, with a total of 40 layers, excluding the anti-reflective film, with a film thickness of about 4 µm.

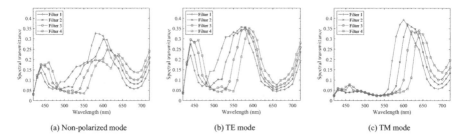

(a) Non-polarized mode   (b) TE mode   (c) TM mode

Figure 12.5. *Spectral transmittance characteristics of deposited filter array. For a color version of this figure, see www.iste.co.uk/funatomi/computational.zip*

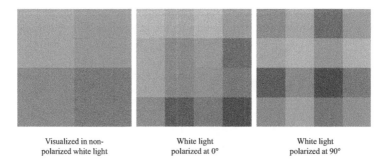

Visualized in non-   White light   White light
polarized white light   polarized at 0°   polarized at 90°

Figure 12.6. *Visual appearance of filter array. For a color version of this figure, see www.iste.co.uk/funatomi/computational.zip*

The spectral transmission characteristics of the deposited filter array are shown in Figure 12.5. Here, the non-polarized mode is the transmittance of non-polarized light, and the TE and TM modes are the transmittance of 0 and 90° linearly polarized light with respect to the lattice direction. For example, the spectral transmission characteristics of Filter 1 in the TE mode are calculated from the average of all Filter 1 transmittances in 100×100 pixels, assuming that the transmittance is equal to that of a filter with $0°$ linear polarization incident light on a Filter 1 of $0°$, and that the transmittance is equal to that of a filter with $45°$ linear polarization incident light on a Filter 1 of $45°$. As shown in Figure 12.5, the anisotropy of the microlattice results in different transmission characteristics in the TE and TM modes. Furthermore, by changing the lattice pitch in Filters 1 to 4, we can shift the peak transmittance.

To compare the appearance of the filter array, a simulation of the RGB components of the transmission characteristics is shown in Figure 12.6. Here, uniform white light is transmitted through the filter array and then converted by the XYZ color matching functions and sRGB color gamut to reproduce the sRGB component under a D65 light source. The incident light is $0°$ linearly polarized, which means that the horizontal linearly polarized light is incident uniformly over the whole area, regardless of the lattice angle of the filter array. Looking at the non-polarized incident light of Figure 12.6, it can be seen that the transmission spectrum for non-polarized incident light is determined by the product of the transmission characteristics of Figure 12.5 regardless of the lattice angle. However, the appearance of Figure 12.6 is different for 0 or $90°$ linearly polarized incident light than for non-polarized incident light, indicating that the observed values of each pixel differ depending on the linearly polarized component of the incident light.

It can be seen that a multispectral polarization filter array with different spectral and polarization characteristics for each pixel can be realized by using photonic crystals. However, the transmission characteristics of the prototype filter array are far from those desired for ideal filter arrays. Specifically, the prototype filter array exhibits a wide bandwidth and multiple peaks, and an extinction ratio between 0 and $90°$ polarizations that depends on the wavelength sometimes being 1:1. To address this deficiency, we would typically take the approach of setting the film thickness, material and optical system to make an ideal bandpass filter (or an ideal linear polarization filter). Since the examples presented in this section are not optimally designed for transmission characteristics, it is possible that better characteristics can be obtained by improving the microstructure. However, the pursuit of ideal attenuation characteristics, as mentioned above, may limit the observed wavelength range, decrease the amount of transmitted light due to an increased film thickness, increase the complexity of the equipment and make it more difficult to manufacture the photonic crystal, thus compromising the convenience of the photonic crystal.

In the next section, we introduce a method to solve this problem by demosaicking. Even for filter arrays with apparently poor transmission characteristics, such as Figure 12.5, if there is a certain degree of independence in the transmission characteristics between pixels, demosaicking can be replaced by a linear inverse problem (the so-called deconvolution problem), and thus, it is possible to recover multispectral polarization images over a wide range of wavelengths.

12.3. Generalization of imaging and demosaicking with multispectral polarization filter arrays

Since capturing with filter arrays inherently lacks image information, a demosaic is generally performed to interpolate the missing information. For example, in a Bayer color filter array, the resolution of green is 1/2, and red and blue are 1/4, so each channel is demosaicked such that the number of pixels is the same as that of the imager using the observed pixels around the missing pixels. Similarly, in the case of MSFA or polarization filter arrays, the estimation accuracy can be improved by focusing on the spatial similarity of the same wavelength or same polarization component, as well as performing an interpolation taking into account the similarity between different wavelengths and polarizations as proposed in Miao et al. (2006), Gao and Gruev (2013), Shinoda et al. (2016), Zhang et al. (2016), Ahmed et al. (2017) and Mihoubi et al. (2017).

In the case of an (ideal) multispectral polarization filter array consisting of a bandpass filter and a wire grid, demosaicking corresponds to filling in simple missing pixels so that the interpolation can be performed by simply extending the conventional demosaicking methods of MSFA or polarization filter arrays. However, in the case of the photonic crystal filter array introduced in the previous section, which has multimodal transmission characteristics or different extinction ratios for different wavelengths, a snapshot grayscale image will be a mixture of various wavelength bands and polarization components for each pixel, and a simple interpolation of missing pixels cannot be applied.

However, capturing with a multispectral polarization filter array can be represented by a simple linear model, regardless of its transmission characteristics. Since demosaicking is equivalent to solving a linear inverse problem, a multispectral polarization image for a desired wavelength, polarization angle and spatial resolution can be recovered from any filter array with any multimodality or extinction ratio, once the solution of the ill-posed problem is determined (note that "recovery" here does not guarantee a complete recovery of the original incident light but means that, like demosaicking with a Bayer CFA, a candidate solution can be derived with errors from the true incident light).

An overview of imaging and demosaicking by a multispectral polarization filter array is shown in Figure 12.7. The light incident on a camera with a filter array is assumed to be a four-dimensional multispectral polarization image, and the image taken by a single exposure is called a mosaicked image. The camera is simply assumed to consist of a monochrome imager and a filter array. In this case, a pixel position (x, y) with a filter array can be considered to be a linear diattenuator with a θ-slanted horizontal reference axis (TE mode) and different extinction ratios for each wavelength. In the example of Figure 12.7, the extinction ratio is low at the peak wavelength in the TM mode because both TE and TM modes have the peak at the same wavelength, but it is relatively high at the wavelength where the transmittance is zero in the TM mode. If the photonic crystal acts as an ideal bandpass filter and linear polarizer, the transmittance for the TM mode is zero, and the bandpass characteristics resemble those for the TE mode. In other words, the overview of Figure 12.7 is a model of a general system that can be applied to any multispectral polarization filter array with any characteristics. In the following, the details of the formulation are explained.

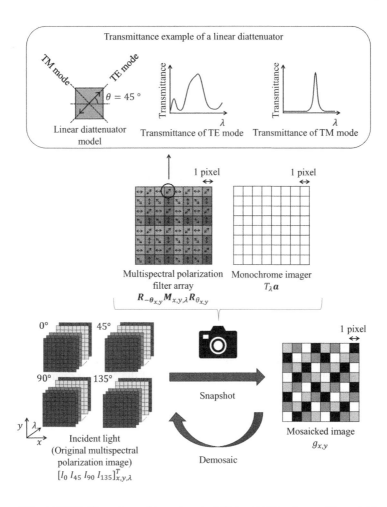

Figure 12.7. *Overview of imaging and demosaicking by multispectral polarization filter array. For a color version of this figure, see www.iste.co.uk/funatomi/computational.zip*

First, we define the first three Stokes parameters for the incident light at a pixel position (x, y) and a wavelength λ as $\boldsymbol{s}_{x,y,\lambda} = [s_0 \ s_1 \ s_2]^T_{x,y,\lambda}$. The relationship between the Stokes parameters and polarization intensity $\boldsymbol{I}_{x,y,\lambda} = [I_0 \ I_{45} \ I_{90}]^T_{x,y,\lambda}$ (i.e. the intensity of the multispectral polarization image) is

$$\boldsymbol{s}_{x,y,\lambda} = \begin{bmatrix} 1 & 0 & 1 \\ 1 & 0 & -1 \\ -1 & 2 & -1 \end{bmatrix} \boldsymbol{I}_{x,y,\lambda}$$

$$= \boldsymbol{P}\boldsymbol{I}_{x,y,\lambda}, \quad\quad\quad [12.1]$$

where I_{135} is abbreviated since $I_{135} = I_0 + I_{90} - I_{45}$. In the imaging architecture using a filter array and monochrome imager, the signal measured intensity by an imager $g_{x,y}$ is

$$g_{x,y} = \sum_{\lambda} T_{\lambda} \boldsymbol{a} \boldsymbol{R}_{-\theta_{x,y}} \boldsymbol{M}_{x,y,\lambda} \boldsymbol{R}_{\theta_{x,y}} \boldsymbol{s}_{x,y,\lambda}, \quad\quad\quad [12.2]$$

where T_{λ} is the sensitivity at λ for the imager, $\boldsymbol{a} = [1 \ 0 \ 0]$ is the extraction vector of s_0 for the imager, $\boldsymbol{R}_{\theta_{x,y}} \in \mathbb{R}^{3 \times 3}$ is the rotation matrix that depends on the lattice angle θ at (x, y), and $\boldsymbol{M}_{x,y,\lambda} \in \mathbb{R}^{3 \times 3}$ is a matrix composed of the upper-left 3×3 components of the Mueller matrix of the filter at (x, y, λ). The right side of equation [12.2] is expressed as follows:

$$g_{x,y} = \sum_{\lambda} T_{\lambda} \begin{bmatrix} [m]^{0,0}_{x,y,\lambda} \\ [m]^{0,1}_{x,y,\lambda} \cos 2\theta_{x,y} - [m]^{0,2}_{x,y,\lambda} \sin 2\theta_{x,y} \\ [m]^{0,1}_{x,y,\lambda} \sin 2\theta_{x,y} + [m]^{0,2}_{x,y,\lambda} \cos 2\theta_{x,y} \end{bmatrix}^T \boldsymbol{s}_{x,y,\lambda}, \quad\quad [12.3]$$

where $[m]^{i,j}_{x,y,\lambda}$ is the ith row and jth column element of $\boldsymbol{M}$ at (x, y, λ). Since the photonic crystal filter array corresponds to a linear diattenuator and has a different extinction ratio for each wavelength, each element of the Mueller matrix of the filter is

$$[m]^{0,0}_{x,y,\lambda} = \frac{[U]^{TE}_{x,y,\lambda} + [U]^{TM}_{x,y,\lambda}}{2}, \quad\quad\quad [12.4]$$

$$[m]^{0,1}_{x,y,\lambda} = \frac{[U]^{TE}_{x,y,\lambda} - [U]^{TM}_{x,y,\lambda}}{2}, \quad\quad\quad [12.5]$$

$$[m]^{0,2}_{x,y,\lambda} = 0, \quad\quad\quad [12.6]$$

where $[U]_{x,y,\lambda}^{mode}$ is the TE or TM mode transmittance for the filter at (x, y, λ) and corresponds to the transmittance of Figure 12.5. Thus, equation [12.3] can be expressed as follows:

$$
g_{x,y} = \frac{1}{2} \sum_\lambda T_\lambda \left[\begin{array}{c} [U]_{x,y,\lambda}^{TE} + [U]_{x,y,\lambda}^{TM} \\ \left([U]_{x,y,\lambda}^{TE} - [U]_{x,y,\lambda}^{TM}\right) \cos 2\theta_{x,y} \\ \left([U]_{x,y,\lambda}^{TE} - [U]_{x,y,\lambda}^{TM}\right) \sin 2\theta_{x,y} \end{array} \right]^T s_{x,y,\lambda}
$$

$$
= \frac{1}{2} \sum_\lambda T_\lambda M'_{x,y,\lambda} s_{x,y,\lambda}. \tag{12.7}
$$

Therefore, the captured pixel value $g_{x,y}$ is determined by the sensitivity of the imager, the TE and TM mode transmittance for the filter, the lattice angle for each filter and the Stokes parameters for the incident light. This linear model has been previously described for a focal plane array (Powell and Gruev 2013), but their model does not consider the spatial and spectral correlations for demosaicking. Therefore, we assume that this measurement model is independent at each pixel and can express the three-dimensional elements (x, y, λ) as one-dimensional data. We define the pixel size as (X, Y), the number of the measurement spectral bands as L, and $g = \begin{bmatrix} g_{0,0} & g_{0,1} \cdots g_{X-1,Y-1} \end{bmatrix}^T \in \mathbb{R}^{XY}$, $s = \begin{bmatrix} s_{0,0,0}^T & s_{0,0,1}^T \cdots s_{X-1,Y-1,L-1}^T \end{bmatrix}^T$ $\in \mathbb{R}^{3XYL}$ and $I = \begin{bmatrix} I_{0,0,0}^T & I_{0,0,1}^T \cdots I_{X-1,Y-1,L-1}^T \end{bmatrix}^T \in \mathbb{R}^{3XYL}$ as a column vector. g can be expressed as follows:

$$
g = TM's
$$

$$
= TM'P'I
$$

$$
= HI, \tag{12.8}
$$

where $T \in \mathbb{R}^{XY \times XYL}$, $M' \in \mathbb{R}^{XYL \times 3XYL}$ and $P' \in \mathbb{R}^{3XYL \times 3XYL}$ are

$$
T = \frac{1}{2} E_{XY} \otimes \begin{bmatrix} T_{\lambda_0} & T_{\lambda_1} & \dots & T_{\lambda_{L-1}} \end{bmatrix}, \tag{12.9}
$$

$$
M' = \mathrm{diag}\left(M'_{0,0,\lambda_0}, M'_{0,0,\lambda_1}, \dots, M'_{X-1,Y-1,\lambda_{L-1}} \right), \tag{12.10}
$$

$$
P' = E_{XYL} \otimes P, \tag{12.11}
$$

where $E_n \in \mathbb{R}^{n \times n}$ is the identify matrix, $\otimes$ is the Kronecker product, and $\mathrm{diag}(X)$ is a diagonal matrix whose diagonal elements are X.

Since g and H given in equation [12.8] are known values after snapshot imaging, demosaicking corresponds to the inverse problem of equation [12.8] as

$$
\hat{I} = Wg. \tag{12.12}
$$

Here, the Wiener estimation (Pratt 1978) is applied for demosaicking to minimize the l_2-norm of the error between the original and estimated multispectral polarization images as:

$$\text{argmin}_W \left\langle ||\boldsymbol{I} - \hat{\boldsymbol{I}}||_2 \right\rangle \ \text{s.t.} \ \hat{\boldsymbol{I}} = \boldsymbol{W}\boldsymbol{g}. \tag{12.13}$$

The norm term in equation [12.13] is expressed as follows:

$$||\boldsymbol{I} - \hat{\boldsymbol{I}}||_2 = \text{Tr}\left[\boldsymbol{R}_I - 2\boldsymbol{R}_I\boldsymbol{H}^T\boldsymbol{W}^T + \boldsymbol{W}\boldsymbol{H}\boldsymbol{R}_I\boldsymbol{H}^T\boldsymbol{W}^T\right], \tag{12.14}$$

where $\text{Tr}\left[\cdot\right]$ denotes the trace of the matrix, and $\boldsymbol{R}_I \in \mathbb{R}^{3XYL\times3XYL}$ is the autocorrelation matrix of $\boldsymbol{I}$. Therefore, the matrix $\boldsymbol{W}$ is

$$\boldsymbol{W} = \boldsymbol{R}_I\boldsymbol{H}^T(\boldsymbol{H}\boldsymbol{R}_I\boldsymbol{H}^T)^{-1}. \tag{12.15}$$

$\boldsymbol{R}_I$ can be calculated as a first-order Markov model (Shinoda et al. 2017).

Various types of images can be obtained from the demosaicked multispectral polarization image $\hat{\boldsymbol{I}}$. In addition, since a simple inverse model is used for demosaicking, various images can be obtained from a snapshot grayscale image by multiplying only one constant matrix. For example, a non-polarized multispectral image $\hat{\boldsymbol{I}}_{msi} \in \mathbb{R}^{XYL}$ (corresponding to the first Stokes parameter) is

$$\hat{\boldsymbol{I}}_{msi} = \boldsymbol{W}_{msi}\,\boldsymbol{g}, \tag{12.16}$$

$$\boldsymbol{W}_{msi} = \left(\boldsymbol{E}_{XYL} \otimes \begin{bmatrix} 1 & 0 & 1 \end{bmatrix}\right)\boldsymbol{W}. \tag{12.17}$$

Similarly, a non-polarized RGB image $\hat{\boldsymbol{I}}_{rgb} \in \mathbb{R}^{3XY}$ is

$$\hat{\boldsymbol{I}}_{rgb} = \boldsymbol{W}_{rgb}\,\boldsymbol{g}, \tag{12.18}$$

$$\boldsymbol{W}_{rgb} = \left(\boldsymbol{E}_{XY} \otimes (\boldsymbol{C}\boldsymbol{T}\boldsymbol{L})\right)\boldsymbol{W}_{msi}, \tag{12.19}$$

where $\boldsymbol{C} \in \mathbb{R}^{3\times3}$ is the conversion matrix from XYZ to RGB, $\boldsymbol{T} \in \mathbb{R}^{3\times L}$ is the XYZ tristimulus function, and $\boldsymbol{L} \in \mathbb{R}^{L\times L}$ is the diagonal matrix of the standard illuminant.

There are various advantages to representing imaging with photonic crystal filter arrays as a linear model and demosaicking it with its linear inverse problem. The first is that demosaics can be generalized regardless of the transmission characteristics of the filter array. As long as the photonic crystal is a linear diattenuator, it is possible to apply equation [12.8] no matter how the transmission characteristics vary, and even when an ideal bandpass filter or polarization filter is used, demosaicking is possible simply by changing the TE and TM mode characteristics of equation [12.8]. Furthermore, we introduced the Wiener estimation as a demosaicking method in this section, which can be computed as the product of a single matrix. In particular, since the similarity of signals in all spatial–spectral–polarization axes is used, the entire visible spectrum can be reconstructed to some extent, and it has the advantage of allowing selective reconstruction of components of arbitrary wavelengths and polarization angles while reducing the number of operations.

12.4. Demonstration

In this section, we present the results of reconstructing various types of images from a single captured image using a photonic crystal multispectral polarization filter array.

We present the results of capturing images of a smartphone screen with the prototyped camera (Shinoda et al. 2018). The photonic crystal filter array is mounted on a monochrome imager, and the imager is mounted on a commercial USB camera housing (ARTRAY, ARTCAM-150P5-WOM) to construct a snapshot multispectral polarization camera. The pixel pitch of the imager is 4,650 nm/pixel, the pattern of the filter array is as described in 12.2, and the deposition area is 100×100 pixels. The captured subject is the red frame area of the LCD screen of the smartphone (Google, Nexus 5) shown in Figure 12.8. The light emitted from the LCD is linearly polarized, but it is impossible to judge the degree of polarization from the appearance of the image taken with an RGB camera. The contents are displayed with a white background and different colored ball patterns.

The demosaicking results are shown in Figure 12.9. The demosaicking reconstructs 420–720 nm at 10 nm intervals ($L = 31$ band) and assumes a D65 light source for the RGB reproduction. "Captured image" is the grayscale image obtained by the imager through the filter array, and all other images are derived from the "captured image". The "RGB non-polarized" and "RGB 0 deg." images show a slight decrease in colorfulness, but the magenta, cyan and yellow colors are reproduced, indicating that the spectral and spatial structures are reproduced over a wide wavelength range in visible light. Furthermore, the "RGB 90 deg." recovered image confirms that the light emanating from the smartphone's LCD screen is darker than "RGB 0 deg.". Because of its structure, the LCD screen has a linear polarization filter on its surface, which cannot be determined by ordinary RGB imaging (Figures 12.8 and Figure 12.9), but it is possible to visualize differences in the polarization components from "RGB 90 deg." in Figure 12.9. The second line in Figure 12.9 shows monochrome images of the reconstructed images at 400 nm–700 nm. Comparing "400 nm" and "600 nm", the cyan ball is slightly darker at "600 nm", and the blue component is stronger than the green component. Furthermore, "500" and "700 nm" display 90°, and therefore, the overall image is dark.

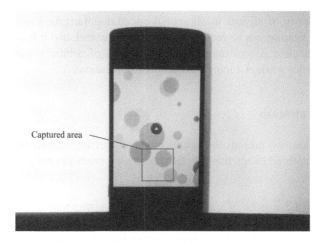

Figure 12.8. *Captured area in the demonstration. For a color version of this figure, see www.iste.co.uk/funatomi/computational.zip*

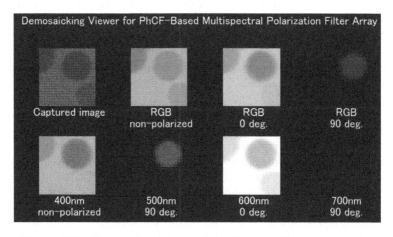

Figure 12.9. *Captured and demosaicked images. For a color version of this figure, see www.iste.co.uk/funatomi/computational.zip*

However, although it is desirable that all the screens at 90 deg. are reproduced in black, it can be seen that the yellow ball is slightly brightly restored at "RGB 90 deg." and "500 nm 90 deg.". This is due to the fact that the separation performance of the polarization components around 500 nm is reduced due to the transmission characteristics of the filter array, and the Wiener estimation, which minimizes the estimated square error, is used for demosaicking. Filter arrays made of photonic crystals have multiple peaks and different extinction ratios depending on the wavelength, and the signal separation performance during reconstruction varies depending on the wavelength band and polarization angle. Furthermore, the Wiener estimation assumes a first-order Markov model for the relationship between the signal values for the different components, so it does not act to conserve rapid signal changes between components. For example, the estimation is based on the assumption that there is some degree of correlation between the signal values at 0 and 90° polarization, and the estimation performance depends on the wavelength; hence, it may not be possible to represent a rapid change in the signal between polarization angles. These can be improved by designing a more rigorous filter array lattice structure and improving the demosaicking algorithm such that the estimation error is minimized while maintaining the gradient in the spatial–spectral–polarization axes. Especially for spectral polarization reconstruction, the loss ratio of the captured signal to the estimated target is higher than for RGB reconstruction, so methods such as l_1-norm minimization considering the sparsity of the signal (Zhang et al. 2014; Zhao and Yang 2015), total variation minimization (Aggarwal and Majumdar 2016) and tensor nuclear norm minimization (Chen et al. 2019) are considered to be effective.

12.5. Conclusion

In this section, we introduced multispectral polarization filter arrays and the restoration process for capturing multispectral polarization images with a single exposure, using a photonic crystal filter array as an example. Although some of the spatial–spectral–polarization information for an object is essentially missing, imaging with a filter array is an effective method for practical applications, since a spectral–polarization image can be captured with a single camera and a single exposure.

In particular, in this section, we introduced a method to create multispectral polarization filter arrays using the optical anisotropy of photonic crystals to deposit different microlattice structures for each pixel. Photonic crystals are characterized by the ability to realize multispectral and polarization filter arrays simultaneously in multilayers, which allows for unification of materials and manufacturing processes compared to manufacturing with different devices (e.g. color resist and wire grids). Furthermore, by treating demosaicking as a linear inverse problem, multispectral/polarization/RGB images can be recovered simultaneously because spectra with a wide wavelength range of visible light can be recovered. Although there are still some problems regarding the image quality of reconstructed images due to the constraints of transmission properties and demosaics, filter arrays are an effective approach for future multispectral and polarization imaging.

12.6. References

Aggarwal, H.K. and Majumdar, A. (2016). Hyperspectral image denoising using spatio-spectral total variation. *IEEE Geosci. Remote Sens. Lett.*, 13(3), 442–446.

Ahmed, A., Zhao, X., Gruev, V., Zhang, J., Bermak, A. (2017). Residual interpolation for division of focal plane polarization image sensors. *Opt. Express*, 25(9), 10651–10662.

Chen, Y., He, W., Yokoya, N., Huang, T.-Z. (2019). Weighted group sparsity regularized low-rank tensor decomposition for hyperspectral image restoration. *IEEE International Geoscience and Remote Sensing Symposium*, pp. 234–237.

Fu, C., Arguello, H., Sadler, B.M., Arce, G.R. (2015). Compressive spectral polarization imaging by a pixelized polarizer and colored patterned detector. *J. Opt. Soc. Am. A*, 32(11), 2178–2188.

Gao, S. and Gruev, V. (2013). Gradient-based interpolation method for division-of-focal-plane polarimeters. *Opt. Express*, 21(1), 1137–1151.

Gruev, V., Perkins, R., York, T. (2010). Ccd polarization imaging sensor with aluminum nanowire optical filters. *Opt. Express*, 18(18), 19087–19094.

Jia, J., Barnard, K.J., Hirakawa, K. (2016). Fourier spectral filter array for optimal multispectral imaging. *IEEE Trans. Image Process.*, 25(4), 1530–1543.

Junger, S., Tschekalinskij, W., Verwaal, N., Weber, N. (2011). Polarization and spectral filter arrays based on sub-wavelength structures in CMOS. *SENSOR+TEST Conferences*, pp. 161–165.

Kawashima, T., Miura, K., Sato, T., Kawakami, S. (2000). Self-healing effects in the fabrication processes of photonic crystals. *Appl. Phys. Lett.*, 77(16), 2613–2615.

Miao, L., Qi, H., Ramanath, R., Snyder, W. (2006). Binary tree-based generic demosaicking algorithm for multispectral filter arrays. *IEEE Trans. Image Process.*, 15(11), 3550–3558.

Mihoubi, S., Losson, O., Mathon, B., Macaire, L. (2017). Multispectral demosaicing using pseudo-panchromatic image. *IEEE Trans. Image Process.*, 3(4), 982–995.

Monno, Y., Kikuchi, S., Tanaka, M., Okutomi, M. (2015). A practical one-shot multispectral imaging system using a single image sensor. *IEEE Trans. Image Process.*, 24(10), 3048–3059.

Ohtera, Y., Onuki, T., Inoue, Y., Kawakami, S. (2007). Multichannel photonic crystal wavelength filter array for near-infrared wavelengths. *IEEE J. Lightwave Technol.*, 25(2), 499–503.

Ohtera, Y., Kurniatan, D., Yamada, H. (2011). Design and fabrication of multichannel Si/SiO$_2$ autocloned photonic crystal edge filters. *Appl. Opt.*, 50(9), 50–54.

Powell, S.B. and Gruev, V. (2013). Calibration methods for division-of-focal-plane polarimeters. *Opt. Express*, 21(18), 21039–21055.

Pratt, W.K. (1978). *Digital Image Processing*. Wiley.

Shinoda, K., Hamasaki, T., Kawase, M., Hasegawa, M., Kato, S. (2016). Demosaicking for multispectral images based on vectorial total variation. *Opt. Rev.*, 23(4), 559–570.

Shinoda, K., Yanagi, Y., Hayasaki, Y., Hasegawa, M. (2017). Multispectral filter array design without training images. *Opt. Rev.*, 24(4), 554–571.

Shinoda, K., Ohtera, Y., Hasegawa, M. (2018). Snapshot multispectral polarization imaging using a photonic crystal filter array. *Opt. Express*, 26(12), 15948–15961.

Sony Semiconductor Solutions Group (2018). *Image Sensor: Polarization Image Sensor* [Online]. Available at: https://www.sony-semicon.co.jp/e/products/IS/polarization/technology.html.

Tu, X. and Pau, S. (2016). Optimized design of N optical filters for color and polarization imaging. *Opt. Express*, 24(3), 3011–3024.

Zhang, H., He, W., Zhang, L., Shen, H., Yuan, Q. (2014). Hyperspectral image restoration using low-rank matrix recovery. *IEEE Trans. Geosci. Remote Sens.*, 52(8), 4729–4743.

Zhang, J., Luo, H., Hui, B., Chang, Z. (2016). Image interpolation for division of focal plane polarimeters with intensity correlation. *Opt. Express*, 24(18), 20799–20807.

Zhao, Y.-Q. and Yang, J. (2015). Hyperspectral image denoising via sparse representation and low-rank constraint. *IEEE Trans. Geosci. Remote Sens.*, 53(1), 296–308.

Zhao, Y., Zhang, L., Zhang, D., Pan, Q. (2009). Object separation by polarimetric and spectral imagery fusion. *Comput. Vis. Image Underst.*, 113(8), 855–866.

List of Authors

Yuta ASANO
Digital Content and Media Sciences
Research Division
National Institute of Informatics
Tokyo
Japan

Ying FU
Beijing Institute of Technology
China

Takuya FUNATOMI
Division of Information Science
Nara Institute of Science and
Technology
Japan

Nathan HAGEN
Department of Optical Engineering
Utsunomiya University
Japan

Felix HEIDE
Princeton Computational
Imaging Lab
Princeton University
USA

Adrian JARABO
Graphics and Imaging Lab
University of Zaragoza
Spain

Antony LAM
Mercari, Inc.
Tokyo
Japan

Daisuke MIYAZAKI
Department of Intelligent Systems
Hiroshima City University
Japan

Hajime NAGAHARA
Institute for Datability Science
Osaka University
Japan

Thanh-Trung NGO
Hanoi University of Science
and Technology
Vietnam

Ko NISHINO
Graduate School of Informatics
Kyoto University
Japan

Takahiro OKABE
Department of Artificial Intelligence
Kyushu Institute of Technology
Fukuoka
Japan

Adithya Kumar PEDIREDLA
Dartmouth College
Hanover
USA

Jérémy Maxime RIVIERE
Realistic Graphics and Imagining
Imperial College London
UK

Imari SATO
Computational Imaging and
Vision Lab
National Institute of Informatics
Tokyo
Japan

Yoichi SATO
Institute of Industrial Science
The University of Tokyo
Japan

Kazuma SHINODA
Graduate School of Regional
Development and Creativity
Utsunomiya University
Japan

Kenichiro TANAKA
Vision and Imaging Group
Ritsumeikan University
Shiga
Japan

Rin-ichiro TANIGUCHI
Kyushu University
Fukuoka
Japan

Florian WILLOMITZER
Wyant College of Optical Sciences
University of Arizona
Tucson
USA

Yinqiang ZHANG
Next Generation Artificial
Intelligence Research Center
The University of Tokyo
Japan

Index

Printed and bound by CPI Group (UK) Ltd, Croydon, CR0 4YY

27/10/2024

14580320-0003